Understanding Lasers

IEEE PRESS Understanding Science & Technology Series

The IEEE PRESS Understanding Series treats important topics in science and technology in a simple and easy-to-understand manner. Designed expressly for the nonspecialist engineer, scientist, or technician as well as the technologically curious—each volume stresses practical information over mathematical theorems and complicated derivations.

Other published and forthcoming books in the series include:

Understanding the Nervous System
An Engineering Perspective
by Sid Deutsch, Visiting Professor, University of South Florida, Tampa
Alice Deutsch, President, Bioscreen, Inc., New York

1993 Softcover 408 pp   IEEE Order No. PP0291-5   ISBN 0-87942-296-3

Understanding Telecommunications and Lightwave Systems
An Entry-Level Guide
by John G. Nellist, Consultant, Sarita Enterprises Ltd.

1992 Softcover 200 pp   IEEE Order No. PP0314-5   ISBN 0-7803-0418-7

Ideas for future topics and authorship inquiries are welcome. Please write to IEEE PRESS: The Understanding Series.

Understanding Lasers
An Entry-Level Guide

Second Edition

Jeff Hecht
Contributing Editor
Laser Focus World

 IEEE
PRESS

IEEE PRESS Understanding Science &
Technology Series

The Institute of Electrical and Electronics Engineers, Inc., New York

This book may be purchased at a discount from the publisher
when ordered in bulk quantities. For more information contact:

IEEE PRESS Marketing
Attn: Special Sales
PO Box 1331
445 Hoes Lane
Piscataway, NJ 08855-1331
Fax: (908) 981-8062

Printed in the United States of America
10 9 8 7 6 5 4 3 2

ISBN 0-7803-1005-5
IEEE Order Number: PP0354-1

Library of Congress Cataloging-in-Publication Data
Hecht, Jeff.
    Understanding lasers : an entry-level guide / by Jeff Hecht. —
2nd ed.
        p. cm.
    IEEE order number PP0354-1" — T. p. verso.
    Includes bibliographical references and index.
    ISBN 0-7803-1005-5
    1. Lasers. I. Title.
TA1675.H435   1993
621.36'6—dc20
                           93-24203
                           CIP

To my parents, Laura B. Hecht and George T. Hecht, for their love and support over a lifetime.

Contents

Preface

"FOR CREDIBLE LASERS, SEE INSIDE"

Shortly after the laser was invented, a now-defunct Sunday newspaper supplement published an article on "The Incredible Laser." In breathless prose, it promised a future of laser weapons straight out of science fiction. Arthur L. Schawlow, a laser pioneer with a whimsical turn of mind who later shared the 1981 Nobel Prize in Physics, posted a copy of the article on the door of his laser laboratory at Stanford University, adding: "For credible lasers, see inside."

I sometimes suspect more nonsense has been written about lasers than about any other "space age" technology of the last half of the twentieth century. Even the technologically unsophisticated have a general idea of what electronics and spacecraft can do. Ask them about lasers, however, and they may start spouting fantastic visions of lasers as deadly ray guns, or as magical devices that can do almost anything with light. These problems have eased somewhat with the spread of lasers in supermarket checkout systems and compact disc players, and the shift of military research away from high-energy laser weapons. However, they are being replaced with marketing hype that attaches the word "laser" to everything from sailboats to sporty cars. This book is about real-world lasers—to borrow the line from Arthur Schawlow's door, "for credible lasers, see inside." What you will learn about is the exciting world of real lasers—how they work and what they can do.

Back in the 1960s, one irreverent observer called the laser "a solution looking for a problem." There was truth to that assessment, because scientists learned how to make lasers work before engineers learned how to put them *to* work. Now that has changed. Lasers are a part of our lives in more ways than we realize. Listen to a compact disc audio recording, and you hear crystal-clear music read by a laser beam. Make a long-distance phone call, and chances are excellent that your voice travels most of the way via laser beam. Go to the supermarket, and a laser beam will read the prices at the checkout counter. Elsewhere, lasers are doing hundreds of other jobs, in places ranging from hospitals and research laboratories to factories and construction sites.

This book will introduce you to real-world lasers and how they work. It is arranged somewhat like a textbook. Each chapter starts by saying what it will cover, ends by reviewing key points, and is followed by a short multiple-choice quiz. We start with a broad overview of lasers. In the second chapter, we review key concepts of physics and optics that are essential for you to understand lasers. You should skim this even if you have a background in physics, especially to check basic optical concepts. The third and fourth chapters cover key concepts of how lasers work. In the fifth chapter, we take a quick look at all the accessories that help lasers do their jobs better. Try to master each of these chapters before going on to the next and those that follow.

The sixth through ninth chapters go into more detail on the workings of specific types of lasers. We've given special attention to the hottest of the commercial technologies, semiconductor lasers, in Chapter 8, which has been extensively rewritten for this new edition. Two of the most exciting lasers still in the laboratory—free-electron and X-ray lasers—are covered in Chapter 9. Chapters 10 and 11 give a brief overview of laser applications. The Glossary and Index are included to help make this book a useful reference after you have studied it.

To keep this book a reasonable length, we concentrate on lasers and their workings. We cover optics and laser applications only in brief, but after reading this book you may want to study them in more detail.

I met my first laser over two decades ago, and have been writing about laser technology since 1974. I have found it fascinating, and I hope you will, too.

Finally, I owe thanks to people who helped make this book possible. I've benefited in many ways over the years from the help of my friends and colleagues in laser journalism: Howard Rausch, who got me into this crazy business; Jeff Bairstow, Carole Black, Herb Brody, Yvonne Carts, Jim Cavuoto, Bob Clark, Richard Cunningham, Tom Farre, Bonnie Feuerstein, Tom Higgins, Breck Hitz, Teri Kwidzinski, Heather Messenger and Mike Wenyon. At IEEE Press, Dudley Kay found a home for the idea and Karen Miller managed the minor miracle of converting a mass of annotated computer output into a finished book.

J. H.

Understanding Lasers

Introduction & Overview

ABOUT THIS CHAPTER

This chapter will introduce you to lasers. It will give you a basic idea of how they work, how they are used, and what their important properties are. This basic understanding will serve as a foundation for the more detailed descriptions of lasers in later chapters.

LASERS IN FACT AND FICTION

The laser sits near the top of any list of the greatest inventions of the last half of the twentieth century. Together with the satellite, the computer, and the integrated circuit, it is a symbol of "high technology." Like the other technologies, lasers affect our lives in many ways, and are growing steadily in importance. Laser technology is both fascinating in itself and an important tool in fields from medicine to communications. However, few people outside the field understand it well.

Arthur Schawlow posted his sign distinguishing the "incredible" lasers of public misconception from the "credible" lasers of his laboratory in 1962. Over the years, he has given away many copies of the poster, and it has amused many students, laser engineers and scientists. Yet the basic point remains true. Too many people confuse real lasers with science-fictional ray guns or death rays. Military researchers have spent immense sums trying to develop laser weapons, but so far none have reached the battlefield. After years of research, the Pentagon backed away from plans to use high-power lasers to defend against nuclear attack. The only realistic uses of laser weapons in the near future are to attack the targets most vulnerable to light—electro-optical sensors and the human eye.

Real laser technology may not be as spectacular, but it is important and fascinating. Lasers can send signals through miles of fiber-optic cable, print computer output, read printed codes in the supermarket, diagnose and treat disease, cut and weld materials, and make ultraprecise measurements. With laser light you can record three-dimensional holograms, project bright light-show images, spot flaws in a centuries-old painting, or play crystal-clear digital music recorded on a compact disc. This book will teach you how this real laser technology works.

WHAT IS A LASER?

The word laser was coined as an acronym, for *l*ight *a*mplification by the *s*timulated *e*mission of *r*adiation, and those words have special meanings. They tell us that laser light is special light. Ordinary light, from the sun or a light bulb, is emitted spontaneously, when atoms or molecules release excess energy by themselves, without any outside intervention. Stimulated emission is different, because it occurs when an atom or molecule that is holding excess energy is "stimulated" to emit that energy as light. We'll come back to look more closely at stimulated emission in Chapter 3.

Albert Einstein first suggested the possibility of stimulated emission in a paper published in 1917. However, for many years physicists thought that atoms and molecules always were much more likely to emit light spontaneously, and that stimulated emission thus always would be much weaker. It was not until after World War II that physicists began trying to make stimulated emission dominate. They sought ways for one atom or molecule to stimulate many others to emit light, amplifying it to much higher powers.

The first to succeed was Charles H. Townes, then at Columbia University. He did not work with light at first, but with microwaves at much longer wavelengths, building a device he called a *maser*, for *m*icrowave *a*mplification by the *s*timulated *e*mission of *r*adiation. He thought of the key idea in 1951, but the first maser was not completed until a couple of years later. Before long, many other physicists were building masers, and trying to find how to stimulate emission at even shorter wavelengths.

The key concepts emerged about 1957. Townes and Arthur Schawlow, then at Bell Telephone Laboratories, wrote a long paper outlining the conditions needed to amplify stimulated emission of visible light waves. At about the same time, similar ideas crystallized in the mind of Gordon Gould, then a 37-year-old graduate student at Columbia, who wrote them down in a series of notebooks. Townes and Schawlow published their ideas in a scientific journal, *Physical Review Letters*, but Gould filed a patent application.

Well over decades later, people still argue about who deserves credit for the laser concept. Townes and two Soviet maser pioneers, Nikolai Basov and Aleksander Prokhorov, shared the 1964 Nobel Prize in Physics for their pioneering work on "the maser/laser principle." Schawlow eventually shared the 1981 Nobel Physics Prize for research done with lasers. For many years, Gould had to settle only for credit for coining the word "laser" in his notebooks. However, after nearly two decades of legal struggles and comparative obscurity, he finally received four patents worth millions of dollars on important aspects of laser technology. The last of those patents was issued in May 1988, over 30 years after Gould sat down to write his notebooks.

Publication of the paper by Townes and Schawlow stimulated many efforts to make lasers. Schawlow and others went to work at Bell Laboratories. Gould took his ideas to a small Long Island company, TRG Inc., which used them in a research proposal to the Department of Defense. The Pentagon brass was enthusiastic about the idea, but not about Gould's brief involvement with a Marxist study group a decade earlier, so they wouldn't grant Gould the security clearance he needed to work on the project. Scientists at many other government, industry and university laboratories also tried to make lasers.

The winner of the great laser race, on May 16, 1960, was a dark horse, Theodore Maiman. He had decided that synthetic ruby was a good laser material, even though Schawlow had publicly said it wouldn't work. Maiman's managers at Hughes Research Laboratories in Malibu, California had told him to stop wasting his time trying to make a ruby laser. But Maiman continued, and by putting mirrors on each end of a ruby rod and illuminating it with a bright flashlamp, he made the world's first laser. A replica is shown in Fig. 1-1.

Many other types of lasers followed, but Maiman's ruby laser was typical in many ways of what we expect a laser to be. It emitted light in a narrow, tightly concentrated beam. The light was at a single wavelength—694 nanometers (1 nm = 10^{-9} meter) at the red end of the

Fig. 1-1 Theodore Maiman and Irnee J. D'Haenens with a replica of the world's first laser, which they made at Hughes Research Laboratories in 1960. (Reprinted from Hughes Research Laboratories, courtesy of AIP Neils Bohr Library).

visible spectrum. The light waves were coherent, all aligned with each other and marching along in step.

Maiman's laser emitted light in short, intense pulses, and many other lasers also operate in pulsed mode. However, other lasers emit steady beams. Power levels can span a wide range. Some lasers emit less than a milliwatt (0.001 watt) of light; others emit steady beams of many kilowatts. The highest laser powers—trillions of watts—are achieved in ultrashort pulses lasting only about a billionth of a second. Figure 1-2 shows a sampling of commercial lasers.

We think of lasers as emitting visible light, but the basic principles behind the laser are valid at many other wavelengths. We already saw that Townes produced the first stimulated emission of

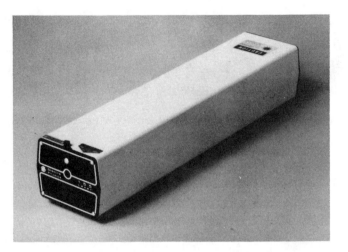

(A) A milliwatt helium-neon laser. (Courtesy of Melles Griot.)

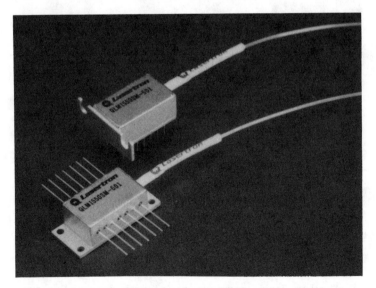

(B) A semiconductor laser packaged for fiber-optic communications.
(Courtesy of Lasertron Inc.)

Fig. 1-2   Representative commercial lasers.

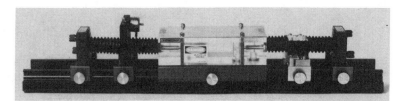

(C) A multiwatt solid-state laser. (Courtesy of Lasermetrics Inc.)

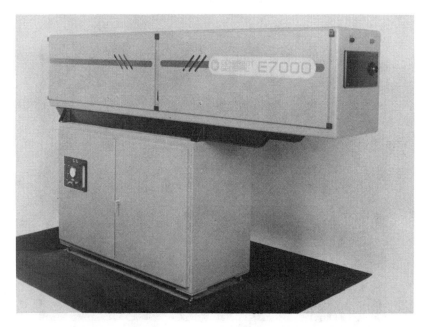

(D) A carbon-dioxide laser designed for industrial materials working.
(Courtesy of Coherent Inc.)

Fig. 1-2 (continued)   Representative commercial lasers.

microwaves. Many lasers emit infrared and ultraviolet light, and researchers now are working on X-ray lasers.

Maiman's first laser was small enough for him to hold in one hand, but modern lasers come in many different sizes. The most common lasers are semiconductor types, the size of a grain of salt, used in compact disc digital audio players and fiber-optic communications. Government laboratories have built monstrous lasers that fill the interiors of good-sized buildings. Other lasers come in many sizes between, including pen-sized laser pointers, benchtop laboratory lasers, and materials-working lasers packaged so they look like heavy machine tools.

We'll learn more about these properties of lasers in later chapters. For now, it is most important to remember that the types of lasers are many and varied.

HOW LASERS ARE USED

In the early 1960s, it seemed that every physicist and engineer wanted to get his or her hands on a laser and play with it. They aimed lasers at just about everything they could find. They shot so many holes in razor blades that for a while laser power was informally measured in "gillettes." Yet there were few practical applications, and for a time the laser seemed to be, in the words of one wag, "a solution looking for a problem."

That is no longer true. Lasers are now used for many purposes in many fields. A sampling of laser applications is listed in Table 1-1; Chapters 10 and 11 describe some of those uses in more detail.

We can get a better feeling for how lasers are used by breaking down major applications into families. One large family of applications uses lasers as highly directional light bulbs for information processing, such as reading, writing, or sending signals. Another uses lasers as precise light sources for a wide variety of measurements. Another puts laser light to work for medical treatment. Laser energy also can be used to change things, from marking symbols on parts to welding thick sheets of steel. Military agencies use lasers for weapon control and simulation, and are studying lasers for use as weapons. Other applications rely on other special properties of laser light, such as the way its light waves are aligned, or the fact that it consists of only a single color.

TABLE 1-1 A sampling of laser applications

Information Handling

Fiber-optic communications
Laser printers for computer output
Optical computing and signal processing
Playing digital audio compact discs
Playing videodisks
Reading and writing computer data on optical disks
Reading printed bar codes for store checkout and inventory control

Measurement and Inspection

Detecting flaws in aircraft tires
Exciting fluorescence from various materials
Illuminating cells for biomedical measurements
Measuring concentrations of chemicals or pollutants
Measuring small distances very precisely
Measuring the range to distant objects
Projecting straight lines for construction alignment and irrigation
Studies of atomic and molecular physics

Medicine

Bleaching of port wine stain birthmarks and certain tattoos
Clearing vision complications after cataract surgery
Dentistry
Laser surgery
Reattaching detached retina
Shattering of stones in the kidney and pancreas
Treatment of diabetic retinopathy to forestall blindness

Materials Working

Cutting, drilling and welding plastics, metals and other materials
Cutting cloth
Drilling materials from diamonds to baby-bottle nipples
Engraving wood
Heat-treating surfaces
Marking identification codes

(continued)

TABLE 1-1 *(Continued)*

Military

Antisatellite weapons
Antisensor and antipersonnel weapons
Range-finding to targets
Simulating effects of nuclear weapons
Target designation for bombs and missiles
War games and battle simulation

Other Applications

Basic research
Controlling chemical reactions
Displays
Holography
Laser light shows
Laser pointers
Producing nuclear fusion
Separating isotopes

Some low-power lasers may seem to do the same job as a light bulb, but they can do it better. For example, supermarkets use laser scanners under checkout counters to read bar codes on packages. The bar code identifies each product to the store's computer, which recalls the price and prints it out at the terminal. Why use a laser? The red laser light is in a thin, highly directional beam, so it can be focused onto packages at various distances above the counter. The scanner includes a filter that cuts out all light other than the single shade of red emitted by the laser, reducing the likelihood of errors caused by overhead lights. Engineers looked at many other alternatives, and decided the laser was best for supermarket checkout. Non-laser light sources may be used when hand-held wands are run directly over the bar code on the surface, keeping out stray light.

Semiconductor lasers are widely used because they are very small and can produce a few milliwatts of light that can be tightly focused to a very small spot. That is ideal for many uses, ranging from playing music on compact discs to sending signals through optical fibers. What's more, by varying the current passing through the laser, you can control how much light it emits—ideal for communications or writing computer output.

Other lasers produce short, intense pulses of light. If these pulses are focused, they can concentrate a lot of energy on a tiny spot. Such pulses can make tiny marks on a metal surface, or even drill holes, particularly if a series of pulses is focused onto the same spot. Because laser pulses do not get dull, they are very good for drilling hard materials like titanium metal or even diamond. They are also very good for drilling into soft objects like rubber or many plastics, because they don't twist, bend or turn the material. A series of pulses, or a higher-power steady beam, can cut or weld. Physicians can carefully control laser energy to remove diseased tissue without damaging healthy cells. Computers can control cutting and drilling lasers in automated machining centers.

The alignment of laser light waves that makes it coherent is ideal for holography, the recording and reconstruction of three-dimensional images. Holograms make striking artistic displays, projecting three-dimensional images that look so real you want to reach out and touch them. They also serve more practical purposes. Holograms are hard to make without special equipment, so they are placed on credit cards to show that the cards are genuine. You can measure tiny deformations in an object by recording two holograms on the same piece of film, one before a stress is applied and one after. Illuminate the film to produce an image, and you can see where the object moved. This lets inspectors spot flaws in aircraft tires before they are put on a plane, so you can ride more safely.

In later chapters, we will delve more into laser applications. We will also see that new applications are being developed steadily. Government and industrial laboratories are working on laser techniques for tasks ranging from treating cancer to controlling nuclear fusion.

IMPORTANT LASER PROPERTIES

Lasers take many different forms and have many differing characteristics, such as their output wavelength and power level. However, different types of lasers often share some common properties, such as the concentration of their output energy in a narrow beam. We list many important properties of lasers below. Each deserves a bit more explanation.

MAJOR LASER PROPERTIES

- Wavelength(s)
- Output power
- Duration of emission (pulsed or continuous)
- Beam divergence and size
- Coherence
- Efficiency and power requirements

Wavelength(s)

As we will see in Chapter 2, light is made of waves, and the length of those waves, or the wavelength, is a fundamental characteristic of the light. We sense the wavelength of visible light as color (although that sensing is not very precise). Each type of laser emits a characteristic wavelength or a small range of wavelengths. The wavelengths depend on the type of material that emits the laser light, the laser's optical system, and the way the laser is energized. Laser action can produce infrared, visible and ultraviolet light; "masers" produce microwaves by essentially the same process. If you want a particular wavelength, your choice is usually limited to no more than a few types of lasers. Table 1-2 lists a sampling of important laser types and their wavelengths in nanometers.

Most lasers are called *monochromatic* (single-colored) and nominally emit only one wavelength. Actually, they emit a range of wavelengths, but the range is narrow enough to be considered a single wavelength for most purposes. Some lasers emit light at different wavelengths under different conditions. For example, the helium-neon laser is best known for its red output at 632.8 nm. However, the same gas mixture can be used with different optics to emit green light at 543 nm, or to emit invisible infrared light at wavelengths as long as 3393 nm.

Some lasers can emit two or more wavelengths at once, called *multiline* operation. In some cases, the wavelengths are close together. The argon laser emits several wavelengths between 450 and 530 nm, with the strongest lines at 514.5 and 488 nm. In others, they are widely separated, as in helium-neon. Some lasers emit many wavelengths in a limited range; one example is the carbon-dioxide laser, with dozens of wavelengths between 9 and 11 µm (9000 to 11,000 nm) in the infrared.

TABLE 1-2 Major wavelengths of important commercial lasers

Type	Wavelength
Argon-fluoride excimer	192 nm
Krypton-fluoride excimer	249 nm
Xenon-chloride excimer	308 nm
Nitrogen gas (N_2)	337 nm
Organic dye (in solution)	320–1000 nm (tunable)
Helium-cadmium	325, 442 nm
Argon-ion	275–303, 330–360, 450–530 nm
Krypton-ion	330–360, 420–800 nm
Helium-neon	543, 632.8, 1150 nm
Semiconductor (GaAlInP family)	630–680 nm
Titanium-sapphire	680–1130 nm (tunable)
Ruby	694 nm
Alexandrite	720–800 nm (tunable)
Semiconductor (GaAlAs family)	750–900 nm
Neodymium-YAG	1064 nm; 532, 355, 256 harmonics
Semiconductor (InGaAsP family)	1200–1600 nm
Hydrogen-fluoride chemical	2600–3000 nm
Carbon-monoxide	5000–6000 nm
Carbon-dioxide	9000–11,000 nm (main line 10,600 nm)

As we will see later, the way lasers emit light tends to concentrate energy at specific wavelengths. However, some lasers can emit light over a fairly wide range of wavelengths. The most important examples are lasers in which light is emitted by organic dyes in solution or certain solid-state crystals. They can be tuned in wavelength by adjusting the laser's optical system.

Output Power

Output power measures the strength of a laser beam, which differs widely among lasers. Strictly speaking, power measures how much energy the laser releases per unit time; it is measured in watts and defined by the formula:

$$Power = \Delta(Energy)/\Delta(Time)$$

One watt is defined as a rate of one joule (of energy) per second.

Laser output powers cover a wide range. Some lasers produce beams containing less than a thousandth of a watt (a milliwatt). Others produce thousands of watts (kilowatts).

You cannot make one laser cover that whole range of output powers just by turning a knob. Some lasers can be adjusted over a limited power range, but others are designed to emit at a stable power level. Some types of lasers cannot be scaled to high power levels; one example is the helium-neon laser, which can emit no more than a few dozen milliwatts. Only a few types, notably carbon-dioxide and chemical lasers, can produce thousands of watts in a steady beam. If you need high laser powers, you often don't have much choice among types of lasers, and thus your choice of wavelengths is limited.

Pulsed and Continuous Output

You may think of lasers as emitting a steady beam of light, like a light bulb. Many do, including the red helium-neon laser you are most likely to have encountered in a laser demonstration (or in a supermarket or construction site). However, others emit pulses of light.

Pulses come in various durations and repetition rates. The length or duration of a pulse can range from milliseconds to femtoseconds—or, in scientific notation, from 10^{-3} to 10^{-15} second. The pulses may be repeated once a minute, or may appear thousands or even millions of times in a second. (Because the eye's response is much slower than the laser's, more than a few dozen pulses per second look continuous to your eye. However, electronic detectors can recognize those short pulses.)

We saw earlier that power is the rate of energy emission. If you want to get the total energy in a pulse, you need to know how the power changes with time. A laser may have extremely high peak power during a short pulse, but because the pulse is short, it doesn't contain much energy. To make a simple approximation:

Pulse Energy = Peak Power × Pulse Length

This works reasonably well as long as you're careful as to how you define pulse length. Strictly speaking, to calculate the exact amount of energy, you need to turn to calculus, and integrate the emitted power over time (t):

$$\text{Energy} = \int \text{Power } dt$$

(The $\int$ is the symbol for integration.) You can best think of this as calculating the area under a plot of power vs. time, as shown in Fig. 1-3. For comparison, the figure shows energy calculated by multiplying power by pulse length, which implicitly assumes the pulse is square, and is a useful approximation.

The average power in a pulsed laser beam differs from the peak power; it is a measure of the average energy flow per second. Thus,

Average Power = (Number of Pulses × Pulse Energy)/Time

Peak and average power can both be important quantities, depending on the laser application. Both are measured in watts, but,

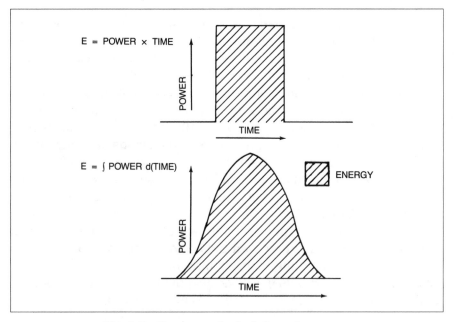

Fig. 1-3   Two ways of calculating pulse energy.

as would be expected, peak power is higher. Pulse energy is measured in joules. The conversion between the two units is

$$\text{Watts} = \text{Joules/Seconds}$$

or

$$\text{Joules} = \text{Watts} \times \text{Seconds}$$

If the pulses come at a constant rate, you can count the number of pulses per second, or repetition rate, giving:

$$\text{Average Power} = \text{Repetition Rate} \times \text{Pulse Energy}$$

Beam Divergence and Size

If you see a laser beam shining through dusty air, it looks as thin as a string or pencil line. However, if you look carefully at the beam, you will see that it spreads out very slowly as it gets farther from the laser. You can see the same effect more readily in the beam from a flashlight or searchlight. This spreading is called *divergence*; it is measured as half the full angle at which the beam spreads.

Laser beam divergence is typically measured in milliradians, or thousandths of a radian, which simplifies calculations because the tangent of a small angle is very close to its size in radians. (You can get away with this simplification for everything but semiconductor lasers.) A radian equals 57.3 degrees; 2π radians equals 360 degrees, or a full circle. Beyond a meter (3 feet) or so from the laser, the beam radius (half the diameter) equals the tangent of the divergence angle times the distance the beam has travelled, as shown in Fig. 1-4. Thus,

$$\text{Radius} = \text{Distance} \times \tan (\text{Beam Divergence})$$

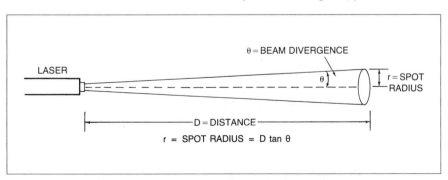

Fig. 1-4   Calculating the size of a laser spot from the beam divergence.

If a 2-milliradian beam travels 10 m, its radius is 0.02 m, and its diameter is 0.04 m. After it has travelled 100 feet, its diameter is 0.4 feet. Not all lasers have such small divergence, but the beams from those that do can be detected at long distances. This high directionality of laser beams is one of their most important properties.

Coherence

Light waves are *coherent* if they are all in phase with one another. Figure 1-5 compares coherent and incoherent light waves. The peaks and valleys of coherent light waves (top) are all lined up with each other. The peaks and valleys of incoherent light waves (bottom) do not line up. Laser light is coherent; light from most other sources, like the sun, an incandescent bulb, or a fluorescent tube, is incoherent.

Take a close look at Fig. 1-5, and you will note something else about coherent light. For light to remain coherent over a long dis-

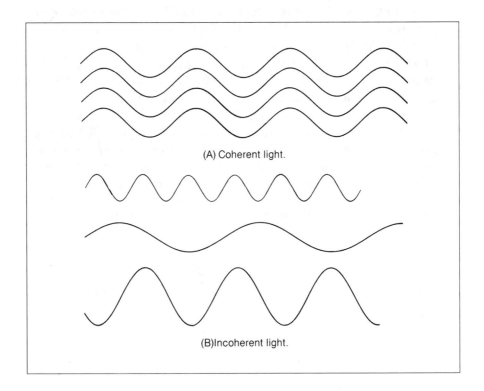

(A) Coherent light.

(B)Incoherent light.

Fig. 1-5   Coherent and incoherent light.

tance, all the waves must have the same wavelength. If one light wave is just 0.1% longer than another, the peaks of one wave will line up with the valleys of the other after 500 wavelengths; thus, the light waves will be out of phase and incoherent. Thus, the more monochromatic a laser is, the more coherent it is. Because the wavelengths of visible light are very small (500 nm or 5×10^{-7} m), it must be quite monochromatic to be coherent over a long distance. If visible light is coherent over 500 wavelengths, that distance is only 2.5×10^{-4} m—a quarter of a millimeter. Monochromatic light need not be coherent, but light that is not monochromatic cannot stay coherent over a long distance.

Many laser applications do not require coherent light; the coherence is incidental. However, coherent light is essential for holography and interferometry, for reasons to be described later.

Efficiency and Power Requirements

Wavelength, output power, coherence and pulse characteristics all are important in picking a laser. Two other factors also enter the picture: efficiency and power requirements. Lasers differ widely in how efficiently they convert input energy (usually electricity) into light. Like other light sources, they are not very efficient in generating light, with the best converting up to about 30% of input energy into light. Many types convert as little as 0.01% or even 0.001% of the input energy into light.

Efficiency becomes a more important consideration at higher output powers. It's not a big problem if a 1-mW laser produces a watt of waste heat, because it's easy to dissipate that little heat. However, it would be very hard to dissipate the million watts of waste heat from a 1-kW (1000-W) laser with the same efficiency.

LASERS, PHYSICS, AND OPTICS

The internal operation of lasers depends on the laws of physics. Laser emission depends on how atoms and molecules emit light, and that, in turn, depends upon their internal structure. To understand that internal structure, and how it affects the ways in which lasers emit light, you will need to take the brief introductory tour of modern physics in Chapter 2.

The way lasers work also depends intimately on optical principles. We will cover some laws of optics in Chapter 2, and touch on optics repeatedly throughout the book. There is no escaping from optics. Lasers are optical devices, and to use lasers you need optics, ranging from the mirrors within the laser itself, to lenses and mirrors that focus and direct laser beams.

WHAT HAVE WE LEARNED

- Laser light is produced by stimulated emission.
- The microwave-emitting maser preceded the laser.
- Over 30 years after the laser was invented, people still argue over who deserves credit.
- Theodore Maiman made the first laser by exciting a ruby rod with a flashlamp.
- The laser was once called "a solution looking for a problem," but it has since found many applications.
- Low-power lasers are good light sources for reading and writing.
- Short, intense laser pulses can drill holes or mark surfaces.
- Different types of lasers share many properties.
- Wavelength is a fundamental characteristic of all light.
- Laser light is usually monochromatic.
- Power is the rate at which light energy flows from a laser.
- Lasers emit different power levels.
- Lasers can emit pulsed or continuous light.
- Average power equals pulse repetition rate times pulse energy.
- A laser beam has a characteristic spreading angle (divergence) and diameter.
- Most lasers emit "coherent" light waves that are in phase with each other and have the same wavelength.
- Efficiency and power requirements are important in picking lasers.
- The internal operation of lasers depends on basic laws of physics.

WHAT'S NEXT

In Chapter 2, we will introduce the basic physical concepts behind lasers. Much of the material is basic, and parts may seem repetitive if you have been exposed to physics before. However, you should at least scan through quickly to make sure you understand the key concepts. They will reappear repeatedly in the following chapters.

Quiz for Chapter 1

1. The word laser originated as:
   a. A military codeword for a top-secret project
   b. A trademark
   c. An acronym for light amplification by the stimulated emission of radiation
   d. The German word for light emitter

2. The first laser was made by:
   a. Charles Townes
   b. Theodore Maiman
   c. Gordon Gould
   d. Nikolai Basov and Aleksander Prokhorov
   e. H. G. Wells

3. The first laser emitted:
   a. Pulses of 694-nm red light
   b. A continuous red beam
   c. Pulses of white light from a helical flashlamp
   d. Spontaneous emission

4. Laser light is which of the following:
   a. Coherent
   b. Stimulated emission
   c. Spontaneous emission

   d. Monochromatic
   e. a, b, and d
   f. c and d

5. Which of the following is NOT a laser application?
   a. Printing computer output
   b. Stimulating rainfall from clouds
   c. Making very precise measurements
   d. Playing compact disc audio recordings
   e. Engraving wood

6. The coherence of laser light is important for:
   a. No practical applications
   b. Drilling holes
   c. Getting laser light to pass through air
   d. Holography
   e. None of the above

7. Which important laser emits light in the visible range, 400 to 700 nm?
   a. Argon-ion
   b. Nitrogen
   c. Carbon-dioxide
   d. Neodymium-YAG
   e. Chemical

8. Which is the proper measurement of average power emitted by a pulsed laser?
   a. Energy × Time
   b. Pulse Energy ×
   Repetition Rate
   c. Pulse Energy /
   Repetition Rate
   d. Peak Power ×
   Pulse Length
   e. None of the above

9. How do you calculate the radius of a laser spot at a given distance if you know the beam divergence?
   a. Multiply the beam divergence in degrees by the distance in milliradians.
   b. Divide the beam divergence in degrees by the distance in meters.
   c. Measure it with a ruler.
   d. Multiply the sine of the beam divergence by the distance in meters.
   e. Multiply the power in watts by the beam divergence.

10. What type of light can be coherent?
   a. Spontaneous emission
   b. Monochromatic and in-phase
   c. Narrow beam divergence
   d. Monochromatic only

CHAPTER 2

Physical Basics

ABOUT THIS CHAPTER

Lasers are among the many developments arising from the modern physics that emerged in the early twentieth century. To understand lasers, you need a basic understanding of light, optics, quantum mechanics and some other aspects of physics. This chapter will give you the basic background in modern physics you need to understand how lasers work.

ELECTROMAGNETIC WAVES AND PHOTONS

Early physicists had long and loud debates over the nature of light. Isaac Newton held that light was made up of tiny particles. Christian Huygens believed that light was made up of waves, vibrating up and down in a direction perpendicular to the direction the light travels. Newton's theory came first, but Huygens' theory explained early experiments better, so for a long time it was assumed to be right.

Today, we know that both theories are right—in part. Much of the time, light behaves like a wave. It is called an *electromagnetic* wave because it is composed of both electric and magnetic fields. As shown in Fig. 2-1, both those fields oscillate perpendicular to the direction in which the light wave is travelling, and perpendicular to each other. Light waves are called transverse waves because they oscillate in a direction transverse (i.e., perpendicular) to the direction in which they travel.

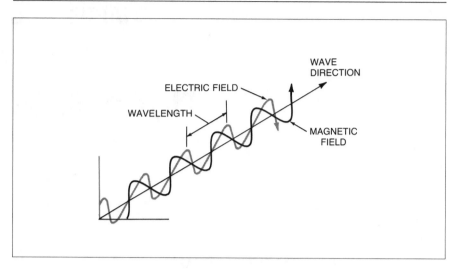

Fig. 2-1   Structure of an electromagnetic wave.

Waves can be identified by certain characteristics. One is wavelength, the distance between successive peaks, illustrated in Fig. 2-1; it is measured in units of length. Another is frequency, the number of wave peaks passing a point in one second. Frequency is measured in cycles (peaks) per second, a quantity called hertz (named after Heinrich Hertz, the 19th-century discoverer of radio waves). Multiply the two units together, and you get the wave's speed in units of length (or distance) per second:

$$\text{Wavelength} \times \text{Frequency} = \text{Speed}$$

The speed of light in a vacuum is a universal constant, about 300,000 km/s or 186,000 miles/s, so this product is always the same for light and other electromagnetic waves. This means that the wavelength is inversely proportional to frequency—the shorter the wavelength, the higher the frequency. We will see later that wavelength and frequency affect the properties of electromagnetic waves; different types of electromagnetic waves have different wavelengths and frequencies.

If we use the standard symbols for wavelength (λ), frequency (ν), and the speed of light (c), we can write simple equations to relate wavelength and frequency (assuming, for the moment, that we're working in a vacuum):

$$\lambda = c/\nu$$

and

$$\nu = c/\lambda$$

At other times, light acts as if it is made up of particles called photons. A photon is a chunk or quantum of energy. Although Einstein's famous $E = mc^2$ equation tells us that mass and energy are equivalent, photons are massless particles, and only massless particles can travel at the speed of light.

Each photon has an amount of energy proportional to the frequency of the light wave, and thus to the wavelength. Photon energy is related to frequency by the formula

$$\text{Energy} = \text{Planck's Constant} \times \text{Frequency}$$

or

$$E \text{ (in joules)} = h\nu = 6.63 \times 10^{-34} J - s \, \nu \text{ (Hz)}$$

where Planck's constant (h) equals 6.63×10^{-34} joule-second; it was named after German physicist Max Planck, who discovered the formula. Combine this formula with the relationship between wavelength and frequency, and you have another expression for wavelength:

$$\text{Wavelength} \times \text{Energy} = \text{Planck's Constant} \times \text{Speed of Light}$$

or

$$\lambda E = hc = 1.99 \times 10^{-25} \text{ joule-meter}$$

As you work with light and other electromagnetic waves, you will come to remember these conversions. For now, the most important lesson is that three descriptions of light waves are equivalent. Each light wave can be described by a photon energy, a wavelength, or a frequency. Different numbers describe the same light. For example, light with a 1-μm wavelength has a frequency of 3×10^{14} hertz and photon energy of 2×10^{-19} joule. Physicists often measure photon energy in units called electronvolts, the energy an electron acquires when it moves through a potential of 1 V. One electronvolt equals 1.6022×10^{-19} joule, so a photon energy of 2×10^{-19} joule equals 1.24 electronvolts, a more convenient unit of measurement.

It is easy to make mistakes in converting units, so it helps to remember these simple rules of thumb:

- The higher the frequency, the shorter the wavelength
- The higher the frequency, the larger the photon energy
- The shorter the wavelength, the larger the photon energy

Because various objects emit or radiate electromagnetic waves (or photons), physicists sometimes call them *electromagnetic radiation*. If the word "radiation" sounds distressingly like something that comes from a leaky nuclear reactor, blame it on the way headline writers tend to throw away adjectives. The word "radiation" covers a very broad range of phenomena, only a few of which are dangerous. Most, like light, infrared radiation, and radio waves, are innocuous parts of the electromagnetic spectrum.

The Electromagnetic Spectrum

We usually think of the spectrum as the rainbow of colors spread out when we pass sunlight through a prism. Those colors are how our eyes see different wavelengths; the sensation of red means longer wavelengths than orange or yellow, for example. Our eyes can see only a narrow range of wavelengths, but the whole electromagnetic spectrum covers a tremendous range, from extremely low frequency waves many miles long to gamma rays a trillionth of a meter long. This range is tabulated in Table 2-1.

In Table 2-1, all frequency and wavelength values are expressed in the same (metric) units to simplify comparison. However, in practice the values of frequency, wavelength, and other quantities such as time and power are expressed in metric units with the standard pre-

TABLE 2-1 **Wavelengths and frequencies of electromagnetic radiation**

Name	Wavelengths (m)	Frequencies (Hz)
Gamma rays*	Under 3×10^{-11}	Over 10^{20}
X-rays*	$3 \times 10^{-11} - 10^{-8}$	3×10^{16} to 10^{20}
Ultraviolet light*	10^{-8} to 4×10^{-7}	7.5×10^{14} to 3×10^{16}
Visible light	4×10^{-7} to 7×10^{-7}	4.2×10^{14} to 7.5×10^{14}
Infrared light	7×10^{-7} to 10^{-3}	3×10^{11} to 4.2×10^{14}
Microwaves	10^{-3} to 0.3	10^{9} to 3×10^{11}
Radio waves	0.3 to 30,000	10^{4} to 10^{9}
Low-frequency waves*	Over 30,000	Under 10,000

*No standard boundaries.

fixes shown in Table 2-2. You probably already know some of these prefixes, but you are likely to discover others as you explore the world of lasers. Virtually everything optical is measured in metric units, and we will follow that practice in this book. Other books often give the wavelength of light in angstroms (Å); 1 Å = 10^{-10} m or 0.1 nm. We will follow the now standard system of metric units that does not include the angstrom. If you encounter angstroms elsewhere, just divide by 10 to convert to nanometers.

The division of the electromagnetic spectrum into various domains is somewhat arbitrary, because different parts of the spectrum were discovered piecemeal. Early physicists discovered infrared and ultraviolet light by making measurements beyond the part of the spectrum visible to the human eye. Heinrich Hertz independently discovered radio waves in the 19th century. X-rays and gamma rays were also discovered independently. Eventually, physicists realized that all those waves were part of the same family of electromagnetic waves. However, some fuzziness remains at the borderlines, especially where little research has been done, such as the border between X-rays and ultraviolet.

Even some boundaries that might seem well defined—the limits of visibility to the human eye—are not rigidly defined. The eye's sensitivity to light drops off gradually with wavelength, not abruptly, especially in the infrared. The limits of visibility are usually defined as 400 to 700 nm (or, equivalently, 4000 to 7000 Å), as shown in Fig. 2-2, but the eye can faintly sense somewhat longer and shorter wavelengths. Thus, some books define the visible range as extending from 380 nm to 750 or 760 nm.

TABLE 2-2 Prefixes used for metric units

Prefix	Abbreviation	Meaning	Number
Tera	T	Trillion	10^{12}
Giga	G	Billion	10^9
Mega	M	Million	10^6
Kilo	k	Thousand	1,000
Deci	d	Tenth	0.1
Centi	c	Hundredth	0.01
Milli	m	Thousandth	0.001
Micro	μ	Millionth	10^{-6}
Nano	n	Billionth	10^{-9}
Pico	p	Trillionth	10^{-12}
Femto	f	Quadrillionth	10^{-15}

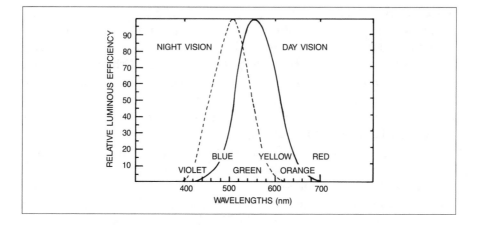

Fig. 2-2   Sensitivity of the human eye to different wavelengths.

Strictly speaking, you might expect the term "light" to apply only to the part of the electromagnetic spectrum sensed by the human eye. However, in practice the definition is broader, and includes much of the infrared and ultraviolet. The reason for the broader definition is that the near-ultraviolet and near-infrared parts of the spectrum behave much like visible light, except for their invisibility to the human eye. The same types of optics used for visible light can also be used in the adjacent parts of the infrared and ultraviolet. Because of this common behavior, this part of the spectrum is sometimes called the "optical" region.

The properties of electromagnetic waves change with wavelength. The farther the wavelength is from the visible, the more differently it behaves. For example, air strongly absorbs ultraviolet wavelengths shorter than about 200 nm. The longest infrared wavelengths (often called "submillimeter" waves because their wavelengths are slightly under a millimeter) are reflected by wire-mesh screens, can travel through waveguides and otherwise behave more like microwaves than visible light. Indeed, if you look at their place in the spectrum, they should behave more like microwaves. Submillimeter waves are a factor of 10 to 100 shorter than microwaves, but a factor of 1000 longer than visible light.

Lasers normally operate at infrared, visible and ultraviolet wavelengths. Similar devices operating at microwave wavelengths are called "masers," for Microwave Amplification by the Stimulated Emission of Radiation. As we saw earlier, Charles Townes invented the maser before anyone had a clear idea of how to build a laser. Recently, scientists at the Lawrence Livermore National Laboratory have pushed the record for the shortest laser wavelength to about 4 nm, in what is called the "soft X-ray" region. However, such laboratory devices only work under unusual conditions and are not yet ready for practical use, as we will see in Chapter 9.

Light Is a Form of Energy

We saw earlier that light and all other electromagnetic radiation are forms of energy. In looking at light energy, it is helpful to think of light as photons, chunks of energy. The higher the frequency or shorter the wavelength, the more energy the photon carries.

Matter absorbs and emits photons. When it absorbs a photon, it gains energy; when it emits or radiates a photon, it releases energy. The amount of energy can be increased by increasing the photon energy, increasing the number of photons, or both. That is, you can get 10 electronvolts of energy from 10 one-electronvolt photons, one 10-electronvolt photon, five two-electronvolt photons, two five-electronvolt photons, or any other combination adding up to 10 electronvolts.

As we will see later in this chapter, the real world is not quite that simple. Materials emit and absorb light most efficiently at specific wavelengths or wavelength ranges. That means that, in practice, you can't count on things absorbing or emitting energy at particular wavelengths—you have to take what you can get. A material may emit 10-eV photons, but not 1-eV photons.

Properties of Light

Some ways materials absorb and emit light energy show the particle side of light's dual personality. Certain materials emit electrons when light strikes them in a vacuum. If you change the wavelength of light, you find that the materials emit no electrons until the wavelength is shorter than a threshold value, which depends on the material. At shorter wavelengths, the number of electrons emitted depends on the light intensity. However, if the wavelength is longer than that threshold level, increasing intensity won't cause the material to emit electrons. This is called the photoelectric effect, and to understand it you need to realize that light energy is packaged as photons. As we saw previously, photon energy increases with frequency (or decreases with wavelength). It takes a certain amount of energy to free the electron from the atom, and the photon can't do the job unless it has at least that much energy. Albert Einstein first explained the photoelectric effect in 1905, laying the groundwork for quantum mechanics. (Interestingly, Einstein received his Nobel Prize in Physics for this work, not for his better-known theory of relativity.)

Another optical phenomenon reveals the wave nature of light. The basic idea is shown in Fig. 2-3, where a bright light illuminates two parallel slits. Light that passes through the slits forms a pattern of bright and dark bands. The explanation for this seemingly mystifying effect—called *interference*—lies in the wave nature of light.

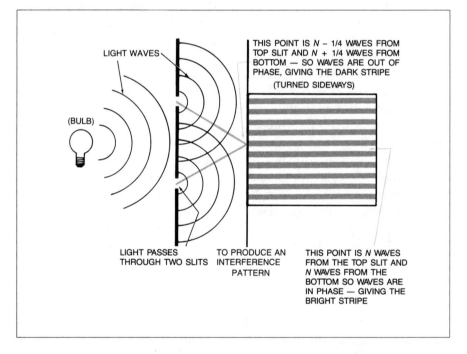

Fig. 2-3   Bright light illuminating two slits causes interference.

When two waves come together, they add in amplitude, as shown in Fig. 2-4. If both have their peaks and valleys at the same point (as in Fig. 2-4A), the two add to form a wave with larger amplitude, producing a bright spot by what is called *constructive interference*. However, if the peak of one light wave comes at the same point as the trough of the other, the negative amplitude of one adds to the

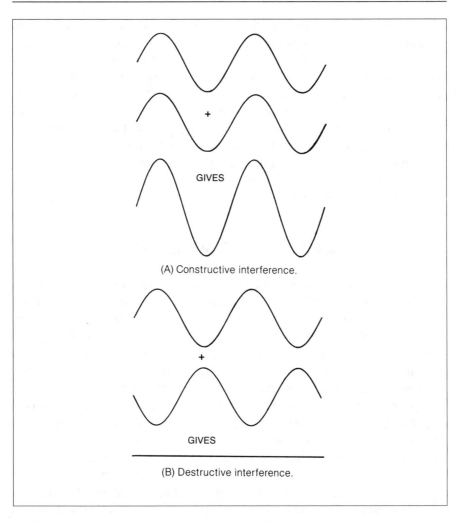

Fig. 2-4   Addition and subtraction of light-wave amplitudes causes interference.

positive amplitude of the other, cancelling it out. If the two waves have equal amplitudes, the result is a null point with zero amplitude. This *destructive interference* gives the dark stripe in the interference pattern. (The energy does not vanish; it just appears somewhere where there is constructive interference.)

How does the two-slit pattern produce the interference stripes shown in Fig. 2-3? Light from one source illuminates the two slits, so identical light waves emerge from each of them. The distance from the slit to the screen where the stripes show is fixed, but the distance light travels to any point on that screen depends on the angle from the slit to that point. Suppose that the central point of the screen is n light waves from each slit and the peaks of the two waves line up with each other, to give constructive interference and a bright stripe. If you move a little to one side, it will be $n + 1/4$ wave from one slit and $n - 1/4$ wave from the other. That means that the two waves will come together 1/2-wave out of phase, and the peaks of one will line up with the valleys of the other, giving a dark stripe.

Interference is a very important property of light and other electromagnetic waves. We will come upon it again and again as we learn more about lasers and optics.

QUANTUM AND CLASSICAL PHYSICS

Einstein's use of photons to explain the photoelectric effect was a key step in the development of quantum mechanics, and quantum mechanics is implicit in our understanding of lasers. Quantum mechanics differs in important ways from the "classical" physics described by Isaac Newton's laws of motion. Classical physics assumes that energy can vary smoothly, so atoms or molecules could have any amount of energy. Quantum mechanics recognizes that energy comes in discrete chunks or quanta, and that atoms, molecules and everything else in the universe can only have discrete amounts of energy. In classical physics, energy can change continuously; in quantum mechanics, it changes in steps. The wave picture of light is classical; the photon picture is quantum-mechanical.

The laser is inherently a quantum-mechanical device, because its operation depends on the existence of light quanta, as we will see in the next chapter. To understand the workings of the laser, we have to look at photons and atoms. In particular, we must look at the quantization of energy levels within atoms. That quantization is a cornerstone of both laser physics and quantum mechanics.

ENERGY LEVELS

In our macroscopic world, energy might seem to vary continuously, like the level of sand in a pail. If we look closely at the sand, we can see that it is made up of many separate grains, and that we can add or subtract only one grain at a time. Likewise, atoms and molecules can only have certain amounts of energy. We call these energy states or levels.

For starters, let's look at the simplest atom, hydrogen, in which one electron circles a nucleus that contains one proton. At first glance, the hydrogen atom looks like a very simple solar system, with a single planet (the electron) orbiting a star (the proton). The force holding the atom together is not gravity but the electrical attraction between the positive charge of the proton and the negative charge of the electron. In a real planetary system, the planet could orbit at any distance from the star. That is not true for the hydrogen atom. The electron can occupy only certain orbits, as shown in Fig. 2-5. We show the orbits as circles for simplicity, but we can't really measure exactly what the orbits look like. Their nominal sizes actually depend on a "wavelength" assigned to the electron, because matter, too, can sometimes act like a wave. The innermost orbit has a circumference of length equal to one wavelength, the next orbit a circumference of two wavelengths, and so on.

If we add energy to our simple planetary system, the planet moves farther from the star. The same happens in the hydrogen atom. As you add more energy, the electron moves to more and more distant

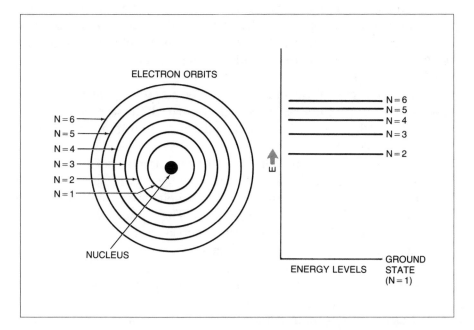

Fig. 2-5 Electron orbits and the corresponding energy levels of the hydrogen atom.

orbits. However, there is a crucial difference between the behavior of a hydrogen atom and our imaginary planetary system. The planet can be at any distance from the sun, or could even fall into it. However, the electron can only occupy certain orbits or energy levels. These energy levels are plotted at the side of Fig. 2-5, with labels indicating the corresponding orbits. The atom is in its lowest possible energy level—the ground state—when the electron is in the innermost orbit, closest to the proton. (Note that the electron does not fall onto the proton.)

The spacing between energy levels in the hydrogen atom is smaller for energy levels higher above the ground state. Eventually, the differences become vanishingly small. If the electron gets too much energy, it escapes from the atom altogether, a process called *ionization*. If we define the energy of the ionized hydrogen atom as zero, we can write the energy of the atom E as a negative number using the simple formula:

$$E = -R/n^2$$

where R is a constant (2.179×10^{-18} joule) and n is the quantum number of the orbit (counting outwards, with 1 the innermost level).

The hydrogen atom is the standard model used to explain quantum mechanics and energy levels because its simplicity shows electronic energy levels clearly. The picture is more complicated in atoms with more electrons.

The energy level occupied by an electron can be specified as a set of quantum numbers. Physicists interpret these as identifying shell, subshell, position in a shell, spin, and other quantities, which we won't explore in detail.

Each electron in an atom must have a unique set of quantum numbers (i.e., occupy a unique quantum state) according to the Pauli exclusion principle. These shells and subshells are usually identified by a number and letter, such as $1s$ or $2p$. The number is the primary quantum number. The letter identifies subshells formed under complex quantum-mechanical rules. Each shell can contain two or more electrons, but each electron must have a unique set of several quantum numbers. In the simplest case of two-electron shells, the two electrons spin in opposite directions.

If atoms are in their ground state, electrons fill in energy levels from the bottom up—that is, from the innermost shell—until the number of electrons equals the number of protons in the nucleus. The primary quantum number of the last ground-state electron corresponds to the row that the element occupies on the periodic table. Table 2-3 shows how electronic energy levels fill up for elements with increasing numbers of electrons, for elements up through strontium (atomic number 38).

TABLE 2-3 Electronic energy levels or orbitals, by increasing energy, up to element 38 (strontium).

1s								
1s	2s							
1s	2s	2p						
1s	2s	2p	3s					
1s	2s	2p	3s	3p				
1s	2s	2p	3s	3p	4s			
1s	2s	2p	3s	3p	4s	3d		
1s	2s	2p	3s	3p	4s	3d	4p	
1s	2s	2p	3s	3p	4s	3d	4p	5s

				Numbers of Electrons in Shell				
2	2	6	2	6	2	10	6	2

These energy levels are the same ones that play a crucial role in chemistry, determining the chemical behavior of the elements and the structure of the periodic table. The elements react with others to fill their outermost electron shell. Atoms with filled outer shells, such as the rare gases neon, helium, krypton, and xenon, tend not to react with other atoms. On the other hand, atoms such as sodium, with a single outer electron, or chlorine, with an outer shell missing just one electron, are highly reactive because of their tendency to lose or gain electrons to have a full outer shell.

The energy levels that electrons occupy in atoms are simple to describe, but there are also many more types of energy levels. Molecules, like atoms, have electronic energy levels, which increase in complexity with the number of electrons and atoms in the molecule. Atomic nuclei themselves have energy levels, although the energies involved are well above those of electrons. Molecules have energy levels that depend on vibrations of atoms within them and on rotation of the entire molecule. All these energy levels are quantized, so atoms or molecules may have many sets of discrete energy levels.

Transitions and Spectral Lines

Nothing would happen in the universe—and we wouldn't be around to wonder why—if atoms and molecules couldn't shift from one energy level to another. The process of making changes or "transitions" between energy levels is critical to laser physics. To look at that process, let's start again with the hydrogen atom.

The electron needs to gain energy to move from the ground state to a higher energy level. Conversely, it must release energy when it drops from a higher level to a lower one. One convenient way (but by no means the only way) for the electron to absorb or release energy is in the form of photons, quanta of electromagnetic energy. The photon energy, the amount of energy absorbed or released, equals the difference in energy between the two energy levels.

Suppose we want to move the electron one step up the energy-level ladder, from the ground state to the first excited level. To do so, we must give the electron exactly as much energy as the energy difference between the ground level and the first excited state. Conversely, for the electron to drop from the first excited level to the ground state, it must release a photon with exactly that same difference of energy between the two levels. In short, the photon energy equals the transition energy. The same is true for any transition: the energy of the absorbed or emitted photon equals the energy difference between the initial and final states. Because the energy associated with each state is constant (except in certain cases that don't matter here), there are only a limited number of possible transition energies. Thus, an electron in a hydrogen atom can emit or absorb light at only certain wavelengths.

Figure 2-6 shows wavelengths corresponding to transitions between the ground state of the hydrogen atom and higher energy levels. The longest wavelength, 121.6 nm, corresponds to the 10.15-electronvolt transition energy between the ground state and first excited level. The next-longest wavelength, 102.6 nm, is the transition between the ground state and second excited state. The wavelengths of transitions to higher energy levels are even shorter, but they eventually reach a limit of 91.2 nm. Light of that wavelength has enough energy to ionize the hydrogen atom, completely removing the electron.

The series of wavelengths in Fig. 2-6 are called the Lyman lines, after American physicist Theodore Lyman, who discovered them in the laboratory. There are similar series of transitions at longer wave-

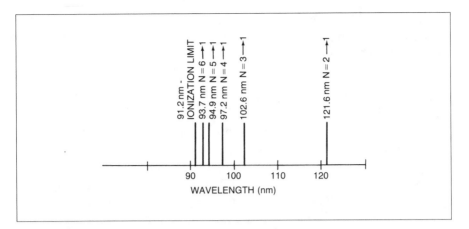

Fig. 2-6   Lyman lines of hydrogen, showing the energy levels involved.

lengths involving the first, second and subsequent excited levels. The
wavelengths (λ) of all these transitions are given by the formula:

$$\text{1/Wavelength} = \text{Constant}\left(\frac{1}{m^2} - \frac{1}{n^2}\right)$$

or

$$1/\lambda = R\left(\frac{1}{m^2} - \frac{1}{n^2}\right)$$

where m is the number of the lower energy level, n the number of the
higher level (the numbers start from 1 for the ground state), and R is
the Rydberg constant, 109,678 cm^{-1}. (The strange units of "per
centimeter" or inverse centimeters give the right dimensions to the
answer, which comes out in centimeters. To convert it to meters,
divide by 100.)

This neat ordering of energy levels is evident only in the hydrogen atom, which has only one electron. Add more electrons, and the
energy-level picture becomes much more complicated. Electrons
interact with each other and with the nucleus, shifting energy levels
slightly, making the energy level structure much more complex
and allowing more transitions between energy levels. The more
transitions, the larger the number of possible spectral lines. Superimpose them all on a single spectrum and they look almost like a set
of random lines.

A further complication is that all transitions are not equally likely. Transition probabilities depend on the differences between the

quantum states. In addition, more atoms are in some states than in others. As a result, absorption or emission is much stronger at some wavelengths than at others.

Types of Transitions

So far we have concentrated on electronic transitions, partly because we picked the hydrogen atom as our introductory example. However, there are other types of energy-level transitions, as shown in Table 2-4, including a variety of electronic transitions. As the table shows, each type has its own energy range. The electronic transitions most important in laser physics are in the middle of the range of possibilities, involving electrons in the outer or "valence" shells of atoms or molecules. These transitions occur at ultraviolet, visible, or infrared wavelengths from about 100 nm in the ultraviolet through to near-infrared wavelengths.

Some electronic transitions occur at longer and shorter wavelengths, but physicists usually classify them in special categories. The shortest wavelengths come from inner-shell electronic transitions in heavy elements, which involve much more energy than outer-shell transitions, and produce X-ray wavelengths. On the other hand, transitions between high-lying electronic energy levels (say, levels 18 and 19 of hydrogen) involve very little energy, putting them deep in the infrared, microwave, or even radio-frequency range; these are called *Rydberg transitions*.

TABLE 2-4 Representative types of transitions and their wavelengths

Transition	Wavelengths	Spectral Range
Nuclear transition	0.0005–0.1 nm	Gamma ray
Inner-shell electronic in heavy element	0.01–10 nm	X-ray
Electronic (Lyman-alpha in H)	121.6 nm	Ultraviolet
Electronic (argon-ion laser)	488 nm	Visible, green
Electronic (H, level 2-3)	656 nm	Visible, red
Electronic (neodymium laser)	1064 nm	Near-infrared
Vibrational (HF laser)	2700 nm	Infrared
Vibrational (CO_2 laser)	10,600 nm	Infrared
Electronic Rydberg (H, level 18-19)	0.288 mm	Far-infrared
Rotational transitions	0.1–10 mm	Far-infrared to microwave
Electronic Rydberg (H, level 109-110)	6 cm	Microwave
Hyperfine transitions (Interstellar H gas)	21 cm	Microwave

Neither Rydberg transitions nor X-ray emission are likely events under normal laser conditions. Normally, very few atoms or molecules are in the high-lying states involved in Rydberg transitions. (Interestingly, some Rydberg lines normally not seen on earth appear as emission lines from interstellar hydrogen, where conditions are greatly different.) Likewise, the conditions needed to break the energetic bond between inner-shell electrons and atomic nuclei are rarely met. However, it is possible to produce extremely high energy concentrations by focusing very powerful laser pulses; and, as we will see in Chapter 9, the development of X-ray lasers now is on the research frontier.

Transitions between nuclear energy levels can produce even higher-energy photons, called gamma rays. In practice, the wavelengths of nuclear and inner-shell-electron transitions can overlap, which has led to some fuzziness in defining the boundary between X-rays and gamma rays. A few researchers are trying to develop gamma-ray lasers, but they face formidable obstacles.

On the other end of the wavelength spectrum are transitions between vibrational and rotational energy levels of molecules. Vibrational transition energies typically correspond to wavelengths of a few to tens of micrometers; rotational transitions have less energy, typically corresponding to wavelengths of at least 100 μm. Laser action can occur on both vibrational and rotational transitions.

Transitions in two or more types of energy levels can occur at once. For example, a molecule can undergo a vibrational transition and a rotational transition simultaneously, with the resulting wavelength close to that of the more energetic vibrational transition. Many infrared lasers emit families of closely spaced wavelengths on such vibrational–rotational transitions.

Remember, in considering transitions, that longer wavelengths correspond to lower energy and shorter wavelengths to higher energy. A visible transition has much larger energy than a rotational transition, even though the rotational wavelength is much larger. A combination of a rotational transition and a vibrational transition thus has a wavelength slightly shorter than the original vibrational transition.

Transition energies or frequencies add together in a straightforward manner:

$$E_{1+2} = E_1 + E_2$$

where E_{1+2} is the combined transition energy, and E_1 and E_2 are the energies of the separate transition. Exactly the same rule holds for frequencies, with ν substituted for the energy. However, wavelengths (λ) of combined transitions add by an inverse rule:

$$1/\lambda_{1+2} = 1/\lambda_1 + 1/\lambda_2$$

or

$$\lambda_{1+2} = 1/(1/\lambda_1 + 1/\lambda_2)$$

Stimulated and Spontaneous Emission

We have seen how an atom or molecule makes a transition to a higher energy level when it absorbs light and to a lower energy level when it releases or emits light. Light absorption can occur when a photon of the right energy is at the right place (the atom or molecule) at the right time (when the atom or molecule is in the lower energy level of that transition). The atom or molecule must be in the right energy state. For example, our simple hydrogen atom will absorb a photon with the transition energy from level 2 to level 3 only if it is in level 2, not if it is in level 1.

Emission is also a complex process, and a critical one for laser physics. Albert Einstein took the first step along the road to the laser when he realized that there could be two types of emission, spontaneous and stimulated, as shown in Fig. 2-7.

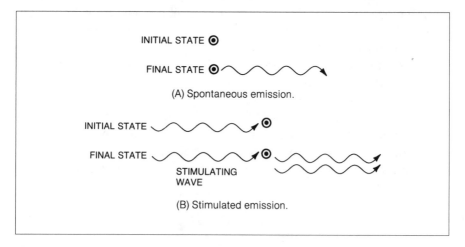

Fig. 2-7   Spontaneous and stimulated emission.

Spontaneous emission (Fig. 2-7A) occurs all by itself and is the source of virtually all light we see in nature from sources including the sun, stars, television sets, incandescent bulbs and fluorescent lamps. An atom or molecule in an energy level above the lowest possible level (or ground state) can drop to a lower level spontaneously, without outside intervention. It can release the excess energy as a photon of light (or in other ways, such as transferring vibrational energy to other atoms, which we won't consider here). Excited atoms or molecules have a characteristic spontaneous emission lifetime (t_{sp}), the average time that they remain in the upper energy level before spontaneously emitting a photon and dropping to the lower level.

Einstein proposed that there could be a second type of emission, stimulated emission. Suppose you had the same excited atom, but this time illuminated it with photons having an energy that exactly matched the transition to the lower state. One of those photons could stimulate the excited atom to emit light on that transition. Looking at the stimulated emission as a wave, it would have precisely the same wavelength and be precisely in phase with the light wave that stimulated it.

The wave viewpoint can help show why stimulated emission has the same wavelength and phase as the light that stimulated it. The light wave stimulates the excited atom or molecule to oscillate at the light-wave frequency, which as we saw earlier equals the energy difference between the excited state and the lower level of the transition. That oscillation amplifies the original light wave, an effect we see as stimulated emission when the stimulated atom or molecule drops to the lower level. (This is still oversimplified from the quantum-mechanical viewpoint, but you really don't want to worry about the details.)

We saw earlier that the word "laser" was coined as an acronym for light amplification by the stimulation emission of radiation. We have seen that, like a laser beam, stimulated emission is all at the same wavelength and in phase—or coherent. But we haven't made a laser beam yet. It took decades for physicists to clear the crucial hurdle needed to amplify stimulated emission.

Population Inversions

The problem with stimulated emission is that it doesn't work very well under what physicists often consider "normal" conditions, thermodynamic equilibrium. At equilibrium, atoms and molecules tend to be at their lowest possible energy level. This isn't precisely the ground state, because they always have some thermal energy at temperatures above absolute zero. However, the tendency of atoms and molecules to drop to lower energy levels creates a problem in what is called *population*—the number of atoms or molecules at each energy level. At thermal equilibrium, the number decreases as the energy level increases, as shown in Fig. 2-8.

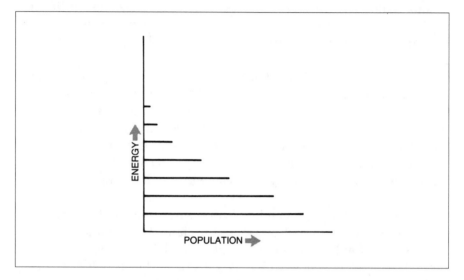

Fig. 2-8   Relative populations of energy levels as a function of energy above the ground state at thermal equilibrium.

The ratio of the numbers of atoms or molecules in states 1 and 2 in thermodynamic equilibrium is given by the equation:

$$N_2/N_1 = \exp[-(E_2 - E_1)/kT]$$

where N_1 and N_2 indicate the numbers of atoms in the two states, E_1 and E_2 are the energies of the two states, k is a constant (the Boltzmann constant) and T is the temperature in Kelvin (the absolute scale). Under normal conditions, this ratio is quite small for transition energies corresponding to optical wavelengths. The Boltzmann constant is 1.38×10^{-23} joule/Kelvin, or 8.6×10^{-5} electronvolt/Kelvin. Plug in the numbers for room temperature (300 Kelvin) and a transition of 1 eV (slightly longer than 1000 nm) and you get a ratio of 1.5×10^{-17}. That means that virtually all atoms or molecules are in the ground state for a visible-wavelength transition at thermodynamic equilibrium.

Why does this make stimulated emission difficult? Let's consider a situation that at first might seem much easier: a collection of atoms with only two energy levels that are in an equilibrium population distribution when there are three times more atoms in the lower level than in the upper level. If one atom drops from the excited state and emits light spontaneously, that light could stimulate emission from another excited atom. However, if it hits an atom in the lower level first, it will be absorbed instead. The problem is that there are three atoms in the lower level for every one in the upper level. Thus, three of every four spontaneously emitted photons hit atoms in the lower state and are absorbed. Only a quarter of the photons encounter an excited-state atom that they can stimulate to emit light, and that stimulated emission is also likely to be absorbed. Thus, stimulated emission doesn't get anywhere.

There is a way to make stimulated emission dominate. If more atoms are in the excited state than in the lower level, photons are more likely to stimulate emission than to be absorbed. Such a condition is called a *population inversion*, because it is the inverse of the normal situation where more atoms are in lower levels than in higher levels. When there is a population inversion, stimulated emission can produce a cascade of light. The first spontaneously emitted photon can stimulate the emission of more photons, and those stimulated-emission photons can likewise stimulate the emission of still more photons. As long as there are more atoms in the upper level than the lower level, stimulated emission is more likely than absorption, and the cascade of photons grows, or is amplified. Once the ground-state population becomes larger, the population inversion ends, and spontaneous emission again dominates.

The problem of producing the population inversion needed for stimulated emission to dominate proved a major hang-up for physicists. Einstein's proposal, made during World War I, was considered mostly a matter of theoretical interest, because the conventional wisdom then said that thermal equilibrium was the normal state of matter. Anything else was unstable and wouldn't last long. The terminology that physicists used then was symbolic of that fact: they called a population inversion a "negative temperature." (The rationale came from working backwards through the Boltzmann equation, which gives a negative value for T if the higher energy level has a larger population.)

It was not until after World War II that physicists began seriously thinking of how to produce population inversions. We'll talk more about that in the next chapter, when we describe more about the workings of the laser.

INTERACTIONS OF LIGHT AND MATTER

So far, we have talked about light in fairly abstract terms. Let's take a break from theory to look at a more practical side of the laser picture: optics and the interactions between light and matter. We can't go into extensive detail here, but it is important to have a general knowledge of these fields to understand how lasers work and how they are used.

We can group objects into three classes according to how they interact with light:

1. Transparent objects (e.g., glass) transmit light

2. Opaque objects (e.g., dirt or rocks) absorb light

3. Reflective objects (e.g., mirrors) reflect light

Nothing is perfectly transparent, opaque, or reflective. Many objects transmit a little light, although we think of them as opaque; hold a page of this book up to a bright light, and some light will pass through. Everything reflects some light, even objects that reflect so little they look black to us. Likewise, everything absorbs some light, even mirrors that look completely reflective. An exceptionally good metal mirror might reflect up to 99% of the incident light, but virtually all the rest is absorbed. Our eyes give us only a rough indication of what is highly reflective (white or shiny) vs. what is dark and strongly absorbing. One reason is that our perception of light or dark depends on the background. The moon reflects only about 6% of the sunlight that strikes it, but it looks bright when we see it against a dark night sky.

To understand how light interacts with matter, we are going to make some simplifying assumptions. For the time being, let's ignore the fact that all materials absorb, reflect and transmit light. Instead, we will group materials in three categories: transparent (or transmissive), absorptive (a more refined version of opaque) and reflective. We will start with the most fundamental quantity that measures the nature of transmissive materials: the refractive index.

Refractive Index

Earlier we saw that the speed of light in a vacuum, usually indicated by the letter c, is a universal constant, defined as precisely 299,792.458 km/s, or roughly 300,000 km/s or 186,000 mi/s. However, light travels more slowly in matter—even matter as tenuous as air. The ratio of the speed of light in vacuum to the speed of light in a material is the refractive index, n_{mat}:

$$n_{mat} = \frac{\text{Speed of Light in Vacuum}}{\text{Speed of Light in Material}}$$

or

$$n_{mat} = c_{vac}/c_{mat}$$

Because the speed of light is faster in a vacuum than in a material, the refractive index is greater than one. Table 2-5 lists the refractive indexes of some common materials.

TABLE 2-5 Refractive indexes of common materials for wavelengths near 500 nm

Material	Index
Air (1 atmosphere)	1.000278
Water	1.33
Magnesium fluoride	1.39
Fused silica	1.46
Zinc crown glass	1.53
Crystal quartz	1.55
Optical glass	1.51–1.81*
Heavy flint glass	1.66
Sapphire	1.77
Diamond	2.43

* Depends on composition. Standard optical glasses have refractive indexes in this range.

We learned earlier that the speed of light in a vacuum equals the wavelength times the frequency. The same holds true in other materials. Frequency remains constant, but the wavelength λ equals the wavelength in vacuum divided by the material's refractive index:

$$\text{Wavelength}_{\text{mat}} = \text{Wavelength}_{\text{vac}} / n_{\text{mat}}$$

or

$$\lambda_{\text{mat}} = \lambda_{\text{vac}} / n_{\text{mat}}$$

This change in wavelength affects how light travels through transparent materials.

Refraction

A ray of light bends as it passes from one transparent medium to another, as shown in Fig. 2-9. The process is called refraction and the degree of refraction depends on the refractive indexes of the two materials. The top material has a lower refractive index, so the wavelength of light is longer in it than in the higher-index material at the bottom. As light waves enter the higher-index material, their wave-

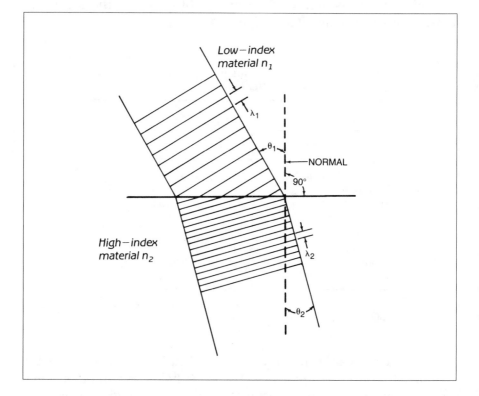

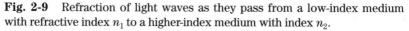

Fig. 2-9   Refraction of light waves as they pass from a low-index medium with refractive index n_1 to a higher-index medium with index n_2.

length shrinks, but they keep in phase (each line represents a wave peak), and they oscillate at the same frequency. This bends the wavefront in the direction shown. If the light wave passes back out into the same low-index material on the other side of the high-index material, the light waves become longer, and the light bends back. You can see how if you turn the figure upside down.

Refraction only occurs at the border between two transparent materials, not within the materials themselves. The amount of refraction is given by measuring the angle between the direction of the light ray and what is called the "normal" angle, perpendicular to the surface where refraction takes place. If light travelling in a medium with refractive index n_1 strikes the surface of a material with index n_2 at an angle of θ_1 to the normal, the light direction in the second material is given by θ_2 in the equation:

$$n_1 \sin \theta_1 = n_2 \sin \theta_2$$

This is known as Snell's law. If the light strikes the surface at an angle θ_1, and you know the refractive indexes of the two materials, this can be rewritten to give:

$$\theta_2 = \arcsin (n_1 \sin \theta_1/n_2)$$

where the arcsin is the angle for which the stated value is the sine.

Suppose, for example, light in water ($n_1 = 1.33$) strikes the surface of crystal quartz ($n_2 = 1.55$) at 30 degrees to the normal. The formula tells us that $\theta_2 = 25.4$ degrees, meaning that the light is bent closer to the normal. If the light was going in the other direction, it would be bent further from the normal. If the difference in refractive index was larger, as when going from air into quartz, the change in angle would be larger.

Something different happens if light goes from a high-index material into a low-index material at a steep angle. Suppose, for example, light in a quartz crystal hit the boundary with air ($n = 1$) at a 50-degree angle to the normal. Plug the numbers into the equation above, and you find that θ_2 equals the arcsin of 1.19. That's impossible because the sine of an angle cannot be greater than one, and your calculator will tell you it's an error. Try an experiment, and you'll find that the light does not escape into the air; it's all reflected back into the quartz, a phenomenon called *total internal reflection*, which is important in fiber optics and in some lasers and optical instruments.

Transparent and Translucent Materials

You may have been confused at some point by the distinction between transparent and translucent materials. Transparent materials are clear, like a window. Translucent materials transmit light, but are cloudy, like wax paper or ground glass. The difference is that trans-parent materials let light pass straight through, while translucent materials scatter the light rays, blurring them so that you cannot see clearly. We won't deal with translucent materials in this book, and will only briefly talk about scattering.

Reflection

Refraction is only one way to redirect light. The other is reflection. The basic law of reflection is simplicity itself:

Angle of Incidence = Angle of Reflection

This means that if light strikes a mirror surface at a 50-degree angle to the normal, it is reflected at the same angle.

There are two types of reflection, and it's important to understand the difference. Mirror-like, or *specular*, reflection is from a surface like a mirror or a piece of window glass that is smooth on the scale of the wavelength of light. Such a smooth surface reflects light back at the angle of incidence, just as we expect a mirror to do. Because the wavelength of visible light is very small, only 0.0004 to 0.0007 mm (400 to 700 nm), a specular surface must be very smooth. However, metal and glass surfaces can be made that smooth and, as we will see later, the surfaces can be coated to enhance their reflectivity.

A surface that is rough on the scale of the wavelength of light reflects light *diffusely*. Each point reflects light at the angle of incidence, but the surface is so rough that the overall effect is to scatter light in all different directions. You can compare diffuse reflection to balls bouncing off a pile of rocks, and specular reflection to balls bouncing off a flat floor. The paper of this book is a good example of diffuse reflection.

All surfaces reflect some light, even those we consider transparent because they transmit the most light. You can see this when it is dark outside, and your windows reflect your household lights. Special antireflection coatings can reduce these reflections.

Mirrors used in high-performance optical systems are reflective on their front surfaces, unlike household mirrors, which have reflective surfaces behind a layer of glass. Such *front-surface mirrors* require more care than household mirrors, because their reflective surfaces are directly exposed to the environment. However, they reflect more light, and are much more accurate optically, because they avoid refraction in the covering layer of glass and the secondary reflection that can occur at the front surface of the glass.

Absorption and Opaque Materials

Perfectly opaque materials, like those that are perfectly transparent or perfectly reflective, are a convenient fiction. What actually happens is that the material absorbs a certain fraction of the light passing through a certain thickness. For example, a material might absorb 60% of the incident light for each centimeter. Thus, after passing through 1 cm of the material, only 40% of the original light would be left; after passing through 2 cm, only 16% (0.4×0.4) would be left, and so on. Mathematically, we can describe this as an exponential equation:

$$I = I_0 \exp^{[-ad]}$$

where I is the intensity of light that has passed through a distance d of the material, I_0 is the initial intensity, and a is the *absorption coefficient* (the fraction of light absorbed per unit length). This explains the well-known fact that thin slices of materials we consider opaque sometimes seem transparent. In fact, all materials have some absorption, although the value of the absorption coefficient may be very small. Likewise—at least in theory—all materials transmit some light, although when the exponential reaches 10^{-20}, the idea of transmission becomes meaningless.

Strictly speaking, absorption is not the only effect that can reduce light transmission. Some light can be scattered as well, like the light scattered from dust in a sunbeam. The light is not absorbed, but instead scattered in different directions so that it is lost from the original beam. Scattering effects are described by an exponential law that looks just like the equation for absorption. In fact, the effects of scattering and absorption together are given by adding one term—s for scattering coefficient—to our equation, giving:

$$I = I_0 \exp^{[-(a+s)d]}$$

This formula is for attenuation, which includes both light absorption and scattering effects. Scattering can make important contributions to losses in laser systems and should not be ignored.

LENSES AND SIMPLE OPTICS

The laws of reflection and refraction show how light is redirected. Refraction and reflection at flat surfaces can be useful, but are limited. If the incoming light is made up of many parallel rays (a handy way to consider laser light), the outgoing rays remain parallel, although they may be going in different directions. Reflection or refraction at curved surfaces can bring light rays together (*focus* them), or spread them out, depending on the curvature. Curved reflective surfaces, like flat reflective surfaces, are called mirrors. Transparent materials with smoothly polished curved surfaces are called lenses.

Lenses have one or two curved surfaces. For the moment, we will consider only the simplest lenses, with surfaces that are flat or with a smooth surface shaped like part of a sphere. Figure 2-10 shows the major types and how they affect a bundle of parallel light rays, such as might come from a laser. If the lens is thicker at the center than at the sides—a *positive lens*—the combined effect of refraction at front and back surfaces bends the light rays so they come together at a *focal point*. If the lens is thicker at the edges—a *negative lens*—it makes the light rays spread out. Each type of lens can have different degrees of curvature on the curved surfaces.

The characteristics of lenses depend on several interrelated parameters, including focal length, surface curvature, size, and the refractive index of the glass. We will describe these parameters for different types of lenses next.

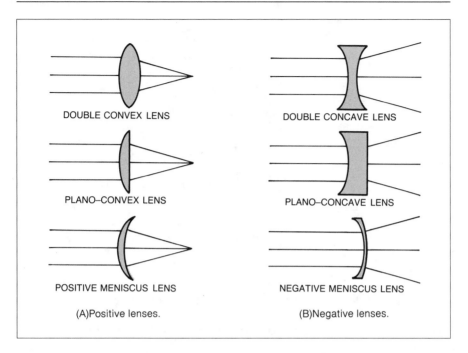

DOUBLE CONVEX LENS

DOUBLE CONCAVE LENS

PLANO–CONVEX LENS

PLANO–CONCAVE LENS

POSITIVE MENISCUS LENS

NEGATIVE MENISCUS LENS

(A)Positive lenses.

(B)Negative lenses.

Fig. 2-10   The six basic types of simple lenses.

Positive Lenses

The most important parameter for any lens is its focal length. For a positive lens, this is defined as the distance from the lens at which it focuses parallel light rays to a point, as shown in Fig. 2-11. The focal length depends on the refractive index of the lens and the curvature of its surfaces. In this case, curvature is defined as the "radius of curvature," the distance from the spherical surface of the lens to the point that is the center of the sphere of curvature. If a lens has surfaces with radii of curvature of R_1 and R_2, and is made of glass with refractive index n, its focal length f is defined by:

$$1/f = (n - 1) \left(\frac{1}{R_1} + \frac{1}{R_2} \right)$$

or

$$f = \frac{1}{(n - 1) \left(\frac{1}{R_1} + \frac{1}{R_2} \right)}$$

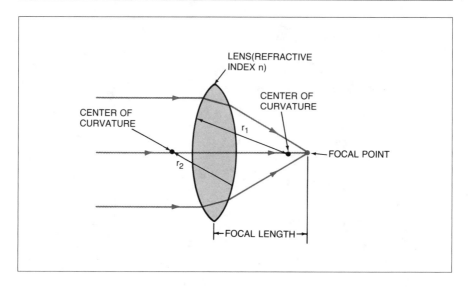

Fig. 2-11　Important parameters of a positive lens.

That formula is for double-convex lenses, with both surfaces curved outward.

If one surface is flat, as in a plano-convex lens, that radius of curvature is taken as infinite, so the value of that $1/R$ is zero. If a surface is concave, or curved inward, as in a positive meniscus lens, that radius of curvature is given a negative sign.

In the real world, light rays from the same object are usually not parallel to each other. Typically, they are spreading out, as shown in Fig. 2-12. A positive lens can also focus these light rays, but in a different way than parallel light rays. Such diverging rays are not all focused to the same point; instead, they form an image farther from the lens than the focal point.

The distance at which the image is formed depends on how far the object is from the lens. If the object is a distance D_o away, the distance of the image D_i is given by:

$$1/D_i = 1/f - 1/D_o$$

$$D_i = \cfrac{1}{\cfrac{1}{f} - \cfrac{1}{D_o}}$$

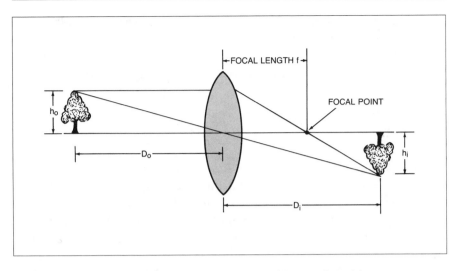

Fig. 2-12 A positive lens forms a real image of an object.

The size of the image also depends on the distance. The ratio of the height of the image h_i to that of the object h_o is defined as the *magnification* ratio. It also depends on the distance between the lens and the object:

$$m = h_i/h_o = D_i/D_o$$

As the formula indicates, the magnification is greater than one if the object is near to the lens and the image is far away—meaning that the image is larger than the object. If the image is nearer to the lens than the object, the magnification is smaller than one, and the image is smaller than the object. If we look at our figure right-side up, we can see that the image is farther from the lens than the object, and thus is magnified. (It would be the other way around if we pretended that the image was the object and vice versa.)

To prevent any confusion, we should stop a minute to explain what we mean by images. Optics specialists speak of two types of images: real and virtual. *Real images* actually exist in space where light rays come together; they can be projected on a screen, a wall, or a piece of paper. These are the types of images produced at the focal point of a positive lens. *Virtual images* are those that you can see with your eyes but can't project onto a screen because the light rays don't actually come together to form them. In a sense, a virtual image is just another view of what you normally see with your eyes, distorted because you are looking through a lens. For example, what you see in a magnifying glass is a virtual image.

Both positive and negative lenses can produce virtual images. However, only positive lenses can bring light rays together to form real images.

Negative Lenses

The same optical laws apply for negative lenses as for positive lenses, but they work in somewhat different ways. The focal lengths of negative lenses are negative. You can get this result mathematically by assigning negative radii of curvature to concave lens surfaces. Its physical meaning is that they don't focus parallel light rays to a point. Pass parallel light rays through a negative lens, and they seem to spread out from a point behind the lens, as shown in Fig. 2-13. The distance from the lens to this point is considered to be the focal length; it is given a negative value because it is on the opposite side of the lens from the focal point of a positive lens. (Optical designers have special sign conventions to keep these things straight, but we won't bother with those.) Because negative lenses don't bring parallel light rays together, they cannot form real images, but they do form virtual images.

The equations giving magnification and image distance are not relevant for negative lenses because they apply only to real images.

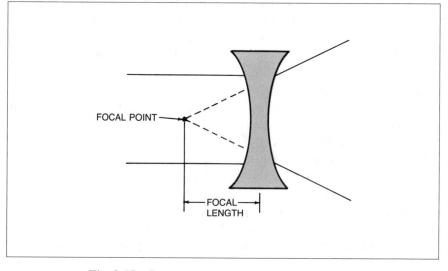

Fig. 2-13   Important parameters of a negative lens.

Mirrors

Mirrors redirect light by reflecting rather than refracting it. A flat mirror does not focus light, but it can redirect light rays. This can be useful in laser systems or other optical systems, where light must be made to go around a corner, or redirected to keep it from escaping an optical system.

Curved mirrors, like lenses, can focus parallel light rays or cause them to spread apart. Mirror surfaces, like lens surfaces, can be either concave or convex. A *concave mirror* seems hollow, like the inside of a bowl. If the surface is spherical (as is usually the case in simple optics), parallel light rays striking it are focused to a point, forming a real image like the one formed by a positive lens. A *convex mirror* is a rounded surface, like the outside of a ball. Parallel light rays striking it are spread out, as they would be by a negative lens, and cannot form a real image.

Like lenses, spherical mirrors have focal lengths, which indicate the distance from the mirror at which parallel rays are focused. The focal length f is defined by the equation:

$$f = R/2$$

where R is the mirror's radius of curvature. To follow the same sign convention as for lenses, we assign a positive radius of curvature to a concave surface and a negative radius to a convex surface. This gives a concave mirror a positive focal length and a convex mirror a negative focal length.

As with positive lenses, the distance of a real image from a concave mirror depends on the object distance. If the object is a distance D_o away from a concave mirror with focal length f, the distance of the real image D_i is given by:

$$1/D_i + 1/D_o = 1/f$$

or

$$D_i = \cfrac{1}{\cfrac{1}{f} - \cfrac{1}{D_o}}$$

Likewise, the magnification m—the ratio of object height h_o to image height h_i—is given by the formula:

$$m = h_i/h_o = D_i/D_o$$

Look back to the formulas for a positive lens, and you will note that they are the same except for a rearrangement of terms. Some optics books use sign conventions that make the terms look different (by assigning a negative sign to distances measured in one direction), but the equations are otherwise the same because the same fundamental thing is happening: light rays are being focused.

Wavelength Effects

We have simplified our discussion of optics so far by ignoring effects of wavelength, but we cannot do that in the real world. Everything changes with wavelength. The refractive index is a function of wavelength. So are the fractions of light absorbed, transmitted, and reflected by an object. Materials that are transparent to visible light, like glass, air, and water, are opaque in parts of the infrared and ultraviolet. Conversely, some things that block visible light transmit light well at other wavelengths.

The details are far beyond the scope of this book, but wavelength effects do have major practical impacts. The variation of refractive index with wavelength causes a prism to bend different wavelengths of light at different angles, so it can spread out the colors of the spectrum. The same variation causes simple lenses to focus different colors of light at slightly different points, a problem called *chromatic aberration.*

Wavelength also affects the choice of optical materials. Silica-based glasses and certain plastics that look transparent to the eye are fine for visible-light lenses, but you need other materials for the infrared and ultraviolet. In fact, you need different materials for different infrared and ultraviolet wavelengths. Some materials strongly absorb certain wavelengths but are comparatively clear at nearby wavelengths.

Further complications come from physical problems with the materials themselves. For example, salts such as sodium chloride or potassium chloride are among the most transparent materials in some parts of the infrared. However, those salts are also so soluble in water that they absorb moisture from the air. Leave them unprotected in a moist room, and eventually a very expensive lens may turn into a salty puddle! Other exotic optical materials are toxic (some contain poisonous thallium) or have other limitations.

Go far enough from the visible region, and air itself becomes a problem. At ultraviolet wavelengths shorter than about 200 nm, air becomes opaque. Scientists call such wavelengths the *vacuum ultraviolet* because they have to perform experiments in a vacuum lest the air absorb the light they're using. Water vapor, carbon dioxide and other air molecules also absorb strongly in parts of the infrared, although not as strongly as in the vacuum ultraviolet.

Optical Complexities

Optics is a complex field, which goes far beyond the realm of the simple lenses and mirrors that we have described so far. If you are going to delve seriously into optics, you will probably want to get a book that deals specifically with that topic.

In this book, we treat optics as an adjunct to laser technology. Thus, we cover the basic elements of optics you need to know to understand lasers, but not the wealth of detail involved in optics *per se*. So far, we have given only a brief introduction to the field. In Chapter 5, we will talk more specifically about optics used with lasers.

WHAT HAVE WE LEARNED

- Light behaves both like waves and like particles (photons).
- Electromagnetic waves can be identified by photon energy, wavelength or frequency. Wavelength is the distance between wave peaks; frequency is the number of wave peaks passing a point each second.
- The speed of light in a vacuum is a universal constant: 300,000 km/s. Wavelength equals the speed of light divided by frequency.
- Photon energy equals Planck's constant (h) times frequency.
- Electromagnetic waves cover a vast range of wavelengths. Visible wavelengths are from 400 to 700 nm.
- Most lasers operate in the visible, ultraviolet or infrared.
- Light is a form of energy emitted and absorbed by matter.
- Materials emit and absorb light at characteristic wavelengths when they make transitions between different energy levels.
- Light waves add in amplitude, causing interference effects.
- Quantum physics recognizes that energy comes in discrete

chunks or quanta. In classical physics, energy levels can change continuously.

- The laser can best be understood by quantum mechanics.
- An electron's energy level is specified by quantum numbers.
- The outer electronic energy levels determine the chemical behavior of elements.
- Molecules have electronic, vibrational and rotational energy levels.
- A transition to a lower level releases energy; absorbing energy causes a transition to a higher level.
- Energy-level structures become more complicated as more electrons are added to an atom.
- All transitions are not equally likely, because of differences in energy-level populations and transition probabilities.
- Electronic transitions can occur at microwave to X-ray wavelengths, but the most important ones are at visible, near-ultraviolet and near-infrared wavelengths. Vibrational transitions typically occur in the mid-infrared; rotational transitions occur at far-infrared or microwave wavelengths.
- Transition energies or frequencies add together simply, but an inverse addition rule is needed for the wavelength.
- There are two types of emission, spontaneous and stimulated. Most light we see in nature is spontaneous emission. Lasers produce stimulated emission.
- Stimulated emission can dominate only if more atoms occupy an upper level than a lower level, a condition called a population inversion. However, at thermodynamic equilibrium more atoms are in lower states than upper ones.
- Objects transmit, reflect, and absorb light.
- The refractive index is a fundamental quantity for all optical materials; it equals the ratio of the speed of light in a vacuum to the speed of light in the material.
- Refraction bends light rays as they pass between transparent media.
- Total internal reflection occurs when light tries to leave a high-index material at a shallow angle, but can't get out.
- The angle of incidence equals the angle of reflection.
- Opaque materials absorb light strongly, according to an exponential law.

- Transparent materials with smoothly polished curved surfaces are called lenses.
- The focal length of a positive lens is the distance from the lens to a point at which it focuses parallel light rays. A positive lens focuses diverging light rays from an object to form a real image. The distance and size of the image depend on the focal length of the lens and the object distance and size.
- You can see virtual images, but you can't project them on a wall.
- Negative lenses spread out parallel light rays and have a negative focal length.
- Curved mirrors can focus light by reflecting it. Concave mirrors can form real images.
- The refractive index and other material properties depend on the wavelength.

WHAT'S NEXT

In the next chapter, we will learn about the internal workings of a generalized laser. Later, we will learn about specific types of lasers, which can differ greatly in detail.

Quiz for Chapter 2

1. A carbon-dioxide laser has a nominal wavelength of 10.6 micrometers. What is its frequency?
   a. 300,000 Hz
   b. 2.8×10^{13} Hz
   c. 1.06 GHz
   d. 2.8×10^{10} Hz
   e. None of the above
2. What is the photon energy for an infrared wave with frequency of 10^{12} Hz?
   a. 10.6 μm
   b. 6.63×10^{-34} joule

   c. 6.63×10^{-22} joule
   d. 10.6×10^{22} joules
   e. About 1 joule
3. What is the wavelength of the infrared wave in Problem 2?
   a. 3×10^{-4} m
   b. 300 mm
   c. 300 km
   d. 300 nm
   e. 10.6 μm
4. What is the metric prefix for 10^{-9}?
   a. Kilo

b. Giga
c. Pico
d. Nano
e. Femto

5. Calculate the wavelength of the transition in the hydrogen atom from the $n = 2$ energy level (the second orbit out) to the $n = 3$ level. This is the first line in the Balmer series of spectral lines.
   a. 121.6 nm
   b. 91.2 nm
   c. 656 nm
   d. 632.8 nm
   e. 900 nm

6. What is the wavelength of a transition that corresponds to the combination of transitions at 500 and 700 nm?
   a. 200 nm
   b. 1200 nm
   c. 600 nm
   d. 292 nm
   e. None of the above

7. At thermodynamic equilibrium and room temperature (300 K), what is the ratio of populations at the upper and lower level of a transition with photon energy of 0.1 electronvolt (i.e., with $E_2 - E_1 = 0.1$ eV)?
   a. 0.0207
   b. −3.9
   c. 1
   d. 0.001
   e. 0.000009

8. What is the ratio in the above problem if the temperature is reduced to 100 K?
   a. 0.0207
   b. 0.27
   c. 1
   d. 0.001
   e. 0.000009

9. Light in a medium with a refractive index of 1.2 strikes a medium with a refractive index of 2.0 at an angle of 30 degrees to the normal. What is the angle of refraction (measured from the normal)?
   a. 50 degrees
   b. 60 degrees
   c. 20 degrees
   d. 17.5 degrees
   e. 15 degrees

10. A material has an absorption coefficient of 0.5/cm, and its scattering coefficient is negligibly small. What fraction of incident light can pass through a 5-cm thickness?
    a. 0.5
    b. 0.1
    c. 0.082
    d. 0.01
    e. None; the material is opaque.

11. A positive lens with a focal length of 10 cm forms a real image of an object 20 cm away from the lens. How

far is the real image from
the lens?

a. 5 cm
b. 10 cm
c. 15 cm
d. 20 cm
e. 25 cm

12. What is the ratio of image
    size to object size for the
    case in Problem 11?

a. 2
b. 1.5
c. 1
d. 0.667
e. 0.5

How Lasers Work

ABOUT THIS CHAPTER

In the last chapter, we learned the basic physics and optics needed to understand how lasers work. In this chapter, we will learn what goes on inside lasers—how they are energized and how they produce beams. Here we will learn about the general principles of laser operation; later on we will learn about laser characteristics and specific lasers.

PRODUCING POPULATION INVERSIONS

In the last chapter, we saw that you need a population inversion to make a laser. If a material is at thermal equilibrium, with more atoms or molecules in the lower level of a transition than in the upper level, absorption will soak up any stimulated emission, because there are more absorbers than emitters. However, if more atoms or molecules are in the upper level, there are more emitters than absorbers. Thus, a photon with the transition energy is likely to encounter an excited state and stimulate emission before it is absorbed. Atoms and molecules have many possible energy levels, but you only need a population inversion on one transition to make a laser.

There are two basic ways to produce such a population inversion. The more obvious one is to excite extra atoms or molecules to the higher state. However, it's also possible to depopulate the lower level, or to pick a system where the lower level is unstable, so there are few or no atoms in the lower level. For a laser to operate continuously, both population of the upper level and depopulation of the lower level are important, because accumulation of too many atoms or molecules in the lower level can end the population inversion and stop laser action.

Excitation Mechanisms

The standard way to produce a population inversion is by putting energy into the laser medium, exciting atoms or molecules to high energy levels. However, it is not enough to deposit energy in a way that heats the material. As long as the material remains in thermodynamic equilibrium, merely heating it always puts more atoms into lower energy levels than in higher states. The ratio of atoms in energy levels 1 and 2 at a temperature T is given by:

$$N_2/N_1 = \exp\left[-(E_2 - E_1)/kT\right]$$

where N_1 and N_2 are the number of atoms in each level, E_2 and E_1 are the energies of the two levels, and k is the Boltzmann constant (1.38054×10^{-23} joule per degree Kelvin). As long as E_2 is greater than E_1, the exponential is negative, and the ratio is below one. Heating the material increases the average energy, but does not make N_2 greater than N_1.

To produce a population inversion, you must selectively excite the atoms or molecules to particular energy levels. This requires some ingenuity, although as we will see later in this chapter, it can occur in nature. (However, if you think only in terms of thermodynamic equilibrium, the only way to make N_2 greater than N_1 in the equation is to make temperature a negative number. This is why, before the advent of the maser and laser, some physicists spoke of a population inversion as a "negative temperature.")

The two most common laser excitation techniques use light and electricity. Both light and electrons can selectively transfer energy to atoms or molecules, exciting them to higher energy levels, as described in more detail for individual lasers in Chapters 6 through 9. This excitation does not have to raise the atom or molecule directly to the upper level of the laser transition. It may excite the light-emitting species to a higher level, from which it can drop to the upper laser level, or it may excite other atoms which transfer the energy to the light-emitting atom or molecule. As we will see later on, the more complex-sounding approaches often make better lasers.

Metastable States and Lifetimes

These excitation techniques only work if atoms or molecules have the right type of energy-level structures. Normally excited states have short lifetimes and release their excess energy in a matter of nanoseconds (billionths of a second) by spontaneous emission. Excited states with such short lifetimes don't last long enough to be stimulated to emit their energy; most of it emerges as spontaneous emission. What is needed is a longer-lived excited state.

Such states do exist. They are called *metastable* because they are unusually stable on an atomic scale, although they may only last for a millisecond (a thousandth of a second) or a microsecond (a millionth of a second). They are very important in laser physics because they make the best kind of upper laser level. Excited atoms and molecules can stay in an metastable excited state long enough to produce significant amounts of stimulated emission.

On a more quantitative basis, laser action requires a build-up of population in the upper laser level. This is possible only if the upper laser is populated faster than it decays, so the population of the upper level exceeds that of the lower level. The longer the spontaneous-emission lifetime, the slower the decay rate—and the easier it is to keep enough atoms or molecules in the upper laser level to get a population inversion and laser output.

Masers and Two-Level Systems

Laser operation will make more sense if we look at some specific examples. We mentioned earlier that Charles Townes made the first maser at Columbia University several years before the first laser. His maser, like the lasers that followed, required a population inversion. However, Townes relied on an unusual trick. He started with a beam of ammonia molecules that contained molecules in both the ground and excited states. Then he separated the excited-state molecules from the rest of the beam and threw away the ground state molecules. Because his final sample held only excited molecules, he had a population inversion and could demonstrate maser action.

That trick does not work for lasers, and it was not long before scientists found other ways to make masers. However, it played an important role in maser and laser development by proving that population inversions could be produced—something some physicists had doubted.

Three- and Four-Level Lasers

Townes's first ammonia maser involved only two energy levels: the upper and lower laser levels. Practical laser systems nominally involve three or four energy levels and can involve more, depending on how the energy is transferred.

The simplest type of energy-level structure is the three-level laser, shown in Fig. 3-1. For simplicity, we assume that all the atoms start in the ground state, which is also the lower laser level. Most are excited to a short-lived high-energy level, then drop quickly to a lower metastable level, which has a much longer lifetime (typically a thousand times or more longer). They accumulate there, producing a population inversion between the metastable state and the depopulated ground state, which are the upper and lower laser levels, respectively. Because this population inversion means that more atoms are available in the upper state for stimulated emission than in the lower

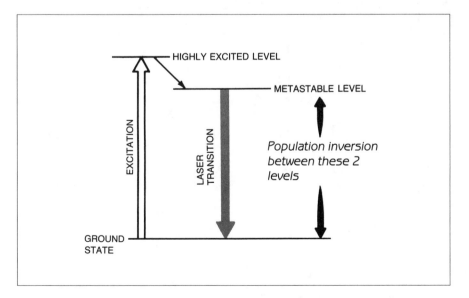

Fig. 3-1 Energy levels in a three-level laser.

level for absorption, stimulated emission can dominate on the laser transition.

Maiman's ruby laser is an important example of this three-level scheme. Although the system works, it is not ideal. One problem is that the ground state is also the lower laser level, so a majority of atoms must be excited to the upper laser level to produce a population inversion. This requires an intense burst of energy—for which Maiman used a bright flashlamp. The population inversion is very difficult to sustain, so three-level lasers operate in pulsed mode.

Most practical lasers involve at least four energy levels, as shown in Fig. 3-2. As in the three-level laser, the excitation energy raises the atom (or molecule) from the ground state to a short-lived highly excited level. The atom or molecule then drops quickly to a metastable upper laser level. The laser transition then takes the atoms or molecules to a lower state which is above the ground state. The atoms or molecules in this lower laser level eventually lose the rest of their excess energy by spontaneous emission or other processes and drop to the ground state.

The key difference is that the lower level is not the ground state. Why is this so important? Because normally most atoms or molecules

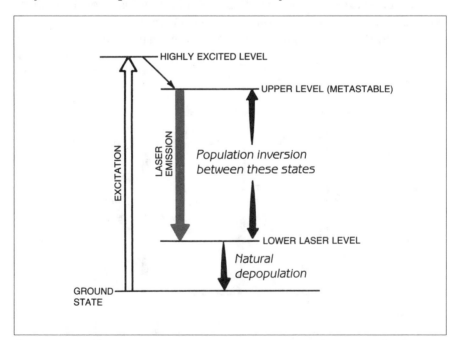

Fig. 3-2 A four-level laser scheme.

are in the ground state. If the ground state was the lower laser level, the only way to produce a population inversion would be to excite most atoms or molecules to the upper laser level. That requires an intense burst of energy, like the flashlamp pulses Maiman used to excite his ruby laser. Suppose, however, that normally only a few atoms or molecules were in the lower laser level—say 1%. Then you could produce a population inversion by exciting only 2% of the atoms to the upper laser level. That is much easier than exciting over half the atoms from the ground state.

Separating the lower laser level from the ground state also brings a more subtle advantage. Atoms or molecules in a four-level laser spontaneously drop from the lower laser level to lower levels or the ground state. If the lower laser level has a much shorter lifetime than the upper level, this steady depletion of the lower level helps sustain the population inversion by avoiding an accumulation of atoms or molecules in the lower level. This lets some lasers produce a continuous beam (often called *continuous-wave* emission). Note that if the lower level has too long a lifetime, a large population can accumulate there, eventually ending the population inversion and stopping laser emission. This happens in a few lasers, limiting them to pulsed operation.

If you look carefully at the operation of real lasers described in later chapters, you will see that the actual energy-level structures are more complex. Excitation is not always to a single high level; it may be to a group of levels, which decay to the same upper laser level. That actually is good news, because it allows excitation over a wider range of energies than if the atom could only be raised to a single specific state. The picture at the lower end of the energy scale also can be more complex, especially where depopulation of the lower laser level is important. The upper and lower laser levels can be much farther above the ground state than indicated in Figs. 3-1 and 3-2, which makes it easier to control their populations but limits overall operating efficiency.

We also have considered only two levels, a single upper level and a single lower level. In most lasers there are more levels. For example, there may be two upper levels with slightly different energies, each leading to one or more separate lower levels. This can produce multiple transitions, which are possible in many lasers. For example, we often think of the helium-neon laser as emitting only on a single red transition, but with suitable optics it also can emit in the infrared or at green, yellow, and orange wavelengths. Other lasers, such as carbon-dioxide, can simultaneously operate on many closely spaced wavelengths because the upper and lower levels are split up into many sublevels.

Another important fact is that the active media in many practical lasers contain more than one species of atoms or molecules. One species may capture the excitation energy efficiently, then transfer the energy to another species to produce a population inversion and laser action. One example is the helium-neon laser, where helium atoms capture energy from electrons passing through the gas and then transfer that energy to neon atoms, generating a population inversion in the neon. In other lasers, another gas may be added to depopulate the lower laser level. In solid-state lasers, the light-emitting atom is normally embedded in a host crystal, which doesn't generate light.

Natural Masers and Lasers

A major reason over four decades passed between Einstein's prediction of stimulated emission and the first working laser was that physicists thought it would be very hard to make a population inversion. They apparently didn't realize that matter could naturally be in a state other than thermal equilibrium. Ironically, it was only after people began building masers and lasers that we discovered natural masers—in outer space. Charles Townes, now at the University of California at Berkeley, has been a leader in studying such "cosmic masers."

Natural masers are gas clouds near hot stars. Starlight excites molecules in the gas to high energy levels, and the molecules drop down the energy-level ladder to a metastable state. If there is a suitable lower laser level, this can produce a population inversion and laser action. Some scientists think similar processes occur in the atmosphere of Mars, generating infrared stimulated emission at the same wavelengths as man-made carbon-dioxide gas lasers.

Cosmic masers produce stimulated emission in the same way as man-made masers and lasers, but they differ from them in important ways. Although they may radiate tremendous amounts of energy, cosmic masers do not produce beams; they emit in all directions, like any cloud of hot interstellar gas. In fact, without special instruments they look just like other gas clouds—which is what astronomers first thought they were. Astronomers did not discover their special nature until they analyzed the light from the gas clouds. Cosmic masers emit strongly at transition wavelengths of certain molecules, such as carbon monoxide. Hot gas emits a broad, continuous spectrum, like the white light from a light bulb.

RESONANT CAVITIES

A population inversion is not all it takes to make a laser. A hot blob of gas with an inverted population, like a cosmic maser, emits light in every direction. The light may be stimulated emission, and it may be at a single wavelength, but it isn't concentrated in a laser beam. If our eyes were sensitive to the right wavelength, that kind of stimulated emission would look like ordinary colored light. To extract energy efficiently from a medium with a population inversion and make a laser beam, you need a resonant cavity that helps build up (or amplify) stimulated emission by feedback—reflecting some of it back into the laser medium.

To see what happens, let's start by looking at the process of amplification.

Amplification and Gain

Stimulated emission can amplify light. One photon with the same energy as a laser transition can stimulate the emission of a cascade of other photons at the same wavelength. You can think of that initial photon as a signal that is to be amplified by stimulated emission. Masers have been used to amplify weak cosmic microwave signals, a job at which they excel because they amplify only a narrow range of frequencies and ignore the background noise. Light amplification by the stimulated emission of radiation (remember the origin of the word laser?) can produce a powerful beam of light.

The workings of laser amplification are shown in Fig. 3-3. The first photon comes from spontaneous emission on the laser transition. When this light wave encounters an atom in the upper laser level, it stimulates it to emit light and drop to the lower laser level. If it encounters an atom in the lower laser level, it could be absorbed. (Stimulated emission and absorption are not automatic—but the probabilities are equal if the photon encounters an atom in the right states.) Remember, though, that we have a population inversion, so

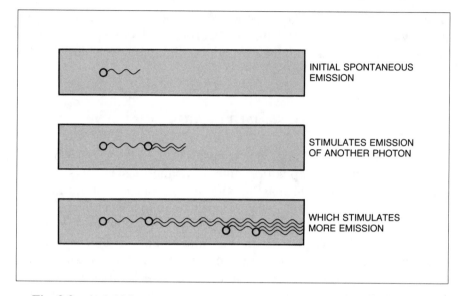

Fig. 3-3   An initial spontaneously emitted photon stimulates the emission of other photons, which stimulate more emission, and so on.

more atoms are in the upper laser level than in the lower level, and the light is more likely to stimulate emission than to be absorbed.

After the first stimulated emission, we have two photons of the same energy in phase with each other. Like the original spontaneous emission photon, each of them is more likely to encounter an atom in the upper laser level than in the lower level. Thus, they, too, are likely to produce stimulated emission. So are the stimulated-emission photons they produce. You can see the trend—the number of photons produced by stimulated emission grows very quickly.

Laser physicists measure the degree of amplification as *gain*, the amount of stimulated emission a photon can generate as it travels a given distance. For example, a gain of 2/cm means that one photon generates two more photons each centimeter it travels. A gain of 0.05/cm means that a photon generates an average of 0.05 stimulated emission photon each centimeter it travels. The result is an amplification factor that to a crude approximation increases like compound interest:

$$\text{Output/Input} = \text{Amplification} = (1 + \text{Gain})^{\text{Length}}$$

Thus, for a gain of 0.05/cm, the amplification factor is (1.05) raised to the power of length (measured in centimeters). Even though this gain may sound small, the total amplification can grow rapidly: for 10 cm the amplification is 1.63; for 20 cm it's 2.65; and for 50 cm, it's 11.5.

That expression actually is a very crude approximation of the degree of amplification. The reason is that the gain term itself depends on how many atoms or molecules are in the upper and lower laser levels. (It also depends on the probability of stimulated emission, and other factors including density and temperature of the laser medium.) The more stimulated emission, the fewer atoms remain in the upper laser level to cause more stimulated emission. The number of atoms or molecules in the upper laser level that can be stimulated to emit light falls, while the number in the lower level that can absorb light rises. This reduces the gain coefficient, and in many pulsed lasers eventually cancels it out and stops stimulated emission altogether.

The gain term is useful in studying lasers, but it is important to understand what it means. The value given for laser gain generally is what is called *small-signal gain*—the gain when the signal being amplified is still weak, and there are plenty of atoms left in the upper laser level. As the signal becomes stronger, laser gain can reach a limit, or *saturate*, because the laser is producing stimulated emission just as fast as it can. The exact increase in stimulated emission with distance travelled through a laser medium is too complex to worry about here. The important point to remember is that the amount of amplification increases sharply with the distance light travels through the laser material.

This increase of gain with distance is one factor that helps concentrate laser output, as shown in Fig. 3-4. If the laser medium is rod-shaped, much stronger output will come from the ends of the rod than from its sides. If there are no mirrors on the rod, as in Fig. 3-4A, this length effect concentrates stimulated emission into an angle θ, defined by an arcsine function:

$$\theta = \arcsin(Dn/2L)$$

where D is the rod diameter, L is its length, and n its refractive index. Some light is lost out the sides, but it is a small amount because it can travel (and experience gain) for only a small distance before escaping from the rod.

Gain and power can be increased by adding a mirror on one end of the laser medium, as in Fig. 3-4B. This reflects light reaching one end of the rod back into it, effectively increasing its length. The light

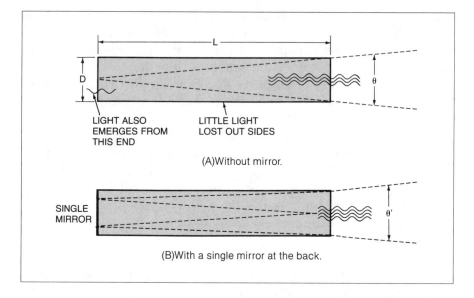

LIGHT ALSO
EMERGES FROM
THIS END

LITTLE LIGHT
LOST OUT SIDES

(A)Without mirror.

SINGLE
MIRROR

(B)With a single mirror at the back.

Fig. 3-4   Geometry alone can concentrate stimulated emission in a rod.

emission angle in such a single-mirror laser is somewhat smaller:

$$\theta' = \arcsin{(Dn/4L)}$$

because of the mirror.

A few lasers with very high gain, notably semiconductor and excimer lasers, can operate in this way. However, some purists try to avoid calling those devices lasers, because they lack the pair of end mirrors that form the optical resonator which helps complete the standard definition of a laser. We won't quibble that much about definitions, but the purists do have a point. It is optical resonators that give lasers many of their special features.

Mirrors, Laser Cavities, and Oscillation

Most lasers have a pair of mirrors at opposite ends of a rod-shaped laser medium. The mirrors form a *laser resonator, cavity,* or *oscillator*, in which the light oscillates back and forth between the mirrors. The idea of an oscillator is important because, like an electronic oscillator, a laser oscillator can generate a signal on its own. (In the laser world, the term amplifier has a separate meaning—it is a laser medium that amplifies, by stimulated emission, a signal from some outside source.)

Why bother building an oscillator when it is not necessary to produce stimulated emission? The reason is that most laser materials have very low gain, so you have to pass light a long distance through them to get much amplification. A lot of amplification is needed because oscillation starts with only a few photons, but it takes a large number of photons to produce observable power. A helium-neon laser must generate 3.2×10^{15} photons to emit 1 mW of red light for one second.

The most practical way to get the light to pass through a long length of the laser medium is to put mirrors on opposite ends of a tube or rod. As the light bounces back and forth between the mirrors, it makes many passes through the laser medium. The amount of stimulated emission grows on each pass through the laser medium until it reaches an equilibrium level.

There is one important point about the laser mirrors that we haven't mentioned yet. One cavity mirror reflects essentially all of the light that reaches it, but the other reflects only some of the light back into the laser cavity. The rest emerges in the laser beam, as shown in Fig. 3-5. If the gain is low, as in a helium-neon laser, the fraction of the light transmitted can be very low, only 1% to a few percent, and most light is returned to the laser cavity to stimulate more emission.

The fact that the output mirror reflects some light and transmits the rest is very important. One of the most common misunderstandings about lasers is the idea that both end mirrors initially reflect all light back into the laser medium; then, at some point, the light magically breaks through one mirror and emerges as a beam. That is not the case. The output mirror always transmits a constant fraction of the light (which emerges as the beam) and reflects the rest back into the laser medium.

Reflectivity of the output mirror is a critical parameter that is selected depending on the gain of the laser medium and the losses within the laser cavity. When a laser is operating in steady or continuous-wave mode, the amplification of a round trip through the cavity must equal the sum of losses and the power escaping from the cavity:

$$\text{Amplification} = \text{Loss} + \text{Output Power}$$

Loss in the laser cavity must be kept as low as possible if laser gain is low, but it cannot be reduced to zero. No mirror can reflect every photon that reaches its surface. Material in the laser cavity also absorbs some light. Suppose, for example, that the light is amplified by 5% as it makes a round trip of the laser cavity, that 1% of the light is absorbed at each mirror (a total of 2%), and that an added 1% is absorbed by the laser gas. Then 2% of the laser power can emerge in the output beam.

As this example implies, laser power is higher inside the cavity than outside. The higher the output mirror transmission, the lower the intracavity power levels. Suppose the output mirror reflects 98% of the light back into the cavity and transmits only 2%, neglecting losses. Then the output power is only 2% of the power within the

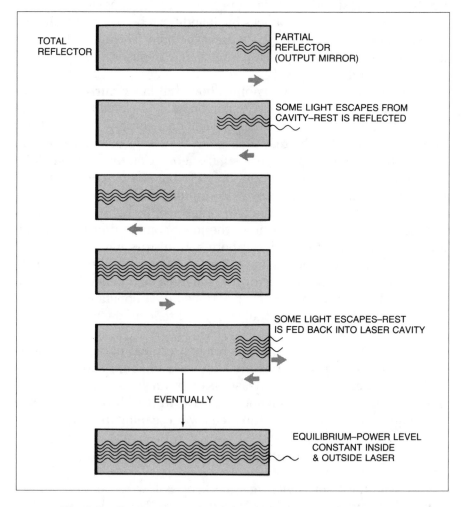

Fig. 3-5 Growth of stimulated emission in a resonant laser cavity.

cavity. If the output beam is 1 mW, the intracavity power is 50 mW. The difference decreases as output mirror transmission increases, but as long as it reflects some light back into the cavity, the power inside the cavity will be higher than in the beam emerging from the laser.

Resonance

There are some subtle implications of the oscillation of light waves back and forth between mirrors in a laser cavity. One is called *resonance*, and depends on the wavelength of the stimulated emission and the length of the laser cavity.

To understand the nature of resonance, we need to turn back to the wave picture of light. In Fig. 3-6, we pretend that light waves are large compared to the length of the laser cavity. (This isn't the case, of course, but light waves are so much shorter than a laser cavity that you couldn't see them if we showed them on a realistic scale.) Recall, as a starting point, that stimulated emission is coherent, so all light waves are in the same phase, and that light waves add in amplitude.

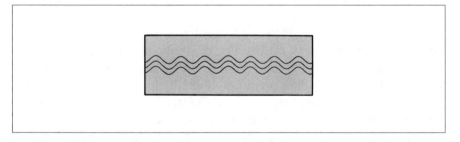

Fig. 3-6   Light waves are resonant if twice the length of the laser cavity equals an integral number of wavelengths.

Figure 3-6 shows what happens when twice the cavity length is an integral multiple of the wavelength (in this case, the cavity is seven wavelengths long). Because all the light waves in the cavity are coherent, they all are in phase. Thus, each wave is at the same phase when it reflects from one of the cavity mirrors. For example, if light starts at a wave peak when it is reflected from the output mirror, it will travel an integral number of wavelengths before it reaches the output mirror again, where it will again be at a peak; so will the light waves stimulated by that wave. Thus, all of them will add in amplitude by constructive interference.

Suppose, however, that twice the cavity length is not an integral multiple of the wavelength. Then a wave will be out of phase with other waves after it has made a round trip. The waves will add in amplitude, but because they are out of phase, destructive interference will reduce their strength.

The result is resonance: light waves are amplified strongly if twice the cavity length is an integral multiple of their wavelength:

$$N\lambda = 2L$$

where N is an integer, λ the wavelength and L the cavity length. Other wavelengths are not amplified strongly, so they die out in the laser cavity.

This might seem to be a very restrictive condition, which could make it hard to build a cavity that would be resonant at a particular wavelength. However, several effects combine to spread laser transitions over a range of wavelengths and make such resonant cavities practical. One is that laser transitions have gain over a range of wavelengths, called the *gain bandwidth*. Fortunately, gain bandwidth is much broader than the range of wavelengths resonant in a laser cavity.

Furthermore, remember that light wavelengths are very much smaller than most laser cavities. For example, 30 cm, a typical round-trip distance in a small helium-neon laser, equals about 475,000 wavelengths of the laser's 632.8-nm red light. Resonance is possible not just at 475,000 waves, but also at 475,001, 475,002, etc. The wavelengths that correspond to those numbers of waves are very close, and several can fit under the gain curve for the helium-neon laser, as shown in Fig. 3-7. Each resonant value of N is called a *longitudinal mode* of the laser, and each has a slightly different wavelength.

PRODUCTION OF LASER BEAMS

We saw that part of the light in the laser cavity emerges through the output mirror as the laser beam, but we glossed over details of producing the beam. Beam characteristics such as size, light distribution, and the rate of spreading or beam divergence depend on the design of the laser cavity and the output optics. Before we look at what influences beam characteristics, we should explain a few laser concepts, including intensity distribution and oscillation modes.

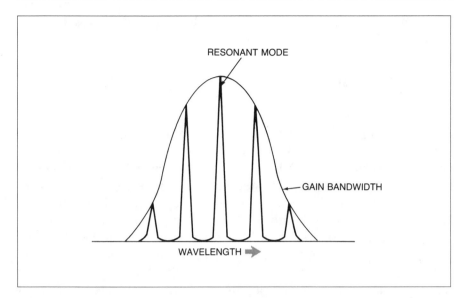

Fig. 3-7   Several cavity resonances can fit within the gain bandwidth of a typical gas laser.

We may think of a laser beam simply as a collection of parallel light rays that form a bright spot on a screen. However, if we drew a line across the beam and measured the light intensity at different points, we would find that the intensity varies. Typically, the beam is brightest in the middle, with intensity dropping off to the sides. It can be hard to define precisely where the beam stops, so normally a cutoff is defined at a certain fraction of the central intensity (often $1/e^2$, where e is the root of natural logarithms).

We saw earlier that lasers oscillate in different longitudinal modes, corresponding to wavelengths which satisfy the equation:

$$N\lambda = 2 \times \text{Cavity Length}$$

where N is an integer and λ the wavelength. Lasers also can oscillate in different *transverse modes*, which manifest themselves in different patterns of intensity across the beam. Those, too, depend on resonator design, and we will look at them in more detail later.

The *beam divergence* is also critical. This angle measures how rapidly the beam spreads far from the laser. As we will see later, if you know beam divergence, you can calculate laser spot size. Normally it, like beam diameter, is measured to points where the beam intensity has dropped to a certain level. It, too, depends on resonator structure.

Types of Resonators

So far we have not said anything about the shape of the mirrors on the ends of a laser cavity, and you probably assumed both are flat. That simple arrangement is only one of the possible resonator configurations shown in Fig. 3-8. In fact, although it seems simple conceptually, a resonator made of two flat mirrors (Fig. 3-8A) can be hard to use. The reason is that if the two mirrors are not precisely parallel, light rays passing back and forth between them would eventually miss one of the mirrors, causing losses that could stop laser oscillation. The misalignment need not be large. If one mirror was only half a degree out of parallel, a light ray striking it from the center of the other mirror in a 15-cm-long laser would be reflected to a point 1.3 mm from the center of the other mirror—and would miss the edge

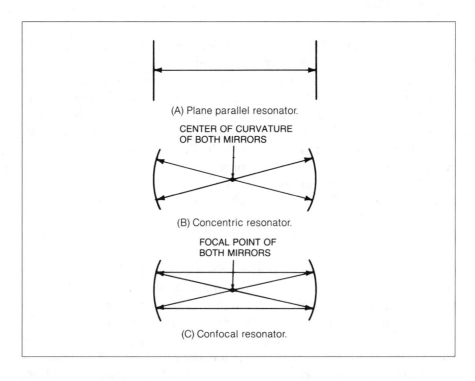

Fig. 3-8   Major laser resonator configurations.

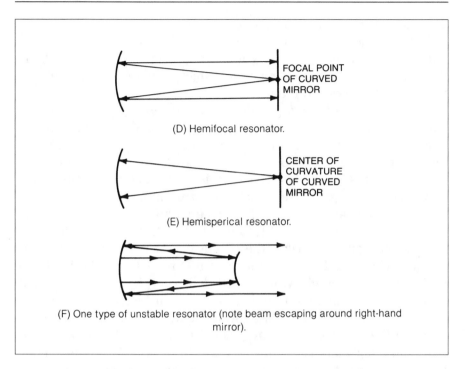

(D) Hemifocal resonator.

(E) Hemisperical resonator.

(F) One type of unstable resonator (note beam escaping around right-hand mirror).

Fig. 3-8 (continued)   Major laser resonator configurations.

of a cavity mirror 2 mm in diameter. Successive reflections would magnify the effects of misalignment and increase the amount of light lost out the sides.

These light-leakage losses can be avoided if one or both of the cavity mirrors are curved, as in most of the designs in Fig. 3-8. Follow the light rays in Figs. 3-8B, 3-8C, 3-8D, and 3-8E, and you can see how the focusing power of the curved mirror keeps light in the laser medium. The curved mirrors reflect light back into the cavity, even if was not emitted precisely along the tube axis. Likewise, minor mirror misalignment doesn't matter, because their curvature focuses the light back toward the other mirror. Such a configuration is called a *stable resonator* because light rays reflected from one mirror to the other will keep bouncing back and forth indefinitely (neglecting losses). Stable resonators are most attractive if the laser gain is low, making loss reduction vital.

On the other hand, if laser gain is high, it is not a problem if some light leaks out the sides of the cavity after repeated reflections. To get the maximum output power, it is more important to extract energy from all the excited media between the cavity mirrors. Look carefully at Fig. 3-8, and you can see that some stable resonators do not pass light rays through the entire volume of the excited medium. If the light rays do not go through a region, they cannot stimulate emission to collect the energy stored there.

One solution to this problem is a different type of resonator, called the *unstable resonator*, shown in Fig. 3-8F. An unstable resonator gets its name because light rays successively reflected from its mirrors eventually drift out to the sides of the laser mirrors. This might make it seem undesirable, and indeed it is for low-gain lasers. However, in high-gain lasers such losses are more than offset by the advantage of collecting energy from a larger volume of laser medium.

You can also see the difference in another way. In a low-gain laser, most light that strikes the output mirror is reflected back into the laser cavity. If the output mirror is 95% reflective, light within the cavity (on the average) will be reflected from the output mirror 19 times before it finally emerges in the output beam. That makes it important to be certain that the ray can reflect back and forth between the mirrors that many times. On the other hand, the output mirror in a high-gain laser may be only 50% reflective, so the average light ray will be reflected once before it emerges in the output beam. This makes alignment of the cavity mirrors much less critical.

A true plane-parallel mirror cavity is neither stable nor unstable, because it neither focuses nor diverges light rays. The mirror cavity's main advantage is the ease of making it, especially for semiconductor lasers, where it is made by cleaving the semiconductor crystal. This causes no problems because semiconductor lasers are very small and have high gain.

We will not go into detail on the theory of laser resonators. It bogs down in the sort of mathematical complexity you're reading this book to avoid, and many of the results are at best arcane. However, laser resonators do (quite literally) shape both the intensity distribution in the output beam and the rate at which the beam diverges.

Intensity Distribution and Transverse Modes

We saw earlier how the length of the laser cavity and the wavelength interact to create longitudinal modes—standing waves along the length of the laser. There are also transverse modes, which determine the pattern of intensity distribution across the width of the laser beam. Transverse mode patterns must meet "boundary conditions," such as having zero amplitude. The mode pattern may have one, two or more peaks in the central part of the beam.

The simplest, or "lowest order," transverse mode is the smooth beam profile with a peak in the middle, as shown in Fig. 3-9. Its shape is that of a mathematical curve called a Gaussian curve, after famed mathematician Karl Gauss. This intensity distribution $I(r)$ as a function of distance from the center of the laser beam (r) is given by

$$I(r) = (2P/\pi d^2) \times \exp(-2r^2/d^2)$$

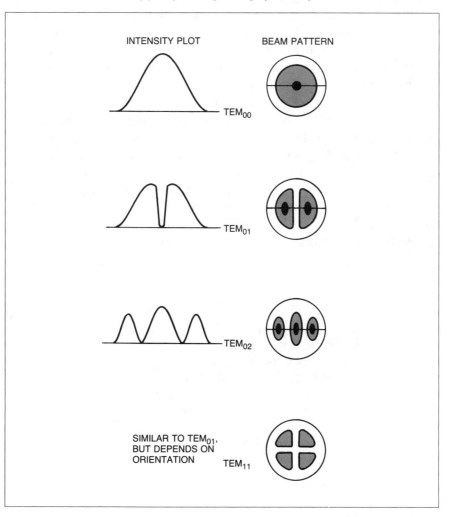

Fig. 3-9 TEM_{00} mode beam and a sampling of other transverse beam modes.

where P is the beam power and d the spot size measured to the $1/e^2$ intensity points. This first-order mode is called the TEM_{00} mode, where the T, E, and M stand for the transverse, electric, and magnetic modes, respectively. Conventionally, the first number is the E mode, and the second the M mode; the two are perpendicular to each other.

As you might expect, there is a large family of TEM_{mn} modes, where m and n are integers. The m indicates the number of zero points, or minima, between the edges of the beam in one direction, and n indicates the number of minima between the edges of the beam in the perpendicular direction. Thus, a TEM_{01} beam has a single minimum dividing the beam into two bright spots, a TEM_{11} beam has two perpendicular minima (one in each direction) dividing the beam into four quadrants, and so forth. The TEM_{00} mode is desirable because it suffers less spreading than higher-order modes. A low-gain laser medium with a stable resonator can readily produce this lowest-order mode; proper design adds losses to suppress the oscillation of higher-order modes. However, some stable-resonator lasers operate in one or more higher-order modes, especially when they are designed to maximize output power.

Unstable resonators have fundamentally different mode structures. You can see the basic reason why in Fig. 3-10, which examines the unstable-resonator example from Fig. 3-8F in more detail. In this simple unstable resonator, the output mirror is a solid metal mirror

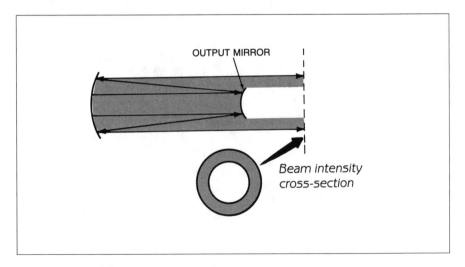

Fig. 3-10   Output pattern from an unstable resonator has a central minimum.

that reflects some light back into the laser medium while the beam escapes around it. Thus, near the laser the beam profile has a "doughnut" cross section—a bright ring surrounding a dark circle where light was blocked by the mirror. If you go far enough from the laser, this intensity distribution averages out to a more uniform pattern.

The details of unstable-resonator modes are even more complex than those of stable-resonator modes, but we can safely skip their many variations because they generally have little practical impact. In Chapter 4 we will learn how transverse modes affect the properties of laser beams.

LASER EXCITATION TECHNIQUES

Our general description of laser physics has given you an idea of how lasers work, but we have glossed over the actual excitation of atoms or molecules to produce a population inversion. We can describe the basic types of laser excitation here; the details depend on the type of laser, and will be covered in Chapters 6 through 9.

Optical Pumping

Optical pumping is the use of photons to excite atoms or molecules. It was the first approach demonstrated, and conceptually it may be the simplest. Remember the simple three- and four-level lasers we described earlier. You can pump them optically by illuminating the laser material with light at the right wavelength to excite the lasing species (atom or molecule) from the ground state to the topmost level. Then it will drop to the metastable upper laser level and, from there, emit light.

Figures 3-1 and 3-2 may have made it seem that the pump light must be a specific wavelength, because they show narrow transitions between isolated energy levels. Fortunately, often this is not the case. The lower and—particularly—the upper levels of the pump transition usually span a range of energies. In practice, there are typically multiple upper levels, which all decay to the metastable upper laser level. This means that the laser can be excited at wavelengths corresponding to any transition between the ground state and those many upper levels. Because of this, many lasers can be optically pumped with a light source emitting a broad range of wavelengths, like a flashlamp.

(Note, however, that narrow-line pumping with another laser may transfer energy more efficiently, and it is required for some lasers to operate.)

One other important fact is that the pump photon must have higher energy—or equivalently shorter wavelength—than the emitted light. This is because the light-emitting species must be raised above the upper laser level from a starting point that is usually below the lower laser level. As we will see later, this is one of many limits on laser efficiency.

Optical pumping can be used with any laser medium that is transparent to the pump light. In practice, it is used most for solid-state crystalline lasers and for liquid tunable dye lasers. The optical pumping energy may be delivered steadily or in pulses, although many optically pumped lasers cannot produce a steady beam.

Electrical Pumping

Most artificial light sources in our lives get their energy from electricity, and lasers are no exception. Electricity is a convenient form in which to transmit energy and, as we shall see, it is also a useful form to use in exciting lasers. (Electricity also provides energy for the light sources used in optical pumping, but here we're talking about directly exciting the laser medium with electricity.)

A fluorescent light is a useful starting point for describing electrically excited gas lasers. An electric current flows through the gas in a fluorescent tube, transferring energy to the atoms and molecules in the gas. The excited atoms and molecules release that excess energy as light. The mercury atoms in ordinary fluorescent tubes actually emit ultraviolet light, which excites a phosphor coating on the inside of the tube to emit visible light.

In a typical small gas laser, an electric current flowing through the tube transfers energy to the gas. That energy excites the light-emitting species to the upper laser level, which then produces stimulated emission. As in fluorescent tubes, a high-voltage pulse initially ionizes the laser gas so it conducts an electric discharge, but a much lower voltage can sustain the current and operate the laser.

Some electrically driven gas lasers produce steady or "continuous-wave" beams as a constant current passes through the gas. Others produce pulses of light after intense electrical pulses pass through the gas. Some high-power lasers are excited by beams of electrons fired into the gas.

Semiconductor Laser Excitation

Electrical currents also power semiconductor lasers, but their operation is so different from that of other electrically pumped lasers that they deserve separate mention here as well as the more detailed description in Chapter 8. The current passing through a semiconductor laser produces a population inversion in a thin layer called the *junction*, where the composition of the semiconductor changes. The voltage levels needed are very small compared to those of gas lasers.

Strictly speaking, what are excited in a semiconductor laser are current carriers, electrons and vacancies called (appropriately enough) *holes*, which could accommodate electrons. The current removes electrons from the crystalline lattice to produce electron-hole pairs, which generate laser light when they recombine, with the electron dropping back down into the hole to form a lower energy state.

Other Energy Transfer Mechanisms

There are many other variations on energy transfer. Some simply involve the capture of electrical or optical energy by one species that transfers it to another. For example, in the helium-neon laser, helium atoms capture energy from an electric discharge in the gas. However, the helium does not lase. The helium atoms have some energy levels close enough to those of neon that they can transfer the excitation energy to neon atoms. Then the neon atoms drop into a metastable upper laser level to produce the population inversion needed to generate light.

There also are pumping techniques quite different from optical and electrical pumping. For example, chemical reactions generate excited species in "chemical" lasers. Atoms and molecules can capture energy produced in nuclear reactions to make nuclear lasers. Free-electron lasers get their energy from a beam of electrons passing through an array of magnets. We will learn more about these unusual types of lasers in Chapter 9.

Energy Efficiency

In describing optically pumped lasers, we pointed out that the pump wavelength is always less than the laser wavelength. This is symptomatic of something true for all lasers: you get less energy out in the beam than you put into the laser. In short, lasers are inherently inefficient. You must give up some energy to convert your input energy into the more orderly form of laser light.

The first energy limitation is simple to see in the diagram of energy levels in a three- or four-level laser. The pump energy raises the laser species from the bottom of the energy scale to the top. The laser transition releases only part of that energy; the rest is lost in other forms. In some cases, the inherent losses from the energy level structure are low. For example, optical pumping at 500 nm may produce laser output at 600 nm. That means that 17% of the pump photon energy does not emerge in the output photon (and does not count losses in producing the pump beam, which may be substantial). In other systems, the laser transition is far above the ground state, so much energy is used just to raise the laser species to a level high enough to emit laser light. As we will see in Chapter 6, this effect puts a very low ceiling—about 0.1%—on efficiency of the argon-ion laser.

Another serious limitation is the fact that excitation is never 100% efficient. The laser medium does not absorb all the pump light or electrical current directed into it.

There also are more subtle limitations. For example, laser action requires a population inversion, so some energy is put into the laser material to generate that inversion. Yet once more atoms or molecules accumulate in the lower level, the inversion no longer exists, but some atoms or molecules remain in the higher energy level, and the energy used to put them there is lost.

These effects combine to limit the overall efficiency of lasers. Some semiconductor lasers can convert about a third of the input energy into output light, and carbon-dioxide lasers can convert several percent or more of the electrical input into light. However, typical overall efficiency (sometimes called "wall-plug" efficiency when measured as power in vs. laser light out) of other lasers is 1% or less. We will look at these limitations in more detail in Chapter 4.

LINE SELECTION AND TUNING

So far, we have said little about the factors that determine what wavelengths a laser emits. As we will see later, some lasers emit only on a single transition, while others emit on two or more. Some, such as the helium-neon laser, can emit on any of several lines, depending upon how they are designed. Several processes are involved in selecting which transitions dominate in any given laser, and several criteria must be met before laser action can occur on any transition.

Atoms and molecules have complex energy-level structures. In general, the more electrons, the more electronic energy levels are possible. Likewise, the larger a molecule, the more vibrational and rotational energy levels are possible. It might seem that the more energy levels, the more transitions are possible, but in practice some transitions are much more likely than others.

It takes a set of transitions with the right set of properties to make a good laser, as shown in Fig. 3-11. These characteristics are inherent in the laser species.

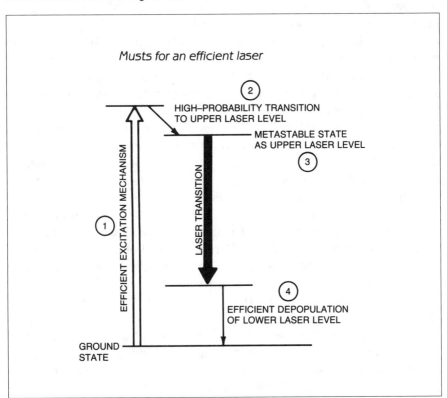

Fig. 3-11   Key transition requirements for a laser.

1. A transition with a high excitation probability is needed to pump the atom or molecule to a high energy level.

2. A high-probability transition is needed from that state to the upper laser energy level to produce a population inversion.

3. A metastable state with low probability of transition to lower energy levels is needed so atoms and molecules will remain in this state (as a population inversion) long enough to be stimulated to emit laser light.

4. (Ideally) A high-probability transition from the lower laser level to a lower state is desirable to depopulate the lower laser level and thus maintain the population inversion.

Meeting these basic requirements is not enough to make a laser, however. The first transition in the sequence works only if there is a practical and efficient way to transfer the excitation energy. The energy can be transferred by light, electrons, collisions with other species in a gas, or other processes—but it must be transferred efficiently enough to generate a population inversion.

Energy-transfer efficiency and transition probability depend on many factors. In a gas laser, for example, these include pressure of the laser gas, concentration of other gases in the laser mixture, temperature, size of the laser tube, and density of the current flowing through the laser medium.

Because energy-level structures are complex, excitation of a laser medium can produce population inversions on two or more transitions, involving one or more metastable levels. This can lead to stimulated emission at two or more wavelengths, which is often not desirable but can usually be controlled.

The nature of laser amplification provides some control over the wavelengths present, because it causes the most likely transitions to dominate the emission. The differences in likelihood of emission need not be great. Even a slight difference in gain—such as between 0.04 and 0.05/cm—is enough to lead to a dramatic difference in the strengths of two transitions originating from the same metastable level.

To understand how this works, let's look at a simple numerical example: two transitions from the same metastable level with gains 0.04 and 0.05/cm. Let's assume they start with the same power, and to simplify our calculations, let's use the simple formula:

$$\text{Power} = (1 + \text{Gain})^{\text{Length}}$$

This neglects the effects of resonator mirrors, losses within the laser medium and limits on possible gain, and does not follow proper rules for handling units of measurement, but it gives useful approximations. Table 3-1 shows how power builds up on the two lines with distance through the laser gas. Power would not be amplified as much as the calculations shown in a real laser, but the small difference in gain would cause the strong line to dominate.

TABLE 3-1 Effect of gain on relative power

	Relative Power		
Distance	Weak Line	Strong Line	Ratio
0.0 cm	1	1	1
1.0 cm	1.04	1.05	1.01
5.0 cm	1.22	1.28	1.05
10 cm	1.48	1.62	1.09
100 cm	50.5	131.	2.59
1000 cm	1.08×10^{17}	1.55×10^{21}	14351.

Cavity Optics

The optics of the laser cavity can also help select the wavelength at which the laser oscillates, although not in the way you might at first think. As long as the laser cavity is much longer than the wavelength, there is no problem finding a wavelength λ that meets the criteria $N\lambda = 2L$. Laser gain occurs over a wide enough range that some wavelength within that band is sure to meet the oscillation criteria.

However, laser cavity optics can suppress transitions by increasing the loss at the corresponding wavelengths. For example, suppose a laser can emit at two transitions, 400 and 600 nm, but you only want it to oscillate at 400 nm. You could use mirrors with different reflectivities at the two wavelengths. If the mirrors reflected light at 400 nm, but transmitted 600 nm of light, the cavity would have higher losses and no feedback at the longer wavelength, so the laser would emit only the shorter wavelength—even if the laser medium had the same gain at both wavelengths.

It's also possible to tune laser emission over a range of wavelengths by putting a prism or diffraction grating into the laser cavity, as shown in Fig. 3-12. Different wavelengths leave the prism or grating at different angles. Only one wavelength is bent at the proper angle to oscillate back and forth between the mirrors in the laser cavity. Shorter and longer wavelengths go off to the sides of the cavity and are lost. As a result, the net gain in the laser cavity is highest at the wavelength selected by the prism or grating. The output wavelength can be tuned across a range of wavelengths (usually limited by the laser material) by turning the grating, prism or mirror. This arrangement is used in tunable lasers of various types.

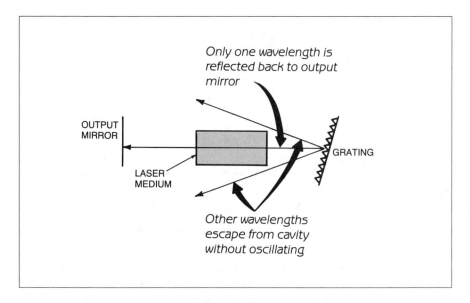

Fig. 3-12   Arrangement for tuning laser wavelength continuously.

WHAT HAVE WE LEARNED

- If the population is inverted, stimulated emission is more likely than absorption.
- Both populating the upper level and depopulating the lower level are important to maintain a population inversion. The upper laser level must have an unusually long lifetime.
- To produce a population inversion, you must excite specific states, not merely heat the material.
- Light and electricity are the types of energy most often used for laser excitation.

- The first maser had a population inversion produced by separating excited molecules from those in the ground state.

- Practical lasers involve three or four energy levels. The three-level laser is not ideal because the ground state is the lower laser level, requiring large excitation energies to produce a population inversion. A four-level laser works better because it requires less energy to invert the population.

- Many laser media contain more than one atomic or molecular species.

- Maser action can occur naturally in gas clouds in space excited by light from hot stars. Cosmic lasers emit light in all directions; they do not generate beams.

- A resonant cavity extracts energy efficiently from a medium with a population inversion.

- Gain measures growth of stimulated emission as light passes through the laser medium. Saturation keeps laser gain from reaching the "small-signal" value that assumes there are no limits on the growth of laser emission.

- Stimulated emission is concentrated along the length of the laser medium even without mirrors in a high-gain medium.

- Oscillation of light between mirrors at the opposite ends of the laser cavity gives the laser many of its distinctive properties.

- Oscillator cavities increase the effective length of the laser medium, producing reasonable powers from low-gain materials.

- The output mirror transmits some light to form the laser beam.

- Round-trip amplification equals loss plus output power in the laser beam.

- Laser power levels are higher inside the cavity than outside. The smaller the output coupling, the higher the intracavity power.

- Light waves are resonant in a laser cavity if twice the cavity length is an integral number of wavelengths, and their amplitudes add to produce constructive interference.

- Beam characteristics depend on the design of the laser cavity and output optics.

- Curved resonator mirrors help confine laser light to the cavity. They work best for low-gain lasers.

- Unstable resonators let light leak around the output mirror; they work best for high-gain lasers.
- Transverse modes manifest themselves as the pattern of light intensity across the laser beam. TEM_{00} is the lowest transverse mode, with intensity rising smoothly to a peak in the middle of the beam. There are a large family of TEM_{mn} modes.
- The pump wavelength must be shorter than the laser output.
- Semiconductor lasers are powered by electrical currents passing through a p-n junction.
- Energy can be transferred from one gas to another in a laser.
- Lasers are inherently inefficient, and suffer from several inevitable loss mechanisms. Laser excitation is never 100% efficient.
- Laser excitation can produce population inversions on multiple transitions. Laser amplification makes the strongest transitions dominant. Cavity optics can suppress oscillation by increasing losses at certain wavelengths.
- A wavelength-selective element in the cavity can tune output wavelength.

WHAT'S NEXT

Now that we have looked at the basic mechanisms behind laser operation, it's time to look at important laser characteristics. Later on, we will look at individual types of lasers.

Quiz for Chapter 3

1. What is the major advantage of a four-level laser over a three-level laser?
   a. More levels to excite atoms to
   b. The lower laser level is not the ground state
   c. More metastable states
   d. No advantage

2. If you put 1 W of laser power into a 10-cm-long oscillator with gain of 0.1/cm, what will be the output power, neglecting losses and saturation?
   a. 1 W
   b. 1.1 W
   c. 2.59 W

d. 10 W

e. 12.6 W

3. What is the emission angle of a laser rod (without mirrors) that is 1 cm in diameter, 1 m long, and has a refractive index of 1.5?

a. 0.005 degree

b. 0.43 degree

c. 1.02 degrees

d. 1.5 degrees

e. None of the above

4. The round-trip loss of helium-neon laser light in a cavity is 2%. The output mirror lets 1% of the light escape in the beam. When the laser is operating in a steady state, what is the round-trip amplification in the laser, measured as a percentage?

a. 1%

b. 2%

c. 3%

d. 5%

e. 99%

5. A helium-neon laser has intracavity power of 100 mW and output power of 1.5 mW. Neglecting cavity losses, what is the fraction of light reflected back into the cavity by the output mirror?

a. 1.5%

b. 50%

c. 90%

d. 98.5%

e. None of the above

6. The helium-cadmium laser has a wavelength of 442 nm. What is the round-trip length of a 30-cm-long cavity measured in wavelengths?

a. 13

b. 300

c. 475,000

d. 1,000,000

e. 1,357,000

7. Which of the following resonator types is not a stable resonator?

a. Plane-parallel

b. Concentric

c. Confocal

d. Hemispherical

e. Hemiconfocal

8. How many internal minimum-intensity points are there in a TEM_{03} mode beam?

a. None

b. 1

c. 2

d. 3

e. 6

9. What type of laser cavity could produce a beam with a central dark spot?

a. Plane-parallel

b. Confocal

c. Unstable

d. Stable

e. Concentric

10. Which of the following factors does not harm laser efficiency?
   a. Atmospheric absorption
   b. Excitation energy not absorbed
   c. Problems in depopulating the lower laser level
   d. Inefficiency in populating the upper laser level

CHAPTER 4

Laser Characteristics

ABOUT THIS CHAPTER

Now that we have seen what goes on inside lasers, it's time to look at what you see when you use a laser. In this chapter, we will learn about important properties of laser light, including coherence, wavelength, directionality, beam divergence, power, modulation, and polarization. We will also learn about the factors that determine laser efficiency.

COHERENCE

Coherence is probably the best-known property of laser light. Light waves are coherent if they are in phase with each other—that is, if their peaks and valleys are lined up at the same points, as shown in Fig. 4-1. Two things are necessary for light waves to be coherent. First, the light waves must start with the same phase at the same position. Second, their wavelengths must be the same, or they will drift out of phase because the peaks of the shorter wave will arrive slightly ahead of the peaks of the longer wave.

As we learned earlier, laser light is coherent because stimulated emission has the same phase and wavelength as the light wave or photon that stimulates it. The stimulated wave, in turn, can stimulate the emission of other photons, which are coherent with both it and the original wave.

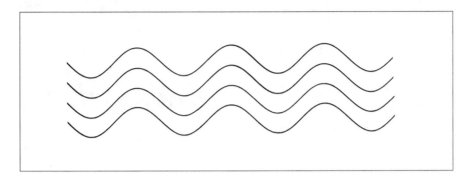

Fig. 4-1 Coherent light waves.

Coherence of Laser Light

The preceding statement might make it sound like all laser light would be perfectly in phase, but this is not the case. Not all photons in a laser beam are descended from the same original photon by stimulated emission, so they don't all start out in phase. In addition, the uncertainty principle causes tiny variations in the wavelengths of the emitted light, which accumulate and become significant after the light goes a long distance. Tiny fluctuations within the laser—such as thermal gradients or vibrations—can also affect one light wave differently than others, degrading coherence. Thus, there is no such thing as "perfect" coherence.

Not all laser light is equally coherent. In fact, the light from some lasers is almost incoherent. The reasons lie inside the lasers themselves. Some lasers emit a broader range of wavelengths than others, and, as we saw above, light waves must have the same wavelength to be coherent. A laser that oscillates in only one longitudinal mode is more coherent than one that emits in multiple longitudinal modes, because the different modes have slightly different wavelengths, which reduce the degree of coherence. Lasers with low gain emit a narrower range of wavelengths than high-gain lasers, so they are more coherent. The uncertainty principle causes the range of wavelengths in a pulse to increase as the pulse length decreases, so the shortest pulses tend to have the broadest wavelength ranges. On the other hand, continuous-wave beams can have the narrowest wavelength range. Add this all together and you find that the most coherent beams come from continuous-wave, low-gain lasers operating in the lowest-order TEM_{00} mode.

Types of Coherence

If you look closely, there are actually two kinds of coherence: temporal and spatial. *Temporal* coherence measures how long light waves remain in phase as they travel (the term "temporal" is used because the degree of coherence is compared at different times). Light waves become incoherent as differences in their optical paths or wavelengths make them drift out of phase. All light has some temporal coherence, but only over a characteristic *coherence length*, which is very close to zero for ordinary light bulbs but can be many meters for lasers. The coherence length depends on the light's nominal wavelength λ and on the range of wavelengths $\Delta\lambda$ emitted:

$$\text{Coherence Length} = \lambda^2/2\Delta\lambda$$

We can convert this equation to calculate coherence length from frequency units, because a laser's bandwidth is often given in frequency terms:

$$\text{Coherence Length} = c/2\Delta\nu$$

where $\Delta\nu$ is the range of frequencies and c the speed of light in vacuum.

We can see what this means for a few representative sources. Light bulbs emit light from the visible well into the infrared. If we assume the range of wavelengths is from 400 to 1000 nm and take an average wavelength of 700 nm, we find that a light bulb has a suitably short coherence length of 400 nm. An inexpensive semiconductor laser has a wavelength of 800 nm and a wavelength range of 1 nm, giving a coherence length of 0.3 mm. An ordinary helium-neon laser has a much narrower linewidth of about 0.002 nm at its 632.8-nm wavelength, corresponding to a coherence length of 10 cm. Stabilizing a helium-neon laser so that it emits in a single longitudinal mode limits the linewidth to about 0.000002 nm, and thus extends the coherence length to about 100 m.

Spatial coherence, on the other hand, measures the area over which light is coherent. Strictly speaking, it is independent of temporal coherence. If a laser emits a single transverse mode, its emission is spatially coherent across the diameter of the beam, at least over reasonable propagation distances.

Coherence and Interference

Light waves must be somewhat coherent to show the interference effects described in Chapter 2. If you superimpose many
incoherent light waves of different wavelengths, interference effects
average out, and they add together to form white light. This is why we
don't see interference effects in everyday things, like scenes illuminated by light bulbs or the sun. However, if the light waves are coherent enough to maintain the same phase relative to each other (even
though they are not all lined up with peaks and troughs at the same
points), adding them together does produce interference effects. At
some points, the total amplitudes cancel, producing a dark zone; at
other points, the amplitudes add together to produce a bright zone.

These interference effects are desirable in many measurement
applications, because they let us measure distance by counting in
units of wavelength. Because light waves are so small, this lets us
measure distance very precisely, a practice called interferometry.
Holography also requires coherent light.

Coherence is not always desirable, however. When coherent light
waves pass through turbulent air, or are reflected from many types
of surfaces, they form grainy patterns called *speckle*. This shifting
speckle is an interference pattern created by slight differences in the
paths that light rays travel. It contains information on the quality of
the air and the surface, but for most practical purposes it is merely
background noise. Such coherent effects make illumination by coherent light inherently uneven, and thus undesirable for many purposes.

LASER WAVELENGTHS

Laser light is usually considered *monochromatic*, meaning single-
colored, but in practice laser light is not perfectly monochromatic.
Some lasers can emit on two or more transitions, and, as we saw in
Chapter 3, stimulated emission generates a range of wavelengths on
each transition. The range of wavelengths (or bandwidth) depends on
the nature of the transition. The amplification that occurs when stimulated emission oscillates within a laser cavity further narrows the
range of wavelengths emitted on each transition, although two or
more wavelengths can oscillate simultaneously in the same cavity.
Thus, lasers normally emit a range of wavelengths, which can be

narrow or broad, and can be changed by adjusting the laser's optics. Nonetheless, the bandwidth of even the broadest-band laser emission is much narrower than that of ordinary light sources such as the sun.

To understand the nature of laser bandwidth, we need to take a closer look at some things that happen inside lasers.

Transitions and Gain Bandwidth

Our earlier discussions may have made a laser transition sound like an abrupt spike at one specific wavelength. In reality, a laser transition is not an abrupt spike, but a curve with a pronounced peak. Figure 4-2 shows two closely related laser characteristics. The broader curve is the net gain on the transition, the amount of amplification possible in the laser medium as a function of wavelength. The narrower one is the intensity of laser output as a function of wavelength. (The laser emission peak has been reduced to fit on the same scale.)

The much sharper peak of the laser emission is caused by amplification and oscillation within a laser cavity. The difference between the gain and emission curves depends on the cavity design. As light passes through the laser medium, stimulated emission is most likely at the wavelengths where gain is highest. Suppose, for example, the gain was 0.1/cm at 500 nm and 0.01/cm at 505 nm. After travelling

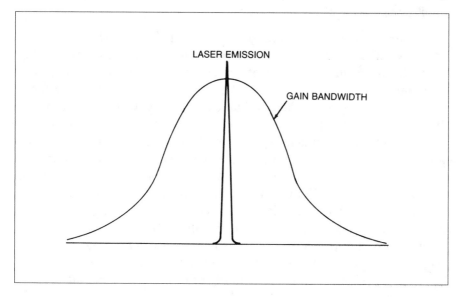

Fig. 4-2   Gain bandwidth and laser emission on a laser transition.

through 100 cm of the laser medium, that 10-to-1 gain ratio alone would produce a 5000-to-1 difference in output power, if the initial powers were equal (which would not be the case because initial power would be lower where gain was weaker). That effect makes the emission curve much sharper than the gain curve.

If the laser cavity optics do not select any particular wavelength, laser intensity will be highest at the wavelength where the gain is highest. However, laser action can occur at any wavelength where gain is enough larger than zero to overcome losses in the laser cavity. Thus, the cavity can select a wavelength away from the peak of the gain curve. As we will see later, this allows tuning or adjustment of the output wavelength.

There is one important qualification to this picture of how laser gain makes the intensity of the emitted light peak sharply with wavelength. The effect is not as strong for transitions between different pairs of energy levels. If two transitions have the same upper energy level, stimulated emission is likely to occur first on the stronger one, depopulating that level and making stimulated emission less likely on the weaker level. However, if they have different upper levels, the upper laser level of the weaker transition will not be affected by stimulated emission on the stronger transition. This allows simultaneous emission on different transitions with different strengths.

Line Broadening

Several effects contribute to the width of a laser's gain curve, and they play different roles in different types of lasers. Some broadening comes from the uncertainties of the quantum world. On a quantum-mechanical level, the likelihood of a transition is a probability function, which is largest at the nominal transition wavelength and drops off sharply—but not precisely to zero—as the wavelength changes.

Interaction of the laser species with other atoms and molecules causes other broadening. In gases, the degree of broadening increases with gas pressure, because increasing the pressure decreases the interval between collisions with other atoms or molecules. These collisions can affect energy levels and energy transfer, in effect "blurring" transitions. Most gas lasers normally operate at low pressures, but in some high-pressure gas lasers this effect can make separate laser emission lines merge into a continuous spectrum. In a solid laser, the transfer of vibrational energy to other atoms can alter emission wavelength.

Another line-broadening mechanism is the inherent motion of atoms and molecules, which can shift wavelength by a process called the Doppler effect. Suppose an atom moves toward you while emitting at a constant frequency ν, which normally corresponds to a wavelength of 500 nm. If it moves toward you 1 nm in the interval between successive wave peaks, the next peak it emits is only 499 nm behind the first. Thus, you see a shorter wavelength (or, equivalently, a higher frequency), sometimes called a "blue shift." If the atom is moving away, the emitted wavelength becomes longer (a "red shift" because it shifts the light toward the red end of the spectrum).

Atoms and molecules in a gas are constantly moving randomly, with an average velocity that depends on temperature and mass. This random motion means that emitted light experiences random Doppler shifts, spreading out the range of emitted wavelengths. For a temperature T and a mass M, the Doppler bandwidth, or broadening caused by this effect measured in frequency units, $\Delta\nu/\nu$, is given by:

$$\Delta\nu/\nu = [5.545\ kT/Mc^2]^{1/2}$$

where $\Delta\nu$ is the change in frequency, ν is the frequency, k is the Boltzmann constant, and c the speed of light in vacuum. This equals a few parts per million for typical gas atoms or molecules. The fractional broadening is the same when measured in units of wavelength.

Wavelength Selection in a Cavity

We also saw, in Chapter 3, that light at many different wavelengths λ could oscillate in a simple plane-parallel mirror cavity of length L as long as they satisfied the equation

$$2L = N\lambda$$

where N is any integer. As long as the laser cavity is much longer than the wavelength (which is the usual case), it's virtually certain that some wavelength given by the equation will fall within the active medium's gain curve. In fact, as shown in Fig. 4-3, two or more oscillation wavelengths usually fall within the laser's gain curve. Each wavelength is a distinct longitudinal mode of oscillation.

We should note that we define wavelength λ as the wavelength in the laser cavity. This is not the laser wavelength in vacuum because

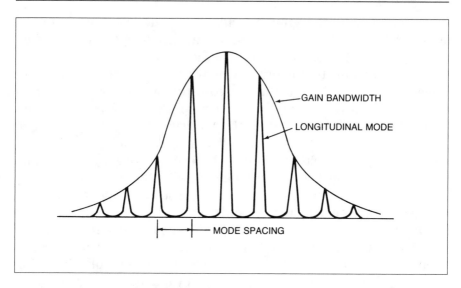

Fig. 4-3   Multiple longitudinal modes fall within the gain bandwidth of a gas laser.

there's something inside the laser cavity—the laser medium, which has a refractive index n. As we saw earlier, the wavelength of light in a material with refractive index n, λ_n, equals the wavelength in vacuum λ divided by refractive index, or λ/n. Thus, to be more accurate, the equation defining possible wavelengths in a laser cavity should read:

$$2nL = N\lambda$$

with n the refractive index in the cavity, L the cavity length, λ the wavelength (in vacuum) and N an integer. This equation is precise only if the laser medium is a single material with uniform refractive index n, but it does indicate how refractive index affects wavelength. As we saw earlier, for a gas the refractive index is very close to 1, but it is close to 1.5 in glass and higher in some other transparent materials.

The actual profile of wavelengths emitted by a laser is the product of the gain profile and the envelope of longitudinal oscillation modes. The wavelengths at the sides of the gain curve tend to be weaker than those in the center, for the reasons we described earlier. The profile in Fig. 4-3 is based on typical emission from a gas laser, which can oscillate in a few longitudinal modes. In some other lasers one central longitudinal mode dominates, with much lower power emitted at one or two peripheral modes.

Single Longitudinal Mode Operation

There are ways to limit laser oscillation to the width of a single longitudinal mode, which is about 1 MHz for helium-neon lasers, or 0.000001 nm (10^{-15} m). The most common technique is to insert into the laser cavity a pair of parallel reflective surfaces that form a resonant cavity, called a Fabry-Perot etalon, which is tilted at an angle to the axis of the laser medium, as shown in Fig. 4-4. The etalon reflects some light away from the cavity mirrors at most wavelengths, increasing cavity loss and suppressing laser oscillation. However, at certain wavelengths interference effects nullify the reflection, eliminating the loss and allowing laser oscillation.

Etalons restrict laser oscillation to wavelengths far enough apart that only one falls under the laser's gain curve. This limits laser oscillation to a single longitudinal mode. The wavelength of the low-loss mode can be tuned by adjusting the spacing between the Fabry-Perot reflectors, or by turning the device in the laser cavity. (Many Fabry-Perot etalons are solid blocks with reflective surfaces, which can be tuned only by turning them.)

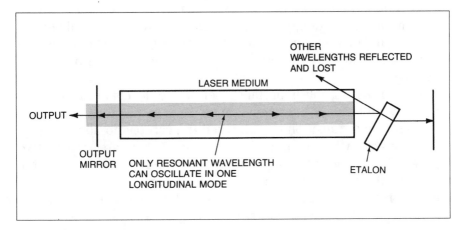

Fig. 4-4   A Fabry-Perot etalon limits a laser cavity to oscillation in a single longitudinal mode.

Laser Wavelength Stability

Laser wavelengths are not absolutely stable. The transition energies do not change but, as we have seen, other factors influence the actual oscillation wavelength. For example, temperature fluctuations may cause thermal expansion of the laser cavity or change the refractive index of the laser medium.

The gradual drift of laser wavelength can change the balance of light emitted at the wavelengths corresponding to different longitudinal modes. In some cases, this can lead to "mode hopping," where the laser shifts quickly from one dominant mode to another, at a slightly different wavelength.

Single- and Multi-Wavelength Operation

We have seen that the nominal "single wavelength" emitted by a laser is not really a single wavelength but a narrow band of wavelengths defined by a single transition. In some cases, a family of closely spaced transitions can emit at a number of closely spaced wavelengths, or allow the wavelength to be tuned continuously through a range or band. In other cases, emission is at several discrete wavelengths, which may involve different upper and/or lower laser levels. The results differ from the user's standpoint.

Two important examples of how transitions can overlap to form a continuous tuning range are the tunable dye laser and the high-pressure carbon-dioxide laser. In later chapters, we will describe them in more detail, but here we will concentrate on their output wavelengths.

In both lasers, the important laser energy levels are broken up into many sublevels. The carbon-dioxide laser is the more straightforward. The main transition occurs when the carbon-dioxide molecule shifts from one vibrational mode to another. (There are actually two such transitions close to one another, one centered near 10.5 μm, the other near 9.4 μm.) While shifting vibrational states, the molecule also changes rotational energy levels. The rotational levels are much smaller than the vibrational transitions, so at low pressures an untuned carbon-dioxide laser can emit many wavelengths, as shown in Fig. 4-5. Each line is a transition between different rotational sublevels during a vibrational transition.

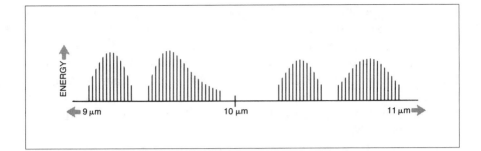

Fig. 4-5 The many wavelengths emitted by a carbon-dioxide laser near 10 μm.

As we mentioned earlier, increasing gas pressure spreads the laser transitions over a wider range of wavelengths. In carbon-dioxide lasers, the broadening is large enough and the transitions close enough that eventually they overlap to form a continuous range of wavelengths.

Even the natural line-narrowing we described above does not prevent these lasers from emitting at multiple wavelengths. It is possible to adjust the cavity optics so that they only allow oscillation at some wavelengths, but that comes at the cost of lower output power, because it limits the number of atoms or molecules from which stimulated emission can extract energy.

Many lasers can emit light on two or more transitions and at widely separated wavelengths. In some, the transitions are between entirely different pairs of energy levels. In others, the transitions may involve some common energy levels. For example, one metastable state of neon can emit on transitions at 543, 632.8 and 3391 nm in helium-neon lasers. The laser cavity optics determine the transition on which such a laser will oscillate.

Laser Wavelength Tuning

At the end of Chapter 3, we mentioned how lasers could be tuned to emit different wavelengths. The basic technique is simple but powerful, and deserves a bit more explanation now that we have talked about gain bandwidth.

Laser cavity optics normally reflect light over a range that is wide compared to the laser's gain bandwidth. Inserting a prism or diffraction grating into the laser cavity effectively narrows the cavity's oscillation bandwidth. The prism or grating directs light of different wavelengths at different angles. As shown in Fig. 3-14, only one narrow range of wavelengths is directed at the right angle to oscillate back and forth within the laser cavity (including the prism or grating). In effect, the mirror or prism increases losses at all other wavelengths by deflecting light out of the laser cavity, thus preventing oscillation at those wavelengths.

Turning the grating, prism or some other optical element in the laser cavity, as we saw earlier, changes the wavelength that can oscillate in the laser cavity. As long as that wavelength remains in the laser's gain bandwidth, the laser can operate. However, if the wavelength falls outside the gain bandwidth, the laser cannot oscillate. Because all laser media have limited gain bandwidths, their tuning ranges are limited. For example, a dye laser might be tunable only between 565 and 615 nm with one dye; to get other wavelengths, you would need other dyes.

As we saw earlier in this chapter, gain varies with wavelength. If you tune laser wavelength, the gain in the laser medium will change. As gain changes, so will output power, so tuning wavelength will alter the power level.

LASER BEAMS AND MODES

Another important characteristic of lasers is that their emission is concentrated in a narrow beam. As we learned in Chapter 3, the nature of this beam depends largely on the laser cavity. Now let's look more closely at beam characteristics outside the laser cavity.

Beam Divergence, Diameter, and Spot Size

We think of laser beams as tightly focused and straight, but they actually spread out slightly with distance. The spreading angle is called *beam divergence*. It is usually measured not in the familiar unit of degrees, but in radians or, more commonly, milliradians—thousandths of a radian. The radian is a unit derived from circular measurements. A full circle equals 2π radians, and one radian equals 57.2958 degrees. One milliradian is 0.0572958 degrees, or 3 arcminutes and 33.36 arcseconds. Radians are convenient because the tangent of a small angle roughly equals its measure in radians. The divergence of most continuous-wave gas lasers is around one milliradian, but it is usually larger for pulsed lasers. (For the few lasers where the beam divergence is much larger, it is often written in degrees.)

Beam divergence from a laser oscillator depends on the nature of the resonator, the size of the output aperture and a process called "diffraction," which sets a theoretical lower limit on divergence.

Diffraction is a consequence of the wave nature of light that we haven't discussed before (although it can be considered an interference effect). It is the spreading out or scattering of light waves as they pass a sharp edge. If we were to consider light waves just as rays, diffraction would seem to bend them as they pass the edge. A better way to understand diffraction is to pretend that each point of the object diffracting light is emitting a new light wave, which spreads out in all directions. Interference effects concentrate the light in the forward direction, but some is scattered in a regular pattern at other angles. In Fig. 4-6, you can see what are called Airy diffraction

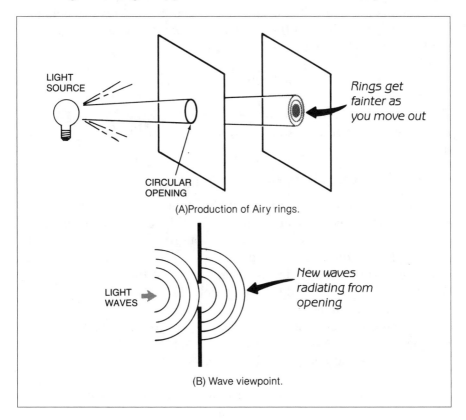

Fig. 4-6 Diffraction of light waves by a circular opening.

rings, formed when light waves pass through a small circular hole. The central bright spot is surrounded by a series of increasingly faint rings.

Our example of light passing through a hole is similar to the way a laser beam emerges from an output mirror (the "hole"). As that implies, a laser beam is subject to the same diffraction effects that spread out ordinary light. Indeed, these diffraction effects determine the minimum divergence and spot size of a laser beam. For TEM_{00} beams the divergence is usually *diffraction-limited* to a theoretical minimum; other beams may spread more rapidly.

Near Field Conditions

Diffraction-caused beam divergence occurs only in what is called the "far field," at comparatively large distances from the laser. In the near field, the beam shows little spreading and remains, essentially, a bundle of parallel light rays.

The distance from the laser over which the light rays remain parallel is sometimes called the *Rayleigh range*. It depends roughly on the beam diameter d and the wavelength λ:

$$\text{Rayleigh Range} = d^2/\lambda$$

For a visible beam with nominal wavelength 500 nm, this range is 2 m if the beam is 1 mm in diameter, and 50 m if the beam diameter is 5 mm.

Far-Field Beam Divergence

Beyond the Rayleigh range, in the laser's far field, beam divergence becomes the critical parameter in calculating beam diameter, as shown in Fig. 4-7. Divergence angle is normally measured from the center of the beam to the edge. The edge itself must be defined because beam intensity drops off gradually at the sides. The usual definition of the beam's edge is the point where intensity drops to $1/e^2$ of the maximum value.

Calculating beam size is a matter of trigonometry. Multiplying distance by the tangent of the divergence angle gives the beam radius, which must be doubled to get diameter:

$$\text{Beam Diameter} = 2 \times \text{Distance} \times \tan(\text{Divergence})$$

or, in more proper mathematical form,

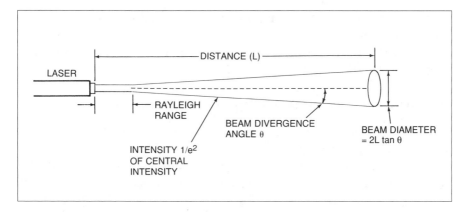

Fig. 4-7   Divergence of a laser beam, exaggerated to make it visible.

$$D = 2 \, (Dist) \tan \theta$$

where D is the beam diameter, $Dist$ the distance from the laser to the spot and θ the divergence angle. If divergence is under about 0.1 radian (6 degrees), you don't have to bother calculating the tangent; the angle in radians is a good enough approximation. That is the case for most lasers, except semiconductor types.

Beam Waist and Divergence

There are several ways to calculate beam divergence. If the resonator mirrors are curved, the beam has a narrow point, or "waist," inside the laser cavity. In the far field, we can approximate the divergence angle with the formula

$$\theta = \lambda / \pi W$$

where θ is the beam divergence, λ the wavelength and W the diameter of the beam waist.

The beam waist diameter itself depends on laser wavelength, cavity length and design. For a simple confocal resonator (with two mirrors both having their focal points at the midpoint of the cavity) of length L, the beam waist diameter W is

$$W = (L\lambda / 2\pi)^{0.5}$$

or about 0.17 mm for a 30-cm- (1-ft) long helium-neon laser emitting at 632.8 nm. (Note that this is smaller than the output spot size, typically about 1 mm for such lasers.) This relation holds for lasers with resonant cavities that produce good-quality beams. The best beams are said to be diffraction-limited, because their divergence is limited by diffraction effects.

Output Port Diameter and Beam Divergence

Although it is helpful to understand these relationships, in practice you will be given a laser's beam diameter and divergence. The diameter is measured at the laser's output port. It can be changed by optical systems that expand or contract the laser beam but still leave the light all going in the same direction. The divergence depends on how intensity varies across the beam as well as on the diameter at the output port, but a good approximation for diffraction-limited beams is:

$$\text{Divergence} = \text{Wavelength/Diameter}$$

or

$$\theta = \lambda/D$$

where D is the diameter of the output optics. A more precise formula is:

$$\theta = K\lambda/D$$

where K is a constant somewhere in the range of 1 that depends on the variation of intensity across the beam.

This formula tells us that the larger the output optics, the smaller the beam divergence. Thus, you need large optics to focus light onto a small spot. Conversely, the shorter the wavelength, the smaller the spot size. In practice, this means that you can't have both a tiny spot diameter and narrow divergence. If you start with a 1-mm-diameter beam from a helium-neon laser, for example, its minimum divergence is about 0.6 milliradian.

Divergence can be larger than these formulas indicate if the beam contains multiple transverse modes, which spread out more rapidly than a single-mode TEM_{00} beam. Lasers without resonant cavities have beams that diverge even more rapidly.

By the way, the reason we're being vague about the value of the constant K is because it changes with the distribution of light across the emitting area or laser beam. If the light is equally bright across the entire opening, its value is 1.22. However, the value is smaller if the intensity drops off smoothly with distance from the center of the beam.

Focusing Laser Beams

Our earlier descriptions of focusing by lenses might have led you to think that they could focus a laser beam down to a perfect point of zero diameter. However, even a perfect TEM_{00} laser beam cannot be focused that tightly. The minimum diameter of a focal spot S formed by a lens of diameter D and focal length f with light of wavelength λ is:

$$S = f \lambda / D$$

(assuming that the laser beam fills the whole lens diameter.) If you're familiar with photography, you may recognize the ratio f/D as the focal ratio, or "f number," of a lens. Thus, for ordinary lenses this formula implies that the smallest focal spot is a few times the wavelength of light. (It is hard to make lenses with focal ratios much smaller than about 2.)

Transverse Mode Effects

In talking about beam diffraction, we have assumed that the beam is in the TEM_{00} mode, where intensity drops smoothly from a peak at the center. However, as we saw in Chapter 3, there are a whole family of TEM_{mn} modes, which can make the beam spread more rapidly.

The crucial factor for far-field divergence is the number of transverse modes (N) the laser emits simultaneously. Far-field divergence increases roughly with the square root of the number of modes ($N^{1/2}$), and the area of the illuminated spot is thus roughly proportional to N.

Longitudinal Modes

We saw earlier that lasers have longitudinal as well as transverse modes. These modes are actually different wavelengths that lie within the laser's gain bandwidth and satisfy the equation for oscillation in a laser cavity, as shown in Fig. 4-3. Each wavelength spike is a longitudinal mode.

Because lasers have both transverse and longitudinal modes,

references to "single" and "multimode" operation can be confusing. In most cases, those terms usually refer to *transverse* rather than longitudinal modes. Lasers oscillating in a single longitudinal mode are often called single-wavelength or single-frequency lasers.

Remember that a laser operating in a single transverse mode can oscillate in two or more longitudinal modes. Indeed, many gas lasers emit a single TEM_{00} mode beam that contains two or more longitudinal modes. As we saw earlier in this chapter, special optics can restrict such lasers to a single frequency.

Polarization

Polarization is the alignment of the electric and magnetic fields that make up a light wave. There are several types of polarization, but the simplest to understand—and most relevant to introductory laser physics—is linear polarization. Light waves that are linearly polarized all have their electric fields aligned in the same direction. One peculiar property of light waves is that they can always be separated into components with linear polarizations that are perpendicular to each other (although those polarizations can change with time). Thus, unpolarized light can be divided into two components with linear polarization, one with a vertical electric field and one with a horizontal electric field.

Normally, most lasers emit unpolarized light. However, adding a component that selects between the two perpendicular polarizations lets a laser generate linearly polarized light. The usual choice is a Brewster window, shown in Fig. 4-8. This is a surface aligned at a critical angle at which different things happen to light waves with different polarization. If the polarization is in the plane of incidence (the plane of the paper in Fig. 4-8), no light is reflected. However, about 15% of the light is reflected if the plane of polarization is perpendicular to the angle of incidence on a glass surface.

Placing a Brewster-angle window at the end of the tube of a low-gain laser makes a critical difference in its operation. The Brewster window makes gain different for light of different polarizations. Light with the higher-gain polarization is amplified more, so only that linear polarization appears in the output beam. Note that designing the laser to produce a linearly polarized output beam does not reduce the output power—it restricts laser oscillation to one polarization, but does not "throw away" the output of the other polarization.

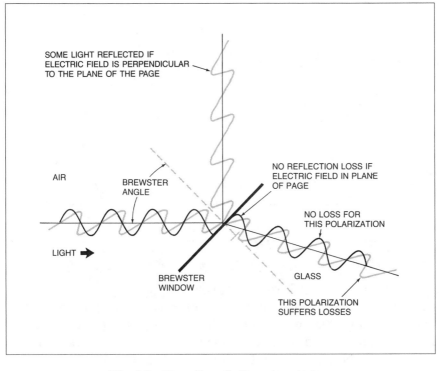

Fig. 4-8   Operation of a Brewster window.

OSCILLATORS AND AMPLIFIERS

So far, we have assumed that all lasers are oscillators with a pair of cavity mirrors, one totally reflective, and an output mirror that transmits a fraction of the incident light which emerges as the laser beam. As we mentioned earlier, you can have a population inversion and stimulated emission without mirrors. In fact, lasers are sometimes used as amplifiers.

The basic arrangement is shown in Fig. 4-9. The laser on the left is an oscillator, with a totally reflective rear mirror and a partly transparent output mirror. The beam from the oscillator passes through another cavity containing the same laser medium, but without mirrors. In our drawing, the oscillator beam makes one pass through the amplifier, stimulating emission to increase output power. Amplifiers are used to increase pulse powers to higher levels than a single-stage oscillator can generate, and to amplify weak signals in fiber-optic communication systems.

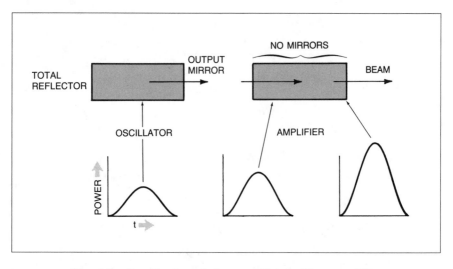

Fig. 4-9   Combination of a laser oscillator with an amplifier.

One functional difference between oscillators and amplifiers is that oscillators can generate their own beams, but amplifiers cannot. However, as we learned earlier, high-gain lasers do not need a highly reflective pair of mirrors. In fact, some types of lasers work well with a totally reflective rear cavity mirror and an output window that reflects only a small fraction of the light back into the laser cavity and transmits the rest in the laser beam.

OUTPUT POWER

From the standpoint of laser users, the power level in the beam is one of the most important laser characteristics. This is a more complex matter than it may seem at first, because it depends upon several things, including variations in power level with time, efficiency of converting excitation energy into laser energy, excitation methods, and size of the laser itself. If we look at each of these factors individually, we also find that we may have to look a bit deeper.

Laser Efficiency

Efficiency is the sort of word that you think you understand until you have to measure it. This is especially true with lasers. The most useful practical measurement is often called *wall-plug efficiency* and measures how much of the energy put into the laser system (through the wall plug) comes out in the laser beam. Unfortunately, this is not always the efficiency that people measure—even when they use the term "overall" efficiency. Sometimes other types of efficiency are given, including "slope" efficiency, "excitation" efficiency and "conversion" efficiency. All those factors enter into the equation for wall-plug efficiency.

Wall-Plug Efficiency

Let's start by looking at all the factors that enter into wall-plug efficiency. For simplicity, we'll consider a gas laser, but the same principles work for other lasers as well. Electrical power enters the laser through a power supply, which converts much (but not all) of the input energy into drive current that is passed through the laser gas. Much (but not all) of that electrical energy is deposited in the gas (energy-deposition efficiency). Much of the deposited energy excites the laser medium and produces a population inversion. Much of the energy in the population inversion is converted into a laser beam.

We could carry that breakdown even further if we wanted, but that should give you the idea. The wall-plug efficiency is the product of the efficiencies of several processes:

       Power Supply Efficiency
×   Energy-Deposition Efficiency
×   Excitation Efficiency
×   Fraction of Energy Available on
       Laser Transition
×   Fraction of Atoms That Emit Laser Light
×   Cavity Output-Coupling Efficiency
=   Overall Efficiency

Plug in some numbers and you can see why lasers have low wall-plug efficiencies. If the power supply, energy deposition, and excitation each are 80% efficient, the laser transition represents 50% of the excitation energy, half the atoms emit laser light, and 80% of the laser energy is coupled out of the cavity in the beam (all of which are optimistic assumptions), then the wall-plug efficiency is $0.8 \times 0.8 \times 0.8 \times 0.5 \times 0.5 \times 0.8 = 0.1024$, or 10%. Let's look briefly at the major elements in the efficiency equation.

Power-Supply Efficiency

No power supply is 100% efficient. In general, the more manipulation of the input electrical energy, the lower the efficiency. Conversion of 120-V alternating-current input to the low direct-current voltages that drive semiconductor lasers is straightforward. Likewise, conversion to steady high voltage to drive a continuous-wave gas laser is efficient. However, losses can be significant if the power supply must generate intense electrical pulses for an electron-beam-driven pulsed laser.

The biggest losses in the power supply category come when converting electricity into other forms of energy. For example, solid-state crystalline and glass lasers require optical excitation. This light is generated by passing electricity through a flashlamp or arc lamp, or using electricity to excite another laser. This conversion process is often inefficient and causes important losses.

Excitation Efficiency

Some energy that enters the laser medium does not excite atoms or molecules to the upper laser level. One reason is that some light or electricity passes through the laser medium unabsorbed. Some absorbed energy does not put atoms or molecules into the right state to emit on the laser transition; for example, it might heat a gas rather than raise atoms to the upper laser level.

Absorption efficiency depends on laser design. The tradeoffs can be subtle. For example, it might seem a good idea to have a lot of laser material to absorb all the energy entering it. However, that would require too much light-absorbing material in some parts of the laser cavity, and that excess material would reduce the population inversion needed for laser action. As such tradeoffs indicate, there is usually an optimum value for absorption efficiency that depends on the properties of the laser material and the way it is excited.

Another limitation on excitation efficiency comes from the fact that not all the energy can be absorbed efficiently. Remember that, in our simple examples of three- and four-level lasers, there was a single pump transition. From a very simplified point of view, the excitation energy must match that transition to excite the laser. Actual requirements aren't that stringent, but the excitation energy must be packaged in the right way to be absorbed. For example, a flashlamp emits light across much of the visible spectrum (and parts of the ultraviolet and infrared), but solid-state crystalline lasers can absorb only some of those wavelengths (at least in ways that produce a population inversion). Some of the remaining energy may heat the laser medium; some of it may not be absorbed at all.

Laser Transition Efficiency

When we described four-level laser transitions, we showed that laser pumping raised an atom or molecule to an excited level, which dropped down into the upper laser level. After laser emission, the atom or molecule again dropped from the lower laser level to the ground state. If you go back and look carefully at Fig. 3-2, you'll note that some energy was lost in the process.

To make the point clearer, let's look at a hypothetical atom in Fig. 4-10 and assign each energy level a value. (For simplicity, we'll use energy units, electronvolts.) Suppose the atom initially absorbs 4 eV to raise it to a short-lived upper level and then drops to the upper laser level at 3 eV. The laser transition takes the atom to the lower laser level at 2 eV, from which the atom drops to the ground state at

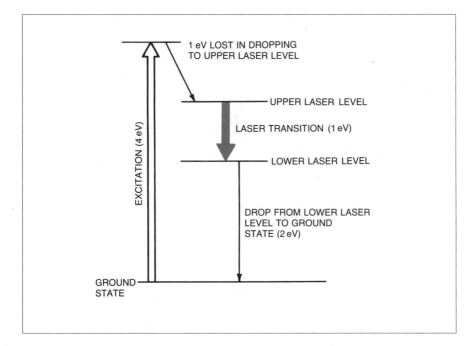

Fig. 4-10 How 4 eV of excitation produces only 1 eV of laser energy.

0 eV. It takes 4 eV to produce a 1 eV laser photon—meaning that the transition is, at most, 25% efficient. Some real laser transitions are better than our example, but many are much worse.

The need to produce a population inversion makes matters even worse, especially in a three-level laser, where the ground state is the lower laser level. For a three-level laser to work, more atoms must be in the upper laser level than in the ground state. Once half the atoms are in the ground state, it won't emit any more laser light, and the energy left in the remaining atoms in the upper laser level is lost.

Matters are not as bad in four-level lasers, where ways exist to get atoms and molecules out of the lower laser level. However, laser action in any system cannot occur until a population inversion is produced. This requires putting enough energy into the laser to pass what is called a *threshold* for laser action, which strongly affects efficiency. Only after the threshold is passed does some input energy emerge as laser light.

We show how a laser threshold works by plotting input energy vs. laser output in Fig. 4-11. No light emerges until the input power

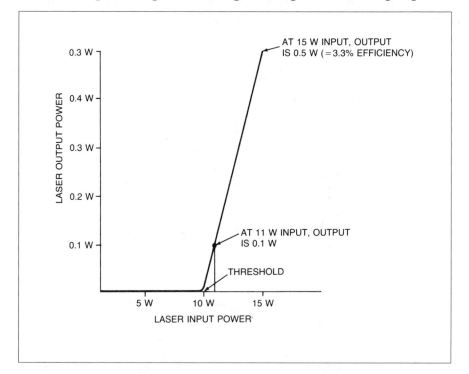

Fig. 4-11 Laser output does not emerge until the input energy has passed a threshold at 10 W. Then it rises with 10% slope efficiency.

passes the threshold, at which point the laser is quite inefficient. However, above the threshold, part of each extra increment of input power becomes output power; the fraction is called the *slope efficiency*. This means that efficiency increases above the threshold. For example, if you need 10 W of input power to pass the threshold, that much input may produce only 0.001 W of output, for 0.01% efficiency. However, if slope efficiency is 10%, adding another watt of input power adds 0.1 W of output power, raising the total power to 0.101 W, for 0.9% efficiency.

Slope efficiency is an important concept, because it shows the fraction of energy above the laser threshold that emerges as output. Note, however, that you should take care in defining slope efficiency —because it, too, can denote wall-plug efficiency or other types of efficiency.

Energy Extraction Efficiency

One of the less obvious limitations on laser efficiency is the efficiency with which energy can be extracted from the laser cavity. There are two related components to this limitation:

1. Geometry of the laser cavity

2. Efficiency of stimulating emission

The geometric problem is that the laser resonator does not collect light energy from the entire excited volume of the laser medium. Figure 4-12 shows the problem graphically. An electric discharge passes through the laser tube, exciting the whole laser medium. However, the resonator mirrors only collect light from the shaded area.

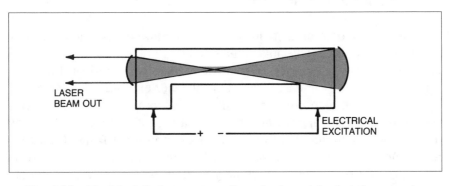

Fig. 4-12   Electrical discharge passes through a laser tube, but the resonator only extracts energy from the gas in part of the tube.

Excitation energy absorbed outside that volume cannot contribute to the laser beam. This problem is more severe with stable resonators; one attraction of unstable resonators is their ability to collect laser energy throughout the laser cavity.

Waste Heat

As we have shown, most energy that goes into a laser does not emerge in the beam. Although semiconductor lasers can exceed 30% efficiency, few other lasers exceed 1%, as shown in Table 4-1.

All this energy does not vanish; most becomes heat, which must be removed from the laser. Convection cooling by air can remove waste heat from low-power lasers, but as power (and waste-heat) levels increase, so do cooling requirements. Some lasers require forced-air cooling with a fan; others require flowing water or some other liquid coolant. (We did not consider the cooling system's energy needs in our rough estimate of efficiency in Table 4-1.)

TABLE 4-1 Efficiency ranges of commercial lasers

Type	Wall-Plug Efficiency
Argon-ion	0.001–0.1%
Carbon-dioxide	5–20%
Copper-vapor	0.2–0.8%
Excimer (rare-gas halide)	1.5–2%
GaAlAs semiconductor	1–over 30%
Helium-cadmium	0.002–0.02%
Helium-neon	0.01–0.1%
Neodymium	0.1–over 10%

Pulsed vs. Continuous Emission

Lasers can emit light steadily or in *continuous-wave* mode, or they can produce pulses. The pulses may be single or repeated at a steady (repetition) rate. The precise characteristics depend on both the physics of the laser and the design of the laser excitation mechanism.

Some lasers can emit only pulses because the lower laser levels fill up during laser operation. This can happen if stimulated emission adds atoms or molecules to the lower excited level faster than they are removed. The lower laser level becomes a bottleneck, with its rapid population increase eventually ending the population inversion. The lower level continues emptying after the end of the self-terminating pulse, so the laser may be ready to produce another pulse quickly, but it cannot emit light steadily.

Excitation methods limit other lasers to pulsed operation. For example, the excimer laser is excited by an electrical discharge flowing through a mixture of gases. That discharge is stable for less than a microsecond, after which the end of the discharge terminates the laser pulse. (Excimer kinetics also limit pulse duration.)

Heat dissipation problems limit other lasers to pulsed operation. One example is neodymium-doped glass, which has thermal conductivity that is too low to dissipate the heat generated if laser action is continuous.

Other lasers can operate in a steady state, generating a continuous beam. This requires a combination of favorable circumstances: a set of energy levels and transitions that allows continuous operation, a stable way to supply energy continuously to the laser medium, and an efficient way to remove waste heat. The higher the power, the lower the efficiency, or the more compact the laser, the more serious the waste-heat problem becomes. Practical ways to remove excess heat include flowing laser gas through the tube, forced-air cooling with a fan, or cooling with a flowing liquid (water or a refrigerant).

In practice, some lasers operate only in pulsed mode because they cannot sustain steady laser emission. Others operate only in continuous-wave because it takes time to establish the right conditions for laser oscillation. Some lasers can operate either pulsed or continuously, depending on operating conditions.

Excitation Techniques

Excitation techniques often determine if lasers operate pulsed or continuously. Laser emission usually occurs almost immediately after excitation starts, and stops almost immediately after the excitation ends. The rise and fall times are not instantaneous, but in pulsed lasers are typically faster than the excitation pulses. Note, however, that it takes time to establish stable operating conditions for some continuous-wave lasers, as well as for pulsed lasers (such as metal-vapor types) that must be heated above room temperature.

The pulse repetition rate depends on the excitation mechanism as well as on internal laser physics. Most gas laser media can dissipate heat and otherwise restore normal conditions faster than a power supply can accumulate enough energy to fire another electrical pulse into the gas. Electrical pulse generation speeds can also limit repetition rates of some flashlamp-pumped solid-state lasers. However, other solid-state lasers are limited by the rate at which heat can be removed from the laser rod. Changing the drive current can modulate low-power semiconductor lasers very fast, at billions of pulses per second, because they are very small and dissipate little power.

Some lasers can operate pulsed or continuous-wave under different conditions. For example, the carbon-dioxide laser emits a continuous-wave beam when excited by a steady discharge at low gas pressure. However, it can produce short pulses if electrical pulses are passed through a gas mixture at much higher pressures near one atmosphere.

Optical and electrical excitation both allow considerable control over pulse characteristics. Even a nominally continuous-wave laser like the helium-neon laser can be made to emit pulses by turning its power supply off and on. However, this can come at a cost in reduced stability, particularly for electrical excitation, and it may be more practical to modulate the beam intensity outside the laser. External modulators effectively block the beam at certain times, switching it on and off, as described in Chapter 5. Pulse characteristics can also be changed by inserting special optical components into the laser cavity. The most important are modelockers, Q switches, and cavity dumpers, also described in Chapter 5.

Limits on Pulse Duration and Bandwidth

The length of time that a laser emits light sets a fundamental limit on the range of wavelengths it can emit. The product of the pulse length (in seconds) times the bandwidth (in hertz) must equal at least 0.441. This is another consequence of the Heisenberg Uncertainty Principle, and it produces what laser specialists call a "transform limit" between bandwidth and pulse length.

This means that the bandwidth of a laser pulse can be no smaller than

$$\text{Bandwidth} = 0.441/\text{Pulse Length}$$

Thus, the longer the pulse, the narrower the minimum bandwidth. Continuous emission does not allow infinitely narrow linewidth, but continuous lasers have the narrowest bandwidth of any laser.

If we turn our formula around, we can see something else—that a laser must have a broad linewidth if it is to generate short pulses. The minimum pulse length is:

Pulse Length (Seconds) = 0.441/Bandwidth (Hertz)

To understand just how broad a bandwidth is required, we should convert the bandwidth measured in hertz to nanometers. First we find the relative bandwidth by dividing the emission bandwidth in nanometers $\Delta\lambda$ by laser wavelength λ:

Relative Bandwidth = $\Delta\lambda/\lambda$

Then we convert wavelength in nanometers into frequency in hertz, ν, using the formula

$$\nu = c/\lambda$$

where c is the speed of light in vacuum (300,000 km/s). Finally, we multiply the frequency by the relative bandwidth (which is a fraction) to find the bandwidth in hertz. Step through this for a laser with 1-nm emission bandwidth at a center wavelength of 500 nm, and you find the minimum theoretical pulse length is 367.5 femtoseconds (367.5×10^{-15} s).

That may sound impressively short, but few lasers fit the bill. They have to emit light over the entire range; most with broad bandwidth emit on a series of lines, corresponding to the longitudinal modes mentioned earlier. The tunable dye lasers described in Chapter 9 have long been used to generate short pulses, but lately some solid-state lasers described in Chapter 6 have become important (notably titanium-doped sapphire). However, even these lasers run out of bandwidth.

To get even shorter pulses, researchers pass light from a laser through special optics, which first stretch the pulse out in length, then stretch its bandwidth out, and then recompress it to an even shorter pulse. This has yielded pulses as short as 6 femtoseconds (6×10^{-15} s).

Power and Energy vs. Time

We can think of the difference between pulsed and continuous-wave lasers as a difference in how their output power varies with time. Before we go on, we should look carefully at these variations.

Figure 4-13 plots the variation of power with time for a generalized laser pulse. The pulse shape is fairly typical, with a Gaussian profile that rises slowly at first, then rapidly to a peak, and then drops off symmetrically. A laser pulse has a characteristic rise and fall time. In practice, rise and fall times are measured between the points where laser power is 10% and 90% of the maximum value, not the 0 and 100% points. In some lasers, there may be a small prepulse (before the main pulse) and/or a long tail dropping off slowly after the main pulse, which are shown as dashed lines in Fig. 4-13. Pulse duration is typically measured as the "full width at half maximum"—meaning the interval from the time the rising pulse passes 50% of the peak power to the time the power drops below that 50% point.

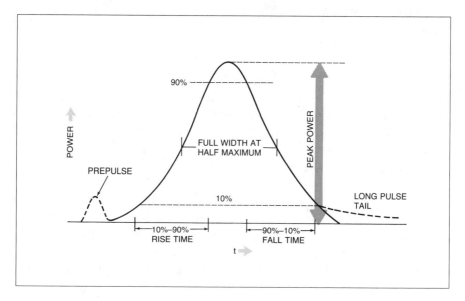

Fig. 4-13   Power variation as a function of time (dashed lines show prepulse and long tail).

The peak power is defined as the highest level of optical power in the pulse. It may be attained only during a tiny fraction of the pulse. However, in practice laser specialists often assume the laser emits peak power for the full duration of the pulse (as measured at half maximum). As we will see, this greatly simplifies calculations of pulse power and energy.

The energy in a laser beam is the power emitted in a given time. If the power is at a constant level:

$$E = Pt$$

where E is energy (in joules), P is power (in watts) and t is time (in seconds). Thus, a 0.001-W laser beam delivers 0.001 joule in 1 s, 0.002 joule in 2 s, and so on. If we were to be rigorous about calculating total energy, we would have to integrate power as a function of time over the appropriate interval:

$$E = \int P(t)\, dt$$

For pulsed lasers, it is useful to know the average level of power that the laser emits over an interval in which the beam is both on and off. This can be calculated by adding together pulse energies and dividing by time, using a summation formula:

$$\text{Average Power} = \sum E/t$$

Alternately, average power can be approximated by multiplying peak power by the "duty cycle"—the fraction of the time the laser is emitting light:

$$\text{Average Power} = \text{Peak Power} \times \text{Duty Cycle}$$

Duty cycle in this case is a dimensionless fraction given by:

$$\text{Duty Cycle} = \text{Pulse Length} \times \text{Repetition Rate}$$

where repetition rate is pulses per second.

To give a concrete example, let's calculate average power of a laser that emits 20 pulses/s, with each pulse lasting 1 μs and having peak power of 100 kW. For simplicity, assume the power is constant during the pulse. The duty cycle is:

$$\text{Duty Cycle} = 10^{-6} \times 20 \text{ pulses/s} = 2 \times 10^{-5}$$

From that figure, we can calculate average power as:

$$\text{Average Power} = (100 \times 10^{3} \text{ W}) \times (2 \times 10^{-5}) = 2 \text{ W}$$

We can get the same value by first calculating pulse energy from peak power:

$$E = (100 \times 10^3 \text{ W})/(10^{-6} \text{ s}) = 0.1 \text{ joule}$$

and then multiplying pulse energy by the repetition rate of 20 pulses/s (usually called hertz) to give:

$$\text{Average Power} = 0.1 \text{ joule/pulse} \times 20 \text{ pulses/s} = 2 \text{ W}$$

Scaling Effects and Power

In Chapter 3, we saw that laser gain was measured as an increment per unit length, for example, 0.02/cm. This means that laser power increases by 2% over the original value after passing through 1 cm of laser medium, or it equals 1.02 of the input value. For a medium of length L, this leads to a formula:

$$\text{Total Power} = (\text{Input Power}) \times (1 + \text{Gain})^L$$

Remember that this formula is valid only for "small-signal" gain, when the power levels are far from the maximum possible from the medium; output power cannot increase without limit. Actual output power is limited by other factors.

The job of raising laser powers to higher levels is called "scaling," and specialists know that each design can scale over only a limited range. They learned that the hard way, after building laser tubes that (counting bends and curves) had gain lengths of up to 750 feet! The designs that work well at low powers often do not work well at much higher powers. Likewise, other designs may make efficient high-power lasers but be of little use at low powers. Some lasers, like helium-neon, work well over only a limited range, from below 1 mW to about 50 mW. Others, like CO_2, can work from under 1 W to tens of kilowatts. However, no one design works for CO_2 lasers over that entire power range. Typically, the lowest-power CO_2 lasers have gas sealed within a tube, but models generating tens of kilowatts require rapid gas flow through the laser cavity.

Energy Storage and Gain

The concept of energy storage in a laser cavity is helpful in understanding how many lasers operate. Suppose, for example, that the active medium has been excited with a flashlamp, but stimulated emission has yet to start. That laser cavity contains stored energy in the form of a population inversion ready to be released as stimulated emission. The amount of stored energy depends on the number of excited atoms or molecules and their excitation energy. It also depends on the excited-state lifetime: the shorter the lifetime, the less energy the laser medium can store.

In practice, the lifetime of the excited state is related to the cross section for stimulated emission, which measures the likelihood of stimulated emission. That cross section, in turn, is related to laser gain. The higher the cross section, the higher the laser gain. Likewise, the higher the cross section, the shorter the lifetime of the upper laser level. This leads to the somewhat paradoxical result that the higher the gain of the laser medium, the less able it is to store energy.

Pause a moment, however, and you can begin to see why energy storage decreases as gain increases. The higher the gain, the more readily the laser transition occurs—and the harder it is to keep energy in the upper laser level.

WHAT HAVE WE LEARNED

- Coherent light is in phase and has the same wavelength.
- Temporal coherence indicates how long light waves remain in phase; spatial coherence is the area over which light is coherent.
- Coherence length equals the speed of light divided by twice the bandwidth.
- Light waves must be coherent to display interference effects.
- Coherent noise generates grainy patterns called speckle.
- Although nominally monochromatic, laser light has a nonzero bandwidth.
- The gain bandwidth of lasers is broader than the range of emitted laser wavelengths.
- Thermal motion of atoms and molecules shifts wavelength by the Doppler effect.
- Multiple longitudinal modes fall within the gain bandwidth of a laser transition.
- Wavelength within the laser cavity depends on the refractive index.

- An etalon can restrict a laser to oscillation in a single longitudinal mode.
- Lasers can emit on many overlapping transitions, or on a set of discrete transitions. Many lasers can emit at two or more widely separated wavelengths on different transitions.
- Lasers can be tuned in wavelength with a prism or diffraction grating that lets only one wavelength oscillate in the laser cavity.
- Beam divergence is observed in the far field; in the near field, laser beams show little spreading. Beam divergence can be calculated from the diameter of the "waist" inside the laser cavity.
- Beam divergence decreases as the diameter of the output optics increases, and increases as the wavelength increases.
- The minimum diameter of a focused laser beam depends on laser wavelength, focal length, and the diameter of the lens.
- Beam divergence increases with the number of transverse modes.
- Lasers oscillating in a single longitudinal mode are often called single-wavelength or single-frequency lasers.
- Addition of a Brewster-angle window can linearly polarize laser output.
- Some lasers are used as amplifiers, without resonator mirrors.
- The most useful measure of efficiency compares laser output to overall input energy entering through the wall plug. Wall-plug efficiency is the product of the efficiencies of several processes.
- The need to maintain a population inversion limits laser efficiency.
- Most energy that enters a laser emerges as waste heat.
- Laser pulse characteristics can be controlled by devices inside or outside the laser.
- Laser pulses have a characteristic rise time, pulse duration, and fall time.
- Energy is calculated by measuring the power received over an interval. If power is constant, this simplifies to the power level times time.

- Average power is the total energy emitted divided by how long energy is collected.
- Laser output power does not depend on length of the medium alone. Scaling is complex and differs among lasers.
- Energy can be stored as an inverted population in a laser cavity.

WHAT'S NEXT

In Chapter 5, we will learn about major laser accessories and how they are used.

Quiz for Chapter 4

1. A semiconductor laser emits 820-nm light with a bandwidth of 2 nm. What is its coherence length?
   a. 822 nm
   b. 0.168 mm
   c. 20 cm
   d. 1 m
   e. 12.8 m

2. A single-frequency dye laser has a bandwidth of 100 kHz (10^5 Hz) and a wavelength of 600 nm. What is its coherence length?
   a. 600 nm
   b. 20 cm
   c. 1 m
   d. 12.8 m
   e. 1.5 km

3. Which of the following contributes to the broadening of laser emission bandwidth?
   a. Doppler shift of moving atoms and molecules

b. Amplification within the laser medium
   c. Coherence of the laser light
   d. Optical pumping of the laser transition
   e. None of the above

4. How many longitudinal modes can fall within a laser's gain bandwidth?
   a. 1 only
   b. 2
   c. 3
   d. 10
   e. No fixed limit, dependent on bandwidth and mode spacing

5. What is the distance over which light from a typical helium-neon laser, with a 1-mm beam diameter and 632.8-nm wavelength, shows no divergence?
   a. No such distance, the

beam starts diverging immediately
b. 0.23 m
c. 1.58 m
d. 6 m
e. 1.5 km

6. A helium-neon laser on the
ground has beam divergence
of one milliradian. What
would its beam diameter be
at the 38,000-km altitude of
geosynchronous orbit?
a. 1 m
b. 38 m
c. 76 m
d. 1 km
e. 76 km

7. Suppose we add a 1-m
output mirror to limit the
helium-neon laser's beam
divergence. What then is the
beam diameter at the 38,000-
km altitude of geosynchronous orbit? (Assume that
$K = 1$.)
a. 1 m
b. 48 m
c. 94 m
d. 1 km
e. 3.8 km

8. What is the smallest possible spot you can make when
focusing a helium-neon laser
beam with a lens having 10-
cm focal length and 2.5-cm
diameter?
a. 632.8 nm
b. 1000 nm
c. 1265.6 nm
d. 2531 nm
e. 2 cm

9. What is the maximum wall-
plug efficiency for a laser
with a power supply that is
70% efficient and energy-
deposition efficiency of 25%,
in which the laser transition
represents 60% of the excitation energy and half the
excited atoms emit laser
light?
a. 1%
b. 5.25%
c. 10%
d. 25%
e. None of the above

10. A pulsed laser generates
500-kW pulses with full
width at half maximum of
10 ns at a repetition rate of
200 Hz. Making simplifying
assumptions about pulse
power, what is the average
power?
a. 1 W
b. 2 W
c. 5 W
d. 10 W
e. 20 W

CHAPTER 5

Laser Accessories

ABOUT THIS CHAPTER

Many optical accessories help lasers perform useful functions. In this chapter, we will briefly describe the most important of those accessories. We will start with classical "passive" optics, such as lenses and prisms, move on to "active" optics, which modulate or otherwise change laser light, and then finally describe some emerging technologies.

ACTIVE VS. PASSIVE OPTICS

We can divide optical devices into active and passive devices. Passive optics always do the same thing to light. A lens, for example, always bends light in the same way, and a mirror always reflects light in the same way. Fiber optics, polarizers, filters, telescopes, and prisms are also passive optics.

Active optics can do different things to light at different times. One example is a modulator, which changes how much light it transmits depending on the signal that drives it. Another example is a rotating mirror that scans a laser beam across a remote surface. The important difference is that active optics are controlled actively, while passive optics do not require external control.

We should warn you that some equipment used with lasers— such as light detectors—does not fall neatly into either category. However, the distinction between active and passive optics can be useful.

CLASSICAL PASSIVE OPTICS

At the end of Chapter 2, we learned how lenses and mirrors refract and reflect light. There we talked only about single, simple lenses and mirrors. However, lenses and mirrors can be combined to make much more useful optics. Thorough coverage of the world of classical optics would require a book in itself, but we can give a brief overview.

Simple Lenses and Mirrors

We can consider most laser beams as sets of parallel light rays, with semiconductor lasers an exception because of their large beam divergence. From the viewpoint of classical optics, parallel light rays are equivalent to light emitted by an object an infinite distance away.

A single positive lens (or a concave mirror) focuses parallel light rays to a small spot at its focal point. This lets you concentrate laser power to a small spot, with a size that depends on the size of the optics and the wavelength of light. The minimum spot size for a lens with focal length f and diameter D, and for a laser wavelength λ is given by the formula we learned in the last chapter:

$$\text{Spot Size} = f\,\lambda/D$$

It is comparable in size to the wavelength.

The same optics that focus parallel light rays to a point can also bend light from a point source to form a parallel beam. (You can see this because optics focus light rays the same way, no matter in which direction they travel.) This is how a flashlight lens or searchlight mirror forms light from a bulb into a narrow beam.

Beam Collimators and Telescopes

Though it may seem amazing today, about two millennia passed between the discovery that lenses or mirrors could focus light and the discovery of the telescope. The ancient Greeks knew about burning glasses, but the telescope was invented about 1600 A.D. It seems doubtful that either discovery was the outcome of a systematic investigation. Legend tells us that children playing with lenses in the shop of Dutch spectacle maker Hans Lippershey discovered the Galilean telescope, which is made of a large positive lens with a smaller negative lens that serves as the eyepiece. Replacing the negative lens with a positive lens makes a more powerful telescope, but turns the image upside down.

Telescopes can do more than just focus light to a point. As shown in Fig. 5-1, they can also expand (or shrink) the bundle of parallel light rays that make up a laser beam. This is important, because it can change the size of a laser beam. A telescope that serves this purpose is called a "beam expander" or "collimator." The same optics can expand or contract the laser beam, depending on which way the beam enters. Although our example shows a refracting telescope (one made with lenses), a reflecting mirror (with a mirror as one element) can also serve as a beam collimator.

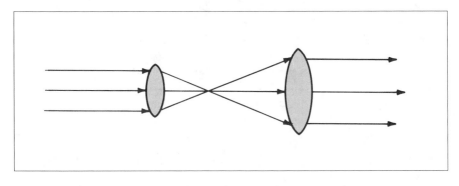

Fig. 5-1 A refracting telescope serves as a beam expander or collimator for a laser beam.

Retroreflectors

We saw in Chapter 2 that a flat mirror reflects an incident beam at an angle equal to the angle of reflection. However, a flat mirror won't make the beam return directly to its source unless the beam strikes it exactly perpendicular to its surface.

You can make light always return precisely back in the direction from which it came by arranging three flat mirrors at right angles to one another, like the corner of a cube, in a *retroreflector*. This device is shown in Fig. 5-2. Some retroreflectors are "hollow," made up of three front-surface mirrors. Others are prisms with reflective rear surfaces.

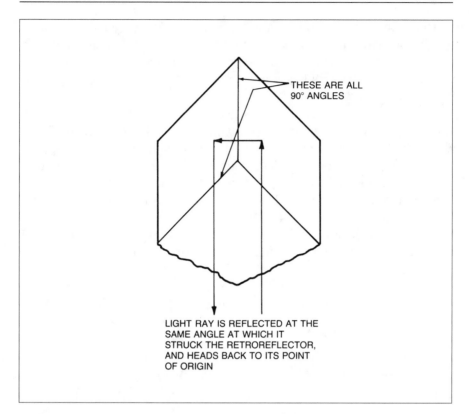

THESE ARE ALL
90° ANGLES

LIGHT RAY IS REFLECTED AT THE
SAME ANGLE AT WHICH IT
STRUCK THE RETROREFLECTOR,
AND HEADS BACK TO ITS POINT
OF ORIGIN

Fig. 5-2 A retroreflector returns light precisely back in the direction from which it came.

Optical Aberrations

So far, we have assumed that simple optics do their job perfectly, bringing light of all wavelengths to a perfect focus. Life is not that kind. Optics suffer several kinds of *aberrations* that spread out the focal spot. The two most important for our purposes are chromatic and spherical aberration.

Chromatic aberration occurs in transmissive optics because the refractive index n is a function of wavelength. Recall from Chapter 2 that a positive lens has a focal length

$$f = \frac{1}{(n-1)\left[\dfrac{1}{R_1} + \dfrac{1}{R_2}\right]}$$

where R_1 and R_2 are the radii of curvature of the surfaces, and n is the refractive index of the glass. This formula indicates that focal length changes with refractive index. The change is not large, but it can be significant. For example, in one common type of optical glass, n is 1.533 at 434 nm in the violet and 1.517 at 656 nm in the red. This means that the focal length in the red is 1.03 times that in the violet, or 3% longer. The effect is strongest near the edge of a lens or optical system, where it may give white objects a red fringe on one side and a blue fringe on the other. You can see this in cheap binoculars, or in thick plastic eyeglass lenses.

Chromatic aberration is not a big problem with lasers that emit only one wavelength. However, compensation may be needed if the laser emits multiple wavelengths, or if the same optical system is used by the human eye. The simplest way to compensate for chromatic aberration is to glue together components made of different glasses, which have refractive indexes that change in different ways with wavelength. Careful selection of glass compositions can largely cancel the variations in refractive index with wavelength, although the correction is not perfect. Lenses with two or more elements designed to cancel chromatic aberration are called *achromatic lenses*.

Spherical aberration occurs because all parts of a spherical lens or mirror do not focus light to the same point. The outer parts have a slightly shorter focal length than the central zone. The difference is larger for more sharply curved surfaces. You can cope with the problem by using only the central part of the lens or mirror, or by picking optics with small curvature (and thus long focal length). An alternative is to make the surface with a nonspherical shape so it focuses all light rays at the same distance, no matter where they strike the surface. Such "aspheric" optics are harder to make than spherical surfaces, but work better, particularly at short focal lengths.

Coatings and Their Functions

Optical surfaces are normally coated with thin-film materials that enhance or suppress the reflection of light. The technology for making these coatings is complex, but the principles are fairly straightforward. We'll describe each basic type separately.

Antireflection Coatings

The theory of electromagnetic waves tells us that some light is always reflected when light passes from one material into a material with a higher refractive index, such as from air into glass. The amount of reflection depends on the polarization of light and the refractive indexes of the materials. If the light leaving a material with refractive index n_1 strikes a transparent material with index n_2 at a normal angle, the reflectance is:

$$\text{Reflectance} = \frac{\left(\frac{n_2}{n_1} - 1\right)^2}{\left(\frac{n_2}{n_1} + 1\right)^2}$$

If the light is entering glass, with $n_2 = 1.5$, from air ($n_1 = 1$), this gives a loss of about 4%.

At different angles, the reflectance differs for light polarized parallel to or perpendicular to the plane of incidence (the plane that contains the incident, transmitted, and reflected light rays). However, the overall dependence on refractive index remains. The larger the refractive index, the larger the losses caused by reflection.

A simple way to reduce this reflective loss is to reduce the difference in refractive index between air and the optical medium. You can't reduce the glass's index, but you can coat it with a lower-index material. The light must pass through two interfaces, but the reflective losses are much lower. For example, if the coating has refractive index of 1.25, the reflectance at the air-coating interface is:

$$\text{Reflectance} = (1.25 - 1)^2/(1.25 + 1)^2 = 1.23\%$$

and the reflection at the coating-glass interface is:

$$\text{Reflectance} = ([1.5/1.25] - 1)^2/([1.5/1.25] + 1)^2 = 0.83\%$$

for a total reflectance loss of about 2%, half the level without the coating. (Total loss is calculated not by adding losses but by multiplying the fractions of light transmitted through the two surfaces, which is one minus the loss, and then subtracting that from one.)

Interference Coatings

Much more complex effects are possible in multilayer *interference coatings*, also called *dielectric coatings* because they are made of "dielectric" materials that do not conduct electricity. These coatings consist of many alternating layers of two different materials, one with a high refractive index and one with a lower index. Layer thicknesses are chosen so that light reflected from one layer constructively (or destructively) interferes with that reflected from other layers, depending on wavelength. This lets the designer control transmission and reflection as a function of wavelength.

Proper selection of the layer materials and thicknesses lets coating designers meet demanding specifications. For example, they can make coatings that reflect or transmit only a narrow range of wavelengths. Such coatings can transmit (or block) a narrow range of wavelengths at a laser line. Thus, they can block white light from overhead lights while transmitting the red line of the helium-neon laser in the laser scanners used for automated checkout in supermarkets.

Interference coatings can give very high peak reflection or transmission (at the desired wavelength). However, they have some practical limitations. One is that their transmission characteristics change when light strikes them at a different angle, and thus "sees" a different layer thickness, causing different interference effects. Tilt an interference coating that strongly reflects one color, and its apparent color will change with the viewing angle.

Optical Filters

Optical filters selectively attenuate some light that reaches them. The degree of attenuation depends on filter design. There are several kinds.

Spectral Filters

Wavelength-selective or *spectral* filters transmit some wavelengths and attenuate others. The degree of attenuation and selectivity vary widely. *Notch* filters transmit only a very narrow range of wavelengths; *bandpass* filters typically transmit a broader range. "High-pass" and "low-pass" filters transmit wavelengths shorter than (or longer than) a certain value.

Typically, interference coatings are used as filters for laser applications because of their good wavelength selectivity and the design flexibility they allow. Because they reflect rather than absorb light, they can withstand reasonable laser powers better than filters that absorb light.

Color Filters

Color filters serve a function similar to spectral filters, in that they transmit certain wavelengths and attenuate others. However, they rely on the absorption of light by colored materials to select the transmitted wavelengths. This is fine for their major applications in photography. However, color filters do not select wavelengths as well as interference filters, and they can be damaged by absorbing high powers they are trying to block.

Neutral-Density Filters

Neutral-density filters reduce or attenuate light intensity uniformly at all wavelengths covered. That is, their absorption is essentially independent of wavelength, and they look grey to the eye.

Attenuation of a neutral-density filter is normally measured as *optical density*, defined as:

$$\text{Optical Density} = -\log_{10}(\text{Transmission})$$

Thus, an optical filter that transmits 1% (0.01) of incident light has an optical density of 2. Reversing the formula gives the fraction of transmitted light:

$$\text{Transmission} = 10^{(-\text{Optical Density})}$$

Spatial Filters

A *spatial* filter transmits one region of a beam and blocks the rest. A simple example and common type of spatial filter is a pinhole in a sheet of black-painted metal. It transmits the center of the laser beam, but not the outer portions. Spatial filters are often used to pick out the central, most uniform part of a laser beam. Slits are also spatial filters.

Optical Materials

In Chapter 2, we saw that the way materials transmit light depends on wavelength. Materials we take for granted as transparent, such as air, water, and glass, are opaque in parts of the infrared and ultraviolet. Other materials that are opaque to visible light are transparent at other wavelengths. Reflectivity also varies with wavelength. As a consequence, different materials are used for optics in different parts of the spectrum.

What we call glass—an impure form of silicon dioxide, SiO_2— remains the most common transparent optical material for wavelengths in and near the visible. It is transparent and durable, and its technology is well-developed. We can make clear glasses with a range of refractive indexes sufficient to meet most optical needs. Quartz, a natural crystalline form of SiO_2 (silica), is used for some optics, as is purified silica. Other optical glasses contain dopants to alter the refractive index.

Clear plastics are used for some visible optics. They are lighter than glass and have a lower refractive index. Their biggest practical advantage is the low cost of making molded plastic lenses. However, internal inhomogeneities can cause high light scattering, which makes plastic lenses unsuitable for many laser applications, although they usually work well enough with white light.

A host of other transparent materials are used in other parts of the spectrum. Figure 5-3 lists some of these materials and shows the wavelengths where they are transparent enough for optical applications.

CYLINDRICAL OPTICS

So far, we have assumed that all lenses and mirrors are radially symmetric. That is, we have assumed that it doesn't matter if you turn them around their centers without tilting them. Light reaching such a lens at the same angle and the same distance from its center is always bent in the same way. Such lenses are fine for most purposes. You don't want a telescope to refract light differently if you rotate the lenses around the telescope axis.

However, there are cases where you do not want circularly symmetric optics. For example, the oval beams from semiconductor lasers diverge more rapidly in one direction than in the direction perpendicular to it. To make the beam circular, you need an optical element that bends the light in one direction but not in the perpendicular direction. Such an element, also shown in Fig. 5-4, is called a *cylindrical lens*, because its surface looks like a piece of a cylinder.

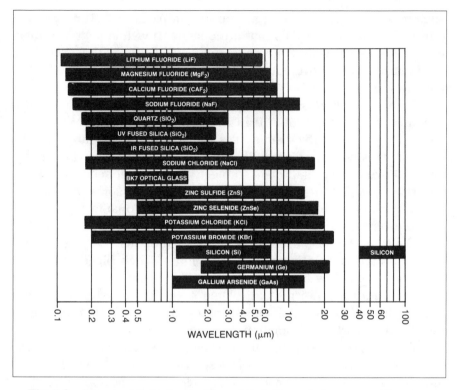

Fig. 5-3 Important optical materials and their transmission ranges. (Courtesy of Janos Technology Inc.)

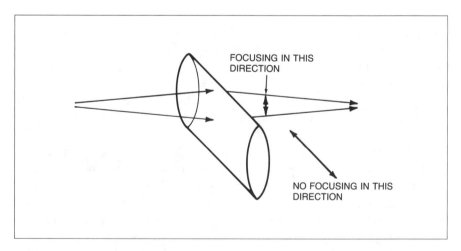

Fig. 5-4 A cylindrical lens focuses light in one direction, but not in the perpendicular direction.

The action of a cylindrical lens depends on how you view it. If you look at a vertical plane in Fig. 5-4, the lens cross section is curved. Thus, the lens focuses light in that direction, bending light rays so they do not diverge as rapidly. However, if you look in a horizontal plane, the cylindrical lens is just a flat piece of glass, with no curvature and no focusing power. Thus, the light rays go straight through, without being bent in that direction.

Cylindrical lenses are not normally used in imaging systems because—as you would imagine—they distort the image. However, they are important in some laser applications because they can collimate the asymmetrical beam from a semiconductor laser to make it look more like the better-behaved beam from a gas laser.

DISPERSIVE OPTICS

Often, it is useful to spread out or disperse different wavelengths of light. Two types of optical components—prisms and diffraction gratings—rely on completely different principles to do the same task.

Prisms

Prisms rely on the variation of refractive index with wavelength. When white light enters a prism, light at the blue end of the spectrum is bent more than that at the red end, because glass has a higher refractive index at blue wavelengths, as shown in Fig. 5-5. If light left the glass through a face parallel to the surface through which it entered, refraction at the exit face would cancel refraction at the entrance, and all wavelengths would emerge at the same angle. However, because the prism's output face is not parallel to the input face, different colors emerge at different angles, forming a spectrum.

Diffraction Gratings

A row of parallel grooves in a piece of glass or plastic can also produce a spectrum because of the way light scatters from the grooves. Light of all wavelengths is scattered at all angles. However, light adds constructively only at one wavelength at many angles; all other wavelengths add destructively, or interfere, reducing their intensity to zero. At the angles where the grating spreads out a

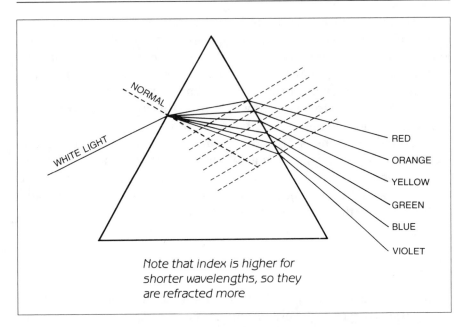

Note that index is higher for shorter wavelengths, so they are refracted more

Fig. 5-5 A prism forms a spectrum by bending different wavelengths at different angles.

spectrum, the result is a gradual change in wavelength with viewing angle, just as from a prism. The more grooves in the grating, the more the grating concentrates light in particular directions. Gratings can be transmissive or reflective, but reflective gratings are more common in optical instruments.

The way a reflective grating scatters light is described by the grating equation:

$$d \, (\sin i + \sin \theta) = m\lambda$$

where d is the groove spacing, i the angle of incidence on the grating (measured from the normal), m an integer indicating the order of the spectrum produced (which we'll explain in a couple of paragraphs), θ the angle of the scattered light to the normal, and λ the wavelength. (Wavelength and groove spacing must be measured in the same units.) If the input beam is at a normal angle to the grating, the equation reduces to:

$$d(\sin \theta) = m\lambda$$

If we know the wavelength and the angle of incidence, we can convert this equation into a form suitable for calculating the emerging angle θ:

$$\theta = \arcsin\{[m\lambda/d] - \sin i\}$$

The factor m is present because a diffraction grating produces more than one spectrum. For $m = 0$, we have undiffracted light reflected directly from (or transmitted directly through) the grating, without any dispersion (mathematically, multiplying by zero cancels the wavelength term). For $m = 1$, we have the first-order or strongest spectrum. At a larger angle, we have the second-order spectrum, formed by the $m = 2$ term. The intensity decreases as the value of m increases, so for most practical purposes the first-order spectrum is best. Note, however, that higher-order spectra are spread out over larger angles because the wavelength is multiplied by m.

Spectroscopes and Monochromators

Diffraction gratings or prisms can separate light of different wavelengths in optical instruments. One such instrument is the *spectroscope*, which projects different wavelengths at different angles, using a prism or grating. The spectroscope does not contain an internal light source.

A *monochromator* contains an internal light source as well as a prism or grating to spread out the spectrum. In its simplest form, the prism or grating spreads the spectrum onto a screen. A movable slit selects one color from the spectrum, and transmits that narrow range of wavelengths to the outside world. The viewer sees the monochromator emitting a narrow range of wavelengths, which is called "monochromatic" although, strictly speaking, it contains more than one wavelength.

FIBER OPTICS

Light waves normally travel in straight lines, but optical fibers can guide them along curved paths. The main use of optical fibers is to carry communication signals over long distances, but they can also be used in other types of optical systems.

Figure 5-6 shows a simplified view of the working of an optical fiber. Light is transmitted by the core, which has a refractive index n_1 and is surrounded by a cladding layer with a lower refractive index n_2. If light in the core strikes the interface with the cladding at a glancing angle, it is reflected back into the core because of the phenomenon of total internal reflection that we discussed briefly in Chapter 2. Light striking at the core-cladding boundary at steeper angles leaks into the cladding layer and can ultimately escape from the fiber itself.

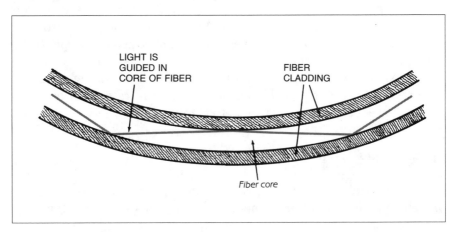

Fig. 5-6   Light guiding in an optical fiber.

We should stress that this is a simplified view of how light travels through optical fibers. (For more details see my book, *Understanding Fiber Optics* [Sams Publishing, 2nd ed., 1993].) The most important point from the user's standpoint is that optical fibers guide light.

Most fibers are made of silica glass that can transmit visible, near-infrared, and near-ultraviolet light. Plastic fibers transmit light best in the visible and are more flexible than glass. However, silica glass fibers have lower loss—particularly at near-infrared wavelengths of 1.0 to 1.6 µm—and thus are preferred for long-distance communications. Longer infrared wavelengths can be transmitted by special-purpose optical fibers made of other materials, such as fluoride glasses or silver halides.

Single fibers are usually used in communications, to carry signals generated by semiconductor lasers or light-emitting diodes (LEDs) to optical detectors, in the same way that wires carry electrical signals. Fibers may be bundled together to "pipe" light into hard-to-reach places (including inside the human body) or to carry images from point to point. Fiber bundles may be made rigid (by melting the fibers together) or flexible (by leaving individual fibers loose in a housing). Fibers or fiber bundles are used to deliver laser light for some applications in medicine and industry. Special-purpose fibers can also be used as sensors.

POLARIZING OPTICS

Polarization indicates the direction of the oscillating electric field that is part of a light wave. As we mentioned earlier, you can divide light into two orthogonal polarizations, with electric fields perpendicular to each other. The simplest way to think of polarization is as a vector in the X-Y plane, as shown in Fig. 5-7. The vector goes from the origin (0,0) to a point on the plane, and measures the amplitude and direction of the electric field. The polarization vector of light is the sum of two perpendicular vectors, one in the X direction, the other

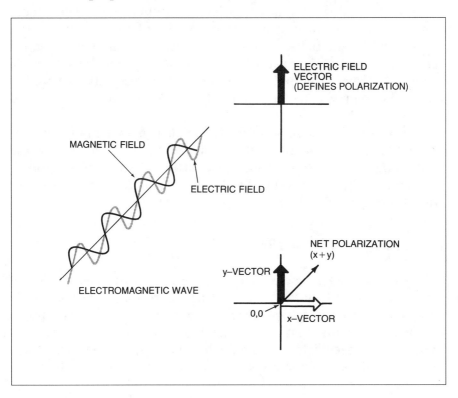

Fig. 5-7   Direction of polarization as a vector.

in the Y direction. Those two vectors are the horizontal and vertical polarization components.

The X and Y directions are not absolute. If you rotate the two axes, you can still break the polarization vector into two perpendicular components aligned along the two new axes. The lengths of the two component vectors will differ, but their sum will remain the same.

Polarizers separate light into its two perpendicular polarization components. Simple dichroic film polarizers (like those used for sunglasses) absorb light polarized along one axis and transmit light polarized along the other axis. For example, a polarizer might transmit light polarized along the Y axis in Fig. 5-7 and absorb light polarized along the X axis. Polarizers used with lasers normally work by other means and separate light into two polarization components. Many use so-called birefringent materials, which have different refractive indexes for light of different polarizations. Others reflect light of certain polarizations.

Light is said to be "linearly polarized" if its electric field is continually oriented in the same plane. If the plane of polarization changes with time but the amplitude is constant, the light is called "circularly polarized", because the polarization vector draws a circle in the X-Y plane. If the polarization vector changes in both direction and length, it draws an ellipse in the X-Y plane, and the light is called "elliptically polarized."

Light is said to be unpolarized if it is the sum of light of all different polarizations. (Random polarization—which might seem to be the same thing—is different, meaning light in which the direction of polarization changes continually in time.) Laser light may be polarized or unpolarized.

Polarization Retarders and Rotators

Because their refractive indexes depend on the polarization of light, birefringent materials can rotate or retard polarization. To understand how this works, consider what happens to the two polarization components if an unpolarized beam hits such a material. The vertically polarized light sees a refractive index n, and the horizontally polarized light sees an index $n + \Delta$, where Δ is a small fraction of n

(for convenience, we'll assume it's a positive number). Because the horizontally polarized light experiences a larger refractive index, it takes longer to travel through the material than the vertically polarized light. This can be useful for certain applications, where it is called "retarding" or "rotating" the polarization.

In practice, polarization retarders are made with a thickness selected to retard one polarization behind the other by a quarter, half, or whole wavelength. Half- or quarter-wave plates can change the polarization of linearly polarized beams in various ways. For example, if the polarization of a linearly polarized input beam is at angle θ from the principal plane of the retarder, the polarization of the output beam is rotated by 2θ. If an input beam is linearly polarized at 45 degrees to the crystalline axis of a quarter-wave plate, the output beam is circularly polarized.

Polarization Analyzers

When polarizers are used in front of instruments to separate the light of one polarization from other polarizations for measurements, they are called "polarization analyzers."

BEAMSPLITTERS

One of the most important optical components used with lasers is the *beamsplitter*. It does what its name says, splitting a beam into two parts. Part of the light is transmitted and part reflected. Normally, the beamsplitter is at an angle to the incident beam, so the reflected light goes off at an angle while the transmitted beam passes right through, as shown in Fig. 5-8. (You need not align the beamsplitter so that the reflected beam is at a right angle to the incident and transmitted beam; this is just a convenient way to draw the beams.)

Beamsplitters come in many types. Some reflect the light of one linear polarization and transmit the light of the orthogonal polarization. Others are insensitive to polarization. Some are multilayer interference coatings, and thus very sensitive to wavelength. Although some beamsplitters divide incident light equally into transmitted and reflected beams, others have different splitting ratios.

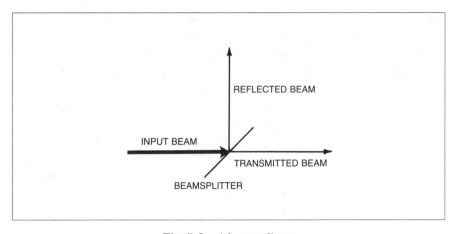

Fig. 5-8   A beamsplitter.

NONLINEAR OPTICS

So far, we have talked mostly about linear optics, which do comparatively simple things to light. However, there also are nonlinear optics, which do more complex things that are impossible with linear optics. The theory of nonlinear optics is the stuff of three-page-long differential equations and Ph.D. theses, but we can give you a basic idea here.

When we talk about linear optics, we mean devices in which the output is a simple function of the input:

$$\text{Output} = A \times (\text{Input})$$

where A is a characteristic of the device. Mathematically, the input and output are time-varying electromagnetic waves. They are easiest to visualize as a cosine function if we ignore the phase of the light wave and consider only how it changes at a single point:

$$\text{Input} = K \cos(\omega t)$$

where K is a constant that gives the amplitude of the electromagnetic wave, ω the angular frequency (2π times the frequency in hertz ν), and t the time. However, the proper description is somewhat more complex:

$$\text{Input} = K (\cos[-i\omega t] + i \sin[-i\omega t])$$

where i is the square root of -1, indicating an imaginary number. Don't be afraid of the imaginary number; it's used to make sure that details we aren't worried about come out right.

We can also write this complex formula in an exponential form:

$$\text{Input} = Ke^{-i\omega t}$$

This formula has exactly the same meaning and is simpler to manipulate mathematically.

In nonlinear optics, the relationship between output and input is more complex, because we have to consider other terms neglected above. Thus, the equation looks like:

$$\text{Output} = A(\text{Input}) + B(\text{Input})^2 + C(\text{Input})^3 + ad\ infinitum$$

where A, B, and C are constants or other expressions that include coefficients representing the strengths of various effects. In practice, terms involving powers higher than the square of the input wave are usually unimportant.

Nonlinear optics is often broadly defined to include effects that do not depend purely upon the input wave. For example, external electric or magnetic fields can interact with a light wave's electromagnetic field to change the phase or polarization. We don't want to worry about the details, but we can give you an overview of how they function.

Harmonic Generation

The simplest nonlinear optical effect is harmonic generation, the production of light waves at integral multiples of the frequency of the original wave. The second harmonic is at twice the frequency of the fundamental wave or, equivalently, at half its wavelength. The third harmonic is at three times the frequency, or a third of the wavelength. In the laser world, the frequencies are very high, so the light waves normally are identified by wavelength, even though the process is still called harmonic generation. Thus, laser specialists say they frequency-double the output of a 1064-nm neodymium laser to get 532-nm light, but they never mention the frequencies involved (2.83×10^{14} and 5.66×10^{14} Hz, respectively).

Harmonic generation makes sense if we look again at the nonlinear relationship of output to input:

OUTPUT $= A(\text{INPUT}) + B(\text{INPUT})^2 + C(\text{INPUT})^3 + ad\ infinitum$

If we substitute $Ke^{-i\omega t}$ for INPUT, we can see that the INPUT^2 term gives a factor of $e^{-i2\omega t}$, which has twice the frequency of the input wave. Thus, we've doubled the frequency. Likewise, INPUT^3 gives a factor of $e^{-i3\omega t}$, which triples the frequency. (You can get similar results with trigonometric identities such as $\cos^2 \omega t = 0.5\ (\cos 2\omega t + 1)$—if you can remember the trigonometric identities.)

We can't forget the wave amplitude, given by K in our simple equation. If INPUT is $Ke^{-i\omega t}$, the INPUT^2 term becomes $K^2 e^{-i2\omega t}$—and depends not on K but on its square. Thus, the amplitude of the second-harmonic wave increases as the square of the field intensity at the fundamental frequency, not just with the field intensity. This means that the second-harmonic output, which is initially low, rises much faster than the input amplitude.

Despite its esoteric origins, harmonic generation is a practical tool in the laser world. There are few good laser materials around, so not all wavelengths are readily available. Harmonic generation makes more wavelengths available. Frequency-doubling of the neodymium laser to give 532-nm green light is the most important example, but there are many others.

Most materials have very small nonlinear coefficients, so only a few materials are usable for second-harmonic generation. Special conditions must be met for the interaction between input and output beams. Because nonlinear effects are small, they usually appear only at high input powers. This makes it easier to generate harmonics from short pulses with high peak power than from continuous-wave beams with a much lower steady power.

Second-harmonic generation is most important in practice, but in some cases the third harmonic may be generated directly. Because the fourth-order nonlinear coefficient is much lower than the second-order coefficient, in practice the fourth harmonic is generated by frequency-doubling the second harmonic.

Sum- and Difference-Frequency Generation

If you take another look at our nonlinear equation, you can see another possible nonlinear effect—generation of light waves at sum and difference frequencies. Suppose the input light is the sum of two waves at different frequencies, ω and ϕ:

$$\text{Input} = K_1 e^{-i\omega t} + K_2 e^{-i\phi t}$$

Then the INPUT^2 term includes a factor $e^{-i(\omega+\phi)t}$, meaning that light is generated at the sum of the two frequencies. There also is a difference-frequency term, with frequency $\omega - \phi$, as well as terms at the two original frequencies.

Like harmonic generation, sum- and difference-frequency generation effects are weak in most materials, and require special conditions. They depend on the product of the two input intensities, $K_1 \times K_2$ in our example. That means that output increases with the product of the input powers.

Parametric Oscillators

In sum-frequency generation, light at frequencies ω and ϕ add together to produce the frequency $\omega + \phi$. Under certain circumstances, you can turn the equation around and use a strong beam at $\omega + \phi$ to generate beams at ω and ϕ.

In a parametric oscillator, pump light at the frequency $\omega + \phi$ enters a nonlinear medium that is placed between two mirrors. The mirrors are reflective at one or both of the frequencies ω and ϕ, so they form a resonant cavity comparable to a laser cavity. The nature and orientation of the nonlinear material determines what frequencies ω and ϕ are produced.

Parametric oscillators can produce light at the higher "signal" frequency we'll call ω, the lower "idler" frequency ϕ, or both. Natural noise processes generate weak light at both ω and ϕ. Mixing this weak light with the pump signal at $\omega + \phi$ generates weak difference-frequency signals at ω and ϕ. Resonance of one or both of the ω and ϕ waves in the cavity enhances the interaction in the nonlinear crystal, thus building up the desired wave, which can be tuned in wavelength by adjusting the parametric oscillator.

Parametric oscillators were developed early in the laser era, but found little practical use because the nonlinear materials then available were very vulnerable to optical damage. They have attracted growing interest in recent years because of new materials with higher damage thresholds, such as KTP (potassium titanyl phosphate). When pumped by suitable lasers, they can provide tunable output in the visible, and at infrared wavelengths to about 5 μm; they can also generate ultrashort pulses in certain operating modes. At this writing, the main interest is in generating tunable light at wavelengths other-wise hard to produce, particularly from about 1.5 to 5 µm.

Raman Shifting

Another important nonlinear interaction is called "Raman shift-ing," after its discoverer, Indian physicist C. V. Raman. It is a small shift in frequency that sometimes occurs when atoms or molecules reflect (or scatter) light. (Normally, light frequencies do not change when they are reflected or scattered.) The shift occurs when a mole-cule or an atom in a crystalline matrix undergoes a vibrational transi-tion at one of its characteristic frequencies while reflecting or scatter-ing a light wave. If the material loses vibrational energy, the Raman shift adds energy to the scattered light wave and decreases its wave-length. If the material gains vibrational energy, it absorbs energy from the light wave, which is scattered at a longer wavelength.

Incident light at any wavelength can experience a Raman shift, but it shows up best at higher power levels. The amount of Raman shift (measured in energy units) is an inherent characteristic of the vibrational transition of the material, and is the same for all input wavelengths. The scattering of one laser wavelength can produce several different Raman-shifted lines.

Raman shifting is a practical way to generate additional wave-lengths from a strong laser line in the visible, near-infrared, or near ultraviolet.

INTENSITY MODULATION

Many applications require changing or modulating power in a laser beam, either to make it vary with time or simply to switch it on or off. This can be done in several ways: by blocking the beam mechanically, by changing the electrical or optical power that drives the laser, or by passing the beam through a device with variable transmission. Let's look briefly at each approach.

Mechanical Modulation

Shutters and beam choppers mechanically block a beam. A shutter, like that on a camera, opens to let the beam through and closes to block light. Shutters are used as safety devices on many lasers, to block the beam when it is not needed or when the laser may not be operated correctly. Control can be manual or automatic.

Shutters operate in a fraction of a second, but that is slow by laser standards. Mechanical beam choppers can modulate a laser beam much faster. A typical example is a rotating disk with holes or slots to let the beam through periodically. You can design shutters and beam choppers to block only part of a beam, but they do not control the fraction of light transmitted precisely (because of uncertainties in the beam position and the distribution of power in the beam), so they are typically used only to turn the beam off and on.

Direct Modulation

We control light bulbs by turning the electricity off and on, and lasers can be controlled in the same way. Direct modulation is often the simplest and cheapest way to control laser power. However, this approach works better for some lasers than for others because of differences in internal operation and response time.

Direct electrical modulation works best for semiconductor diode lasers. The devices are small and respond very quickly to changes in drive current; specially designed lasers can be directly modulated at billions of pulses per second. Only a simple circuit is needed to switch the drive current on and off, although circuit complexity increases with operating speed. Above laser threshold, their output usually increases linearly with drive current. The availability of such a simple modulation technique is one important attraction of semiconductor diode lasers.

Gas lasers are not well-suited for direct electrical modulation. They can be turned off and on, but they take much longer to stabilize—several minutes for some types. Because they need a thousand volts or more across the tube, modulation also is more complex electronically. You might compare the difference between semiconductor and gas lasers to the difference between incandescent bulbs and fluorescent tubes. Incandescent bulbs turn off and on faster, and their light output can be adjusted with a dimmer; fluorescent tubes are harder to ignite and their brightness cannot be adjusted with a dimmer. In practice, gas lasers—as well as crystalline and glass solid-state lasers—are normally modulated externally, or by varying input optical pump energy.

Direct Optical Modulation

We saw earlier that some lasers are pumped optically. Such lasers respond very quickly to changes in the pump light, so you can modulate them by changing the optical input. If a laser is pumped by a flashlamp or pulsed laser, its output rises and falls with pump-light intensity. This does not make direct modulation simple, however; it just shifts the problem of controlling output intensity to a different place—the pump source. Pulsing a flashlamp does produce a laser pulse from a flashlamp-pumped laser, but electronics must pulse the flashlamp (which again is slow by laser standards). Likewise, modulating the pump laser can control the output of an optically pumped laser, but you still have to adjust the pump laser output.

External Electro-Optical Modulation

Often the most practical way to modulate the intensity of a laser beam is with an external device that can be controlled to transmit varying fractions of the beam. One family of external modulators uses an electric field to change the refractive index of a material. The electric field has different effects on the refractive indexes for light with different polarizations, so it effectively rotates the polarization of light in suitable materials. This does not change light intensity, but combining such a material with a polarizer makes an electro-optical modulator.

When we talked about polarization rotators earlier, we learned that the rotation angle depended on the material's thickness and birefringence. For electro-optic crystals, the polarization angle depends not only on the material thickness but also on the electric field. Once the electro-optic material is chosen and cut to size, the only significant variable is the electric field, which thus controls the angle of polarization rotation. (Other factors, including temperature and pressure, can change birefringence slightly.)

Figure 5-9 shows the construction of an electro-optic modulator. First, a polarizing filter gives the beam a linear polarization. Then the crystal rotates the polarization, by an amount that depends on the voltage applied across the electrodes. Then the emerging light must pass through another polarizing filter, with its plane of polarization at the proper angle. Adjust the voltage across the crystal so it rotates the plane of polarization to be parallel to that of the output filter, and the transmitted beam is at its maximum intensity. Change the voltage to rotate it perpendicular to the plane of the filter, and the beam will be blocked. In Fig. 5-9, we see the output intensity somewhat reduced.

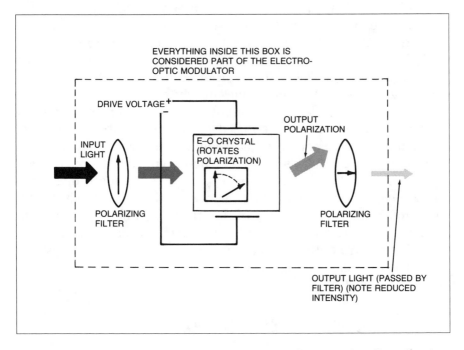

Fig. 5-9   An electro-optic modulator, with an electro-optic cell mounted between crossed polarizers.

There are two types of electro-optic modulators: Kerr and Pockels cells. They rely on subtly different effects. The more common Pockels cell relies on the Pockels effect in crystals, in which the change in birefringence is proportional to a constant (p) times the applied electric field (E), or pE. The constant is small; typical values of p are 10^{-10} to 10^{-11} m/V even for the best materials. The Kerr effect arises from alignment of molecules in a liquid or gas; it changes birefringence by the product of a constant K times the wavelength λ times the square of the electric field E, $K\lambda E^2$. Kerr cells are fast, but they require higher voltages than Pockels cells to rotate the polarization by the same amount.

Both types of electro-optical modulators are expensive, costing hundreds of dollars and up. They also have other limitations, including the need for bias of tens or—more typically—hundreds of volts.

External Acousto-Optic Modulators

A laser beam can also be modulated by interactions with sound waves in a solid. As an acoustic wave passes through a solid, it alternately raises and lowers the pressure, which in turn changes the refractive index. Light waves passing through the material see these variations in refractive index as a series of layers, like the multilayer dielectric coating we described earlier. Alternatively, the refractive-index variations can be seen as lines in a diffraction grating.

Interference and diffraction combine to scatter light from the "acoustic grating" at an angle that depends on the ratio of the wavelengths of light and sound in the medium. The stronger the acoustic waves, the more light is scattered or deflected from the main beam. This can modulate light if we look only at the main beam and throw away the scattered light. (Later, we will see other uses for the deflected beam.) The intensity of the transmitted beam drops as the intensity of sound waves increases—modulating beam intensity.

Like electro-optic modulators, acousto-optic modulators cost hundreds of dollars and up. The two types compete with each other to some degree.

BEAM SCANNERS

Many applications require scanning a laser beam back and forth, much as your eyes scan this page, or as an electron beam scans the screen of a television tube. There are several major approaches.

Perhaps the simplest concept is the rotating mirror. This is a reflective polygon that spins rapidly around a central axis. Each mirror face turns as the laser beam strikes it, scanning the reflected beam across an angle dictated by the rule that the angle of incidence equals the angle of reflection. Each new facet repeats the scan. Holographic scanners are similar in concept, but rely on rotating holographic optical elements to direct the light.

Mechanical motion of mirrors is also the basis of resonant and galvanometer scanners. However, these scanners twist flat mirrors back and forth over a limited angular range, so the laser beam always strikes the same flat mirror surface. As the mirror swings back and forth, it sweeps the laser beam across a line.

As we mentioned in describing acousto-optic modulators, the interaction between a light wave and a sound wave can deflect some light passing through certain materials. In modulators, we use the undeflected beam, but in acousto-optic beam scanners we use the deflected light. The angle at which the deflected light emerges is proportional to the ratio of optical to acoustic wavelengths; the strength of the deflected light depends on the acoustic power. Thus, you can make the light scan a pattern by changing the acoustic frequency driving the scanner. Unlike mechanical scanners, acousto-optic deflectors are solid-state devices with no moving parts, an important attraction for some uses.

It also is possible to make electro-optical deflectors, which, like acousto-optic scanners, are solid-state devices, but depend on electro-optic interactions. However, these are rarely used.

CONTROLLING LASER PULSE CHARACTERISTICS

We saw earlier the importance of how laser output varies with time. It is often useful to concentrate laser energy into short time intervals. Like focusing a broad beam down to a tiny spot, this concentrates the energy in a single place and time. There are three primary tools: Q switches, cavity dumpers, and modelockers.

Q Switches

Every resonant laser cavity has a characteristic quality factor, or Q, that measures internal loss. The higher the Q, the lower the loss.

Loss is not necessarily bad, because a high-loss cavity can store more energy than one with lower loss. Suppose, for example, you block one laser mirror and then excite the laser medium. The excitation can produce a population inversion, but the laser cannot oscillate because the mirror does not reflect light. If you suddenly unblock the mirror, the population inversion will be much larger than needed for laser action. Stimulated emission will quickly drain the stored laser energy from the cavity in a short pulse with peak power much higher than the laser could otherwise produce, as shown in Fig. 5-10. You can think of a Q switch as a device that quickly switches from absorptive to transmissive, suddenly reducing cavity loss.

The length of a Q-switched pulse depends on output-mirror reflectivity R and on the time t it takes laser light to make a round trip through the laser cavity, and is roughly given by:

$$\text{Pulse Length} = \frac{t}{(1-R)}$$

Typical pulse lengths are around 10 ns. The round-trip time equals twice the cavity length L divided by the speed of light in the laser medium, which itself equals the speed of light in vacuum c divided by the laser medium's refractive index n. Put this together, and you have a more useful approximation for Q-switched pulse length:

$$\text{Pulse Length} = \frac{2Ln}{c(1-R)}$$

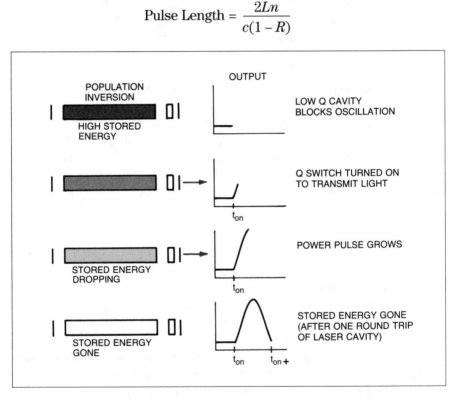

Fig. 5-10 Q-switching generates a laser pulse.

A Q switch can change the cavity Q once to produce a single shot, or repeatedly to generate a series of pulses at a specified interval, depending on how it is driven. Both electro-optic and acousto-optic modulators can serve as Q switches.

Other Q switches incorporate rotating mirrors. When the mirror face is not perpendicular to the laser axis, it does not reflect light back to the output mirror, and the Q is low; when the mirror face rotates perpendicular to the laser axis, the Q is momentarily high, and the laser generates an intense pulse of light.

Not all Q switches require active control. A "passive" Q switch can be built using a dye that normally absorbs light. After each dye molecule absorbs a photon, it stays awhile in an upper energy level, where it can't absorb more light. Strong enough stimulated emission from the laser medium can quickly raise all of the dye molecules to the excited state. This "bleaches" the dye, making it transparent because no molecules are left in a state that can absorb light. The dye's abrupt change from absorbing to transmitting light raises cavity Q by exposing a mirror behind it, and thus generates a Q-switched pulse.

Q-switching is widely used, but not all lasers can be Q-switched. Q-switching works only if the laser can store energy in excited-state atoms or molecules for a period longer than the Q-switched pulse. That is, the excited state must have a longer spontaneous-emission lifetime. Some lasers do not meet this criterion.

Cavity Dumping

Like Q-switching, cavity dumping works by releasing energy stored in the laser cavity; however, the details are different. In essence, a cavity dumped laser normally has two fully reflective mirrors, one on each end of the cavity. The laser normally operates with this high-Q cavity, building up a steady power inside the cavity. However, the mirrors keep the light from escaping.

How do you get the power out? You can put another mirror in the laser cavity that aims the beam out of the cavity, instead of back at the other resonator mirror. This dumps all the energy circulating in the laser cavity in a single pulse, which lasts only as long as the light takes to make a round trip of the laser cavity. This means that the length of a cavity dumped pulse is

$$\text{Pulse Length} = 2Ln/c$$

where L is the cavity length, n the refractive index of the laser medium and c s the speed of light in a vacuum. For a 30-cm gas laser, this would be 2 ns, significantly shorter than a Q-switched pulse.

Cavity dumpers don't actually use pop-up mirrors. Instead, they rely on beam deflectors or other components inside the laser cavity that briefly deflect the beam outside of the laser. Cavity dumping generates lower energy pulses than Q-switching, and it is not used as often. However, lasers that cannot be Q-switched can be cavity dumped.

Modelocking Pulses

The shortest pulses that can be produced in a laser cavity—lasting on the order of a picosecond (10^{-12} s) or less—are generated by a process called "modelocking." The idea, from a theoretical physicist's point of view, is to lock together many longitudinal modes so that a laser simultaneously oscillates in phase on all of them. This control comes from inserting a special optical element in the laser cavity that makes all the modes oscillate together.

It is simpler to visualize the process by imagining that it locks together a clump of photons, not modes. The clump bounces back and forth between the laser mirrors. The modelocking element in the laser cavity "opens up" to transmit light each time the photon clump passes through. Each time the photon clump hits the output mirror, a short pulse escapes. Outside the laser, we see many short pulses, one emitted each time the clump of photons makes a round trip of the cavity. Modelocked pulses are very short, but during those very short pulses the peak power can reach a very high level.

The length of a modelocked pulse depends on the range of wavelengths (or, equivalently, frequencies) emitted by the laser. The wider the frequency range $\Delta \nu$, the shorter the pulse can be. The lower limit on pulse duration is roughly

Minimum Pulse Duration = 0.44/Frequency Range

or

$$t = 0.44/\Delta \nu$$

Pulses of this minimum length are sometimes called "transform-limited," because the relationship is considered a transform between time and frequency domains. In a laser with a broad enough frequency range, modelocking can produce minimum pulses a little longer than 10 femtoseconds (10^{-14} s). The shortest pulses to date, 6 femtoseconds (6×10^{-15} s), were produced by compressing short laser pulses in an external optical system.

Note that the natural repetition rate of a modelocked laser is the time it takes light to make a round trip of the laser cavity. That isn't long—light takes only about 2 ns for a round trip of a 30-cm (approximately 1-ft) laser. Thus, a 30-cm modelocked laser would generate 500 million pulses/s, a 500-MHz repetition rate, much higher than the repetition rates of cavity dumping or Q-switching.

Combining Pulse Controls

It is possible to combine two or more of these pulse control techniques. For example, combining cavity dumping and modelocking can reduce the repetition rate of modelocked pulses, to ease measurement requirements or help isolate effects of successive pulses.

POWER AND ENERGY MEASUREMENT

Special instruments, devices, and terms are used in measuring laser power and energy. While we don't have room to give a comprehensive overview, we can introduce the basic ideas of light detection and measurement.

Light Detection

The most common way to measure light is to first convert it to electricity, and then measure electrical current, resistance, or voltage. Various light detectors can convert optical signals into electronic form. The most common are semiconductor devices, in which incident light raises an electron from the valence band to the conduction band, and also creates a "hole" in the conduction band, which functions as a current carrier. An older type is the *photoemissive* detector, a vacuum tube in which light frees electrons from a metal surface, and those electrons are collected by a positively biased electrode. Photomultipliers are photoemissive tubes with several stages that amplify the electrical signal. (Originally, "photodiode" meant a vacuum photoemissive tube with two electrodes, one emitting electrons and the other collecting them; but the word is now used for semiconductor detectors.) There also are some detectors that monitor light in other ways.

Each type of detector is sensitive to a limited range of wavelengths, depending on its composition. A small sampling is listed in Table 5-1. Many types are available, with the choice depending on wavelength range, sensitivity to light, the speed with which they can respond to signals, durability, operating requirements, and cost.

TABLE 5-1 Wavelength ranges of detectors

Type	Material	Wavelengths (nm)
Photoemissive	Potassium-cesium-antimony	200–600
Semiconductor	Silicon	400–1000
Semiconductor	Germanium	600–1600
Semiconductor	Gallium arsenide	800–1000
Semiconductor	Indium gallium arsenide	1000–1700
Semiconductor	Indium arsenide	1500–3000
Semiconductor	Lead sulfide	1500–3300
Semiconductor	Lead selenide	1500–6000

The electrical output from detectors can be used for measurement or other purposes. For example, electrical signals from detectors in optical communication systems are processed by other electronic circuitry so that they can drive electronic equipment such as computers or telephone switching systems.

Sometimes the term "detector" can be used in a broader sense to include light sensors that do not produce electrical output. Simple examples include photographic film and photochromic materials, in which light changes how the material looks to the eye.

Radiometry, Photometry and Light Measurement

The measurement of light is a specialized area often divided into two fields, radiometry and photometry. Radiometry is the science of measuring the power and energy contained in electromagnetic radiation, regardless of wavelength. Photometry measures similar quantities, but only for light visible to the human eye, with the contribution of each wavelength weighted according to the eye's sensitivity. Thus, photometry ignores infrared wavelengths, which are invisible to the human eye, and counts photons of green light (to which the eye is very sensitive) much more strongly than red or violet light (to which the eye does not respond as strongly). Unfortunately, some people are sloppy with that terminology, but you should remember the difference.

The eye almost never looks directly at laser light (see Appendix A on safety), so virtually all laser measurements are made in radiometric units. Table 5-2 summarizes these units and lists common symbols for them, with brief descriptions of their meanings.

TABLE 5-2 Radiometric units

Quantity and Symbol	Meaning	Units
Energy (Q)	Amount of light energy	Joules
Power (P or ϕ)	Flow of light energy past a point at a particular time (dQ/dt)	Watts
Intensity (I)	Power per unit solid angle	Watts/steradian
Irradiance (E)	Power incident per unit area	Watts/cm^2
Radiance (L)	Power per unit angle per unit projected area	Watts/steradian-m^2

Note that power ϕ (sometimes called radiant flux) measures the rate at which electromagnetic energy flows by a point. Mathematically, it is the derivative or rate of change of the energy Q:

$$\phi = dQ/dt$$

or

$$\text{Power} = d(\text{Energy})/d(\text{Time})$$

The unit for measuring power, the watt, equals one joule/s.

Some commercial instruments also measure power in decibels. The decibel (dB) is a relative unit that gives the ratio of two power levels, P_1 and P_2:

$$\text{Power Ratio (dB)} = 10 \log_{10}(P_1/P_2)$$

Sometimes power is measured in decibels relative to a predefined level, typically 1 mW or 1 μW. A positive number indicates that the measured power level is above the comparison; a negative number indicates that it is lower.

Pulse Duration and Spectral Measurements

Pulse duration and light wavelength are two other important quantities that must sometimes be measured while working with lasers. Both require special measurement instruments.

The simplest way to measure pulse length is to feed the output of an electronic detector into an oscilloscope. However, this runs into instrumental limitations as pulse lengths decrease to the nanosecond realm. Sophisticated sampling oscilloscopes can measure nanosecond pulses accurately, but more complex techniques are needed to measure picosecond and subpicosecond pulses. Such techniques are outside the scope of this book.

The simplest way to analyze the spectrum of light is to spread it out with a prism or diffraction grating. This is adequate for some simple measurements, but when more quantitative measurements are needed, this may have to be combined with electronic techniques. One valuable approach is to spread out the spectrum on a linear array of light sensors and then measure the amount of light reaching each sensor. With proper calibration, the measurements from individual sensors indicate how much light is present at particular wavelengths.

MOUNTING AND POSITIONING EQUIPMENT

Walk into any well-equipped laser laboratory, and the first thing you are likely to notice is a massive table, probably painted flat black, such as the one shown in Fig. 5-11. This behemoth is called an "optical table." On it, you will find an array of special mounting equipment that holds lasers, lenses, and other optical components.

Optical tables and mounts are not optics *per se*, but they are used with optics and lasers. Optical tables and massive linear rails, called "optical benches," serve as foundations. Many are mounted on shock-absorbing legs that isolate them from vibrations in the room. This vibration isolation is essential for holography and sensitive measurements because small effects, such as a person walking through the room or a truck passing on the street outside, can cause vibrations comparable in size to the wavelength of light. The tables are expensive, but alternatives exist if you're working on a low budget in a home or school laboratory. Most books on making your own holograms describe how to isolate optical setups from vibrations by mounting them in a sandbox.

An optical table or bench also serves as a firm foundation. Typically, threaded holes are drilled in the surface of optical tables to accommodate screw-in mounts. The edges of optical rails and benches mate with optical mounts that hold components. The mounts themselves are made to hold standard lenses and optics firmly.

Fig. 5-11   Optical table, with mounts. (Courtesy of Newport Corp.)

In addition to fixed mounts, tables, and optical benches, optical laboratories also use a variety of positioning equipment. These are devices that hold and move optical components. The motion may be driven manually or mechanically, and some equipment incorporates computer controls. Precision is crucial because laser and optical systems can be very sensitive to small misalignments.

The technology of optical mounts and positioning equipment is mundane compared to lasers. You can often live without the best-quality equipment (indeed, if your budget is tight, you may not have any choice). However, good mounting and positioning equipment can make life much easier.

EMERGING TECHNOLOGIES

Some potentially important optical technologies remain in the laboratory. Three of them deserve brief mention because you will be hearing about them in the years to come: integrated optics, adaptive optics, and phase conjugation.

Integrated Optics

The basic idea behind integrated optics is something like that behind integrated electronic circuits. Today's lasers and optics are discrete, each made from a separate chunk of material. Electronics used to be that way, too, with circuits made of separate transistors, diodes, resistors and capacitors. If we could integrate the function of an entire optical system—laser, lenses, mirrors, and modulator, for example—into a single chunk of material, we would have integrated optics. Such integrated optical circuits are attractive for applications, such as communications and signal processing, that require manipulating low-power beams of light.

Researchers have been working on integrated optics for many years, and have achieved modest success. However, they also face some stubborn problems. Important difficulties come from the nature of the materials themselves. For example, you can't make lasers from the materials most useful for making modulators. The interactions that affect how light travels through a "waveguide" in an integrated optical circuit are so weak that the light must travel through a comparatively long channel to be switched between output ports—meaning that integrated optics must be much larger than their electronic counterparts. Although the technology remains promising, it is far from ready for routine use.

Adaptive Optics

One of our implicit assumptions about optics is that they are fixed and unchanging. We think of a mirror or lens retaining the same curvature throughout its usable lifetime—and changing its shape may end its useful lifetime. Suppose, however, that optical components could change their shape to meet changing conditions.

Interest in making adaptive optics comes from the effects of the atmosphere on light. Air currents and turbulence bend light rays. We see this at night as the twinkling of stars. Astronomers with large telescopes find that it spreads starlight, limiting resolution of the largest telescopes to about one arcsecond. Such fluctuations also affect efforts to send beams from high-power lasers through the atmosphere.

The idea of adaptive optics is to bend the surface of a mirror to compensate for atmospheric effects. The changes must be made continually, because the atmosphere changes continually. The optics first sense what is happening in the atmosphere, and then the surface of the mirror is bent to compensate for those atmospheric effects. If this works properly, it can let astronomers focus the image of a star to a point, or military engineers send a high-energy laser beam to a small spot on the target. (Although the Pentagon has paid most of the bills so far, the military problems are the toughest because high-power laser beams themselves generate some side effects that are hard to correct.)

Phase Conjugation

Another idea being studied to correct for atmospheric effects on light propagation is called *phase conjugation*. In a sense, this is a step beyond the retroreflector that we described earlier.

A retroreflector sends light rays back in the direction from which they came. Phase conjugation does somewhat the same thing in a different way. It generates what is called a "conjugate" of an incoming light wave. The conjugate wave is essentially a reversed version of the original wave—the wave is at the same phase, but travelling in a different direction, as shown in Fig. 5-12. This phase-conjugate wave travels in the same path as the incoming wave, but in exactly the opposite direction. If the incoming light came from a small laser spot, the phase-conjugate wave will go back to that same spot.

That may sound exactly like retroreflection. However, it is actually a time-reversed version of the original wave, reconstructing the light's phase and amplitude. Because of the way the phase-conjugate wave recreates the original wave, it automatically compensates for things that happened to the laser beam. In Fig. 5-12, the incoming light experiences perturbations that cause it to spread out. The phase-conjugate wave experiences the same perturbations but in the opposite direction, as if it's going backwards in time. Thus, the spread-out light in the phase-conjugate beam is pushed back together and focused on the laser source.

Phase conjugation remains largely in the laboratory, where researchers are trying to find better materials to generate the conjugate beam. Developers have shown that internal phase conjugation can improve the quality of a solid-state laser beam by compensating for harmful effects inside the laser cavity. Other applications are in development.

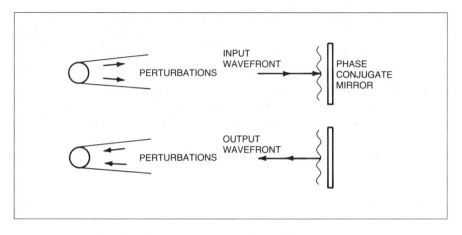

Fig. 5-12   Generation of a phase-conjugate wave.

WHAT HAVE WE LEARNED

- Passive optics always do the same things; active optics can change what they do to light.
- A positive lens or concave mirror focuses parallel light rays to a point.
- Telescopes can expand or shrink the diameter of a bundle of parallel light rays.
- A retroreflector returns an incident beam to its starting point, regardless of the angle at which it strikes.
- Optics suffer from chromatic and spherical aberration.
- Antireflection coatings reduce reflective losses at optical surfaces by reducing the refractive-index differential.
- Multilayer coatings use interference effects to control light transmission and reflection as a function of wavelength.
- Spectral filters transmit some wavelengths and block others.
- Neutral-density filters absorb the same fraction of light at all wavelengths in their operating range.
- Different optical materials must be used at different wavelengths. Glass is useful only in and near the visible range.
- Cylindrical optics refract light in one direction but not in the perpendicular direction.
- Prisms spread light out by wavelength because of the way the refractive index varies with wavelength.

- Diffraction gratings are rows of parallel grooves that scatter light by interference effects to form multiple spectra.
- Fiber optics guide light in a core that has a higher refractive index than the surrounding cladding.
- Polarization is the direction of the electric field in a light wave. We can think of it as a vector. It can be retarded or rotated by birefringent materials.
- Beamsplitters separate an input beam into two parts.
- Nonlinear optical materials can produce light waves at multiples of the input frequency. Only a few materials have high nonlinear coefficients.
- The amplitude of a second-harmonic wave increases as the square of the fundamental intensity.
- Nonlinear interactions can generate light waves at sum and difference frequencies.
- A parametric oscillator can generate tunable light at longer wavelengths than the input, by effects related to sum-frequency generation.
- Raman shifting changes the wavelength of light scattered from certain materials.
- Beams can be blocked by mechanical shutters and beam choppers.
- Direct modulation turns laser excitation off and on to control output power. It works best for semiconductor lasers.
- Electro-optic modulators change polarization to modulate light intensity.
- Interactions with acoustic waves in a solid can modulate a laser beam.
- Mechanical, electro-optic, or acousto-optic devices can scan laser beams.
- A Q switch controls cavity losses to generate short intense pulses. Q-switching works only for laser media that can store energy.
- A cavity dumper "dumps" energy from a laser cavity with totally reflective mirrors.
- Modelocking generates ultrashort pulses by effectively making a clump of photons circulate back and forth through the laser cavity.

- The length of a modelocked pulse depends on laser bandwidth.
- Light is usually converted to electricity so it can be measured.
- Radiometry is the measurement of electromagnetic power at all wavelengths.
- Photometry is the measurement only of light visible to the human eye.
- Power is the rate of flow of electromagnetic energy.
- Adaptive optics, which change their surface shape to correct for changing conditions in the atmosphere, are in development.
- Phase conjugation generates a time-reversed version of an input light wave.

WHAT'S NEXT

In Chapter 6 we will start looking at specific types of lasers. Our first family of lasers will be gas lasers.

Quiz for Chapter 5

1. A double-convex lens with two equal radii of curvature of 20 cm has refractive index of 1.60 at 400 nm and 1.50 at 700 nm. What is the difference between the lens' focal lengths at those two wavelengths?
   a. 400 nm focal length is 3.33 cm shorter
   b. 400 nm focal length is 0.33 cm shorter
   c. 700 nm focal length is 0.033 cm shorter
   d. 700 nm focal length is 1 cm shorter
   e. No difference

2. Silicon has a refractive index of 3.42 at 6 μm in the infrared. What is the reflective loss for 6-μm light incident from air normal to the surface?
   a. 5%
   b. 10%
   c. 20%
   d. 30%
   e. 40%

3. For the silicon sample in Problem 2, what is the reflective loss if a coating with refractive index of 2 is applied on the surface? (Remember that you can't just add the reflective losses. You must multiply the fractions of transmitted light to get the total transmitted

light to assess overall reflective loss.)
a. 5%
b. 10%
c. 15.4%
d. 17.2%
e. 20.5%

4. What type of filter would
   you use to block light from
   a laser, but let other light
   through?
   a. Neutral-density filter
   b. Interference filter
   c. Color filter
   d. Spatial filter
   e. Any of the above

5. At what angle (from the normal) will a diffraction grating with 1-μm spacing scatter 500-nm light if the input
   light is incident at a normal
   angle? Calculate for the first-
   order spectrum ($m = 1$), and
   don't worry about the sign.
   a. 0.5 degree
   b. 5 degrees
   c. 30 degrees
   d. 60 degrees
   e. None of the above

6. What is the wavelength of
   the fourth harmonic of the
   ruby laser (694 nm)?
   a. 2776 nm
   b. 1060 nm
   c. 698 nm
   d. 347 nm
   e. 173.5 nm

7. Raman shifting does what to
   input light?
   a. Modulates its intensity

b. Doubles its frequency
c. Changes its wavelength
   by a modest amount
d. A and B
e. None

8. Which type of laser is the
   simplest to modulate directly by changing its excitation?
   a. Semiconductor
   b. Ruby
   c. Helium-neon
   d. Neodymium-YAG
   e. All equally difficult

9. What is the length of a
   Q-switched pulse from a
   10-cm long neodymium-
   glass laser with a 90%
   reflective output mirror?
   Assume the refractive
   index of glass is 1.5.
   a. 1 ns
   b. 5 ns
   c. 10 ns
   d. 20 ns
   e. 1 μs

10. A laser has a bandwidth of
    4 GHz (4×10^9 Hz). What is
    the shortest modelocked
    pulse it can generate,
    according to the transform
    limit?
    a. 1 picosecond
    b. 10 picoseconds
    c. 30 picoseconds
    d. 100 picoseconds
    e. 250 picoseconds

CHAPTER **6**

Gas Lasers

ABOUT THIS CHAPTER

Gas lasers are one of the three major laser families, and are in many ways the most varied. In this chapter, we will learn how basic laser concepts apply to gas lasers. We will then learn about the most important gas lasers.

THE GAS-LASER FAMILY

The family of gas lasers is large and varied. The first gas laser was demonstrated at Bell Telephone Laboratories by Ali Javan, William R. Bennett, Jr., and Donald R. Herriott in December 1960. It was only seven months after Maiman operated the first laser, which was ruby, a crystalline solid-state laser. Since then, laser action has been demonstrated at literally thousands of wavelengths in a wide variety of gases, including metal vapors, rare gases, and complex molecules.

Gas lasers vary widely in their characteristics. The weakest commercial lasers emit under a thousandth of a watt, but the most powerful emit over 10,000 W. The most powerful continuous laser beams—a couple of million watts—have been generated by experimental military gas laser weapons. Some gas lasers can emit continuous beams for years; others emit pulses lasting a few billionths of a second. Their outputs range from deep in the vacuum ultraviolet—at wavelengths so short they are blocked completely by air—through the visible and infrared to the borderland of millimeter waves and microwaves.

What makes the gas-laser family so large? Researchers say it is partly the ease of testing different gases for laser action in the same glass tube. They pump out the old gas, pump in the new, seal the tube, and excite the gas. Such experiments are so simple that some modest university laboratories ran up impressive lists of discoveries in the 1960s. The ease of experiments led to some accidental discoveries. One example came when William Bridges put argon into a tube while studying a mercury laser at Hughes Research Laboratories in Malibu, California, in the 1960s. He tried to pump the argon out, but enough remained in the tube to generate a visible beam at 488 nm.

Today, semiconductor and solid-state lasers have begun replacing gas lasers for some applications, but gas lasers continue to play important roles in others. Red semiconductor lasers have become common, but gas lasers remain the standard sources of most shorter visible wavelengths. Gas lasers also offer the highest powers in many parts of the spectrum. Although far more semiconductor lasers are sold than any other type, gas lasers account for almost half the dollar volume of laser sales around the world—about $500 million in 1992, according to *Laser Focus World* magazine.

GAS-LASER BASICS

Gas lasers share some common features, as shown in the generic gas laser of Fig. 6-1. The laser gas is contained in a tube with cavity mirrors at each end, one totally reflecting and one transmitting some light to form the output beam. Most gas lasers are excited by passing an electric current through the gas; the discharge usually runs the length of the tube, as shown in Fig. 6-1. Electrons in the discharge transfer energy to atoms or molecules in the laser gas, typically through at least one intermediate step. Then the excited gas emits light, which resonates within the laser cavity and emerges to form the laser beam.

This picture is a very general one, and—as we will see later in this chapter—each gas laser has its own distinct characteristics. However, the picture is useful, and it will help to elaborate on it before describing individual types of gas lasers.

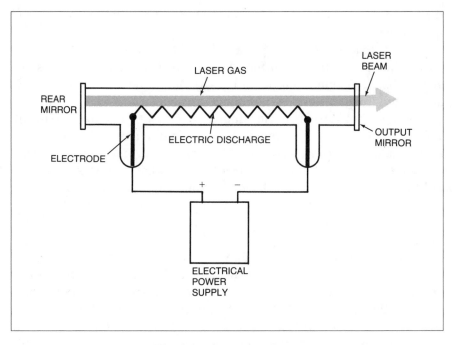

Fig. 6-1 Generic gas laser.

Gas-Laser Media

Many different gases are used in lasers. In developing the helium-neon laser, Javan's group at Bell Labs systematically studied energy levels in neon and other gases. During the great laser boom of the 1960s, some groups simply put different gases into tubes and zapped them with an electric discharge to see what worked. New lasers are still being discovered, and some old types are being developed commercially for the first time, but as laser technology has matured, development has once again become systematic.

There are both obvious and subtle requirements for laser gases. The most obvious is that they must have sets of energy levels suitable for laser action. In practice, energy levels are not enough. For optimum operation, the laser medium must contain a certain mixture of gases at a certain pressure.

Different gases in a mixture serve different functions. For example, carbon-dioxide gas lasers contain helium and nitrogen as well as carbon dioxide. The nitrogen molecules absorb energy from the electric discharge and transfer it to CO_2 molecules. The helium atoms help CO_2 molecules drop from the lower laser level (maintaining the population inversion) and assist in heat transfer. In other lasers, extra gases are added to help absorb energy, transfer heat, or deactivate the lower laser level. The optimum mixture depends not just on the light-emitting species, but also on operating conditions such as the power level, wavelength, and tube design. Thus, all carbon-dioxide lasers don't contain the same proportions of different gases.

Gas pressure is also an important variable, especially because it affects how well the laser gas conducts electricity. Most continuous-wave lasers require a pressure that is a small fraction of one atmosphere to sustain a stable electric discharge. Many pulsed lasers can operate at much higher pressures, sometimes over one atmosphere, because the discharge need not be stable for long. Again, the optimum pressure is not the same for all lasers of the same type; it depends on details of the laser design.

Many gas lasers contain gases present in the atmosphere. Carbon-dioxide and helium-neon lasers are the most common examples, but the less common nitrogen laser is even more striking because air—which is more than three-quarters nitrogen—can serve as the active medium. The gases are not always in the same form as in the atmosphere. For example, the 488- and 514.5-nm lines of argon are emitted by ions—atoms from which one electron has been stripped. In other gas lasers, light is emitted by hot metal vapors that are sometimes ionized as well. Generally, the tubes of such lasers initially contain some solid metal that is partly vaporized to reach the pressure required for laser operation.

Gas Replacement, Flow and Cooling

Many gas lasers normally operate with their tubes sealed like a vacuum tube. However, some gas lasers require periodic fills of new laser gas, while in others the gas flows through the laser tube.

Early gas lasers needed periodic gas replacement because the tubes were not sealed well. Helium atoms, which are very small, can leak out of almost anything, and helium is an essential component of helium-neon lasers. Great strides have been made in glass-sealing technology, and helium-neon lasers are now rated to operate for 20,000 hours or more.

Some sealed gas lasers require periodic replacement of the laser gas because contaminants accumulate and gradually degrade laser action. For example, excimer laser tubes are designed for periodic purging and refilling with fresh laser gas. It is sometimes possible to refurbish tubes of other lasers, like argon, by adding new gas, cleaning the tube, and replacing some elements of the tube.

Gas flows through the tubes of some higher-power gas lasers, both cooling the gas and removing contaminants from the discharge area. Many flowing-gas lasers operate in a closed cycle, but some lasers have an open cycle that requires removing the waste gas. Cooling systems are also needed for some sealed-tube lasers; if cooling requirements are modest, they may be met by a fan, but some high-power lasers require water cooling.

Excitation Methods

The most common way to excite gas-laser media is by passing an electric discharge along the length of the laser tube, as shown in Fig. 6-1. The basic idea is similar to a fluorescent tube. First, a high direct-current voltage ionizes the gas so that it will conduct electricity. Once the gas becomes conductive, the voltage is reduced to a much lower level that will sustain the modest direct current needed to excite the laser medium.

This *longitudinal* excitation is fine for low-power lasers, but it cannot effectively deliver the high current needed for high-power pulsed or continuous-wave lasers. Such lasers are usually excited by a discharge perpendicular, or *transverse*, to the length of the laser tube.

Gas-laser power supplies must convert alternating current from the commercial power grid into different forms to excite lasers. For continuous-wave gas lasers, the power supply must raise the voltage and rectify the current to generate both the trigger voltage needed to ionize the laser gas and the steady voltage needed to operate the laser. A pulsed power supply must deliver high energy in a short interval. Typically, it includes a DC module that charges a capacitor and a fast discharge circuit that applies the electrical pulse from the capacitor across the laser, as shown in Fig. 6-2. A key element is the switch that applies the high voltage across the laser gas. High-voltage, high-current switching technology is difficult, and such switches typically limit the repetition rate of lasers with high pulse energy.

A few gas lasers are powered by zapping the laser medium with a beam of electrons from an accelerator. This technique is rare, and is used only to obtain very high pulsed powers.

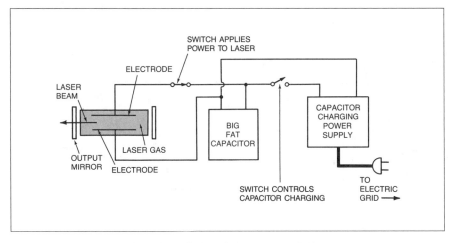

Fig. 6-2   Pulsed laser power supply.

As we will see at the end of this chapter, a few gas lasers are pumped optically, usually by an electrically excited gas laser with a shorter wavelength. Although the overall efficiency is limited, this approach can generate otherwise unobtainable wavelengths.

Tube and Resonator Types

If you go back and look at Fig. 6-1 carefully, you'll note that it's a bit vague about the placement of the mirrors. In practice, the mirrors can be combined with or separate from the windows at the ends of the laser tube. Combining the two can reduce the laser tube cost, but it exposes the mirror coatings to the discharge in the laser tube, which can shorten their life in some lasers. On the other hand, separate mirrors are more expensive, but offer better performance in certain lasers.

The window at the end of the gas tube and the mirror at the end of the laser cavity have distinct functions. The window, like the rest of the laser tube, isolates the laser gas from air, preventing contamination and maintaining the required pressure. The window should have loss as low as possible. The cavity mirrors should have low loss as well, but their main function is to reflect laser light back and forth in the cavity to generate stimulated emission and to couple some light out of the cavity as the laser beam.

One common design approach is shown in Fig. 6-3. The windows at the ends of the laser tube are mounted at what is called Brewster's

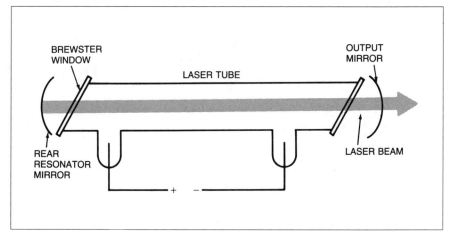

Fig. 6-3   Laser tube with Brewster-angle window and confocal resonator.

angle, which we described in the last chapter. As we indicated then, surface reflection depends on the angle from the normal, the refractive index, and polarization. If light strikes a transparent material with refractive index n from air, no light polarized parallel to the plane of incidence can be reflected if the light strikes at an angle θ_B given by:

$$\theta_B = \arctan n$$

For optical glass with refractive index 1.5, Brewster's angle is about 57 degrees. Some light polarized perpendicular to the plane of the incidence is reflected, leading to higher losses for that polarization, so the orthogonal polarization is amplified. As a result, the laser beam is polarized parallel to the plane of incidence of the Brewster window.

In Fig. 6-3, the cavity mirrors are curved, defining a confocal resonator. As we saw in Chapter 3, this stable-resonator design can provide a good-quality diffraction-limited beam with low divergence. It is common in continuous-wave gas lasers at visible wavelengths, which typically have low gain and require optimized cavities to oscillate.

Different types of cavities are used in high-gain pulsed lasers, notably the rare-gas halide excimer lasers described later in this chapter. Such lasers do not need as careful optimization to oscillate, and they use cavities where only the rear mirror is highly reflective. (The few-percent reflection from an uncoated glass surface provides enough feedback to serve as an output mirror.) This design gives high-gain gas lasers comparatively large beam divergence and diameter. This design also lends itself to oscillator-amplifier configurations, where an external laser cavity without mirrors amplifies output pulses from a master oscillator.

Wavelength and Bandwidth

Transitions in gas lasers, like those in other types, have a well-defined nominal wavelength. However, gas atoms and molecules emit at a broader range of wavelengths because they are continually moving at any temperature above absolute zero. This effect is called Doppler broadening because it depends on the Doppler shift, the change in wavelength (or frequency) of light emitted by something moving relative to the observer. We described it briefly in Chapter 4.

The average speed of gas atoms is proportional to the atomic mass M and the gas temperature T. The usual formula gives the root mean square (the square root of the sum of the squares) $<v>$, a better defined "average":

$$<v> = (3kT/M)^{1/2}$$

where k is the Boltzmann constant mentioned earlier. This velocity is large enough that Doppler broadening accounts for most of the linewidth of many gas lasers. For example, the Doppler width (defined as full width at half maximum) of a typical helium-neon gas laser is about 1.4 GHz, or about 0.0019 nm. Although this is small compared to the 4.738×10^{14} Hz frequency of the laser's 632.8-nm transition, it is much larger than the 1-MHz bandwidth of one longitudinal mode of a typical helium-neon laser cavity. That Doppler width is broad enough to include several longitudinal modes, separated by about 500 MHz.

Doppler broadening occurs in gas lasers because the atoms are not fixed in place as they are in crystalline solid-state lasers or semiconductor lasers. Other effects can broaden the range of wavelengths emitted by those types.

Differences Among Gas Lasers

Although gas lasers share a number of common features, they also differ in many ways. Prominent among them are wavelength and output power, which are the most important laser characteristics for many applications. Table 6-1 lists wavelengths and output power ranges for major commercial gas lasers; the most important types are described in more detail below.

TABLE 6-1 Major gas lasers, by wavelength and power level, grouped under type of transition involved

Type	Wavelength (nm)	(Approximate) Power Range (W)*	Operation
Electronic Transitions			
Molecular fluorine (F_2)	157	1–5 (avg.)	Pulsed
Argon-fluoride excimer	193	0.5–50 (avg.)	Pulsed
Krypton-fluoride excimer	249	1–100 (avg.)	Pulsed
Argon-ion (UV lines)	275–305	0.001–1.6	Continuous
Xenon-chloride excimer	308	1–100 (avg.)	Pulsed
Helium-cadmium (UV line)	325	0.002–0.1	Continuous
Nitrogen	337	0.001–0.01 (avg.)	Pulsed
Argon-ion (UV lines)	333–364	0.001–7	Continuous
Krypton-ion (UV lines)	335–360	0.001–2	Continuous
Xenon-fluoride excimer	351	0.5–30 (avg.)	Pulsed
Helium-cadmium (UV line)	354	0.001–0.02	Continuous
Krypton-ion	406–416	0.001–3	Continuous
Helium-cadmium	442	0.001–0.10	Continuous
Argon-ion	488–514.5	0.002–25	Continuous
Copper-vapor	510 and 578	1–120 (avg.)	Pulsed
Xenon-ion	540	—	Pulsed
Helium-neon	543	0.0001–0.001	Continuous
Gold-vapor	628	1–10 (avg.)	Pulsed
Helium-neon	632.8**	0.0001–0.05	Continuous
Krypton-ion	647**	0.001–7	Continuous
Helium-neon	1153	0.001–0.015	Continuous
Iodine	1300	—	Pulsed
Vibrational Transitions			
Hydrogen-fluoride (chemical)	2600–3000***	0.01–150	Pulsed or CW
Deuterium-fluoride (chemical)	3600–4000***	0.01–100	Pulsed or CW
Carbon-monoxide	5000–6500***	0.1–40	Pulsed or CW
Carbon-dioxide	9000–11000***	0.1–45,000	Pulsed or CW
Vibrational or Rotational Transitions			
Far-infrared	30,000–1,000,000***	<0.001–0.1	Pulsed or CW

*For typical commercial lasers; not indicated for lasers that are not often sold

**Other wavelengths also available

***Many lines in this wavelength range

Gas lasers can operate on electronic, vibrational, or rotational transitions. As we saw earlier, electronic transitions are at near-ultra-violet, visible, or near-infrared wavelengths. Vibrational transitions (which actually include rotational elements as well) are in the infrared, while rotational transitions are at the long-wave end of the infrared and reach into the microwave region at wavelengths longer than 1 mm.

HELIUM-NEON LASERS

The gas laser best known to people outside the laser world is the helium-neon laser. Often called the "He-Ne," its low-power red beam reads labels at supermarket checkout counters, records holograms, and demonstrates how lasers work in countless school laboratories. Although semiconductor lasers are now far more common because of their use in compact disc audio players, their presence is not obvious because they are deep inside. Compact red semiconductor lasers are starting to replace He-Ne lasers in many applications, but most operate at 670 nm, a longer wavelength than helium-neon and one at which the human eye is less sensitive. The red beams of helium-neon lasers remain common.

Physical Principles

The energy levels involved in the helium-neon laser are shown in Fig. 6-4. Electrons passing through a mixture of five parts helium to one part neon excite both species to high energy states, but the more abundant helium atoms collect more energy. As shown in Fig. 6-4, the high-energy helium states have nearly the same energy as 5s and 4s energy levels of neon. (Those combinations of letters and numbers are the spectroscopic notations for electronic energy levels used in Table 2-3. Although they have specific meanings, it is much simpler just to regard them as labels. Some books use other notation.) The energy levels are close enough that the excited helium atoms can transfer energy to neon atoms when they collide, which happens often in the gas. The 5s and 4s energy levels of neon are metastable, so atoms stay in those states for a comparatively long time.

Note that excitation raises neon atoms directly to a high energy level, without stopping at the intermediate levels at the right in Fig. 6-4. The result is a population inversion, with more neon atoms in

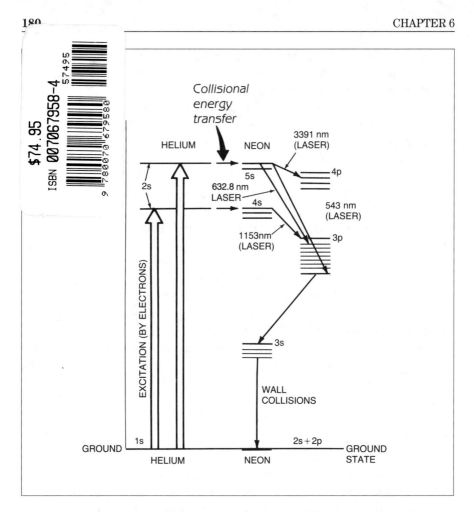

Fig. 6-4 Energy levels and laser transitions in the helium-neon laser shown on a relative scale. (They actually are much higher above the ground state.)

the upper levels than in the lower levels of certain transitions. Several energy levels are involved, and several laser transitions are possible, depending on operating conditions and optics.

The first helium-neon laser operated at 1153 nm in the infrared, but the 632.8-nm red line was discovered soon afterwards and has become the standard helium-neon laser wavelength. Now laser manufacturers offer helium-neon lasers that emit at weaker visible lines, particularly the 543-nm green line.

After dropping to the lower laser level, neon atoms remain high above the ground level. However, they quickly lose that energy and drop through a series of lower energy levels to the ground state. From the ground state they can again be excited to the upper laser level by energy transfer from helium atoms.

Overall gain of the helium-neon laser is very low, so care is needed to minimize laser cavity losses, as we will see below. Overall efficiency is also low—typically 0.01% to 0.1%—because the transitions are far above the ground state. However, the helium-neon laser is simple, practical, and inexpensive; mass-produced sealed-tube versions can operate continuously for tens of thousands of hours. About 10,000 V is needed to ionize the gas, but after ionization a couple of thousand volts can maintain the current of a few milliamperes needed to sustain laser operation.

Laser Construction

Figure 6-5 shows the internal structure of a typical mass-produced helium-neon laser. Note that the discharge passing between electrodes at opposite ends of the tube is concentrated in a narrow bore, one to a few millimeters in diameter. This raises laser excitation efficiency and also helps control beam quality. The bulk of the tube volume is a gas reservoir containing extra helium and neon. Gas pressure within the tube is typically a few tenths of a percent of atmospheric pressure.

Mirrors are bonded directly to mass-produced helium-neon tubes by a high-temperature process that produces what is called a "hard seal." This seal slows helium leakage, which otherwise might limit laser lifetime. The mirrors must have low loss because of the laser's low gain. The rear cavity mirror is totally reflective. The output mirror reflects most incident light back into the laser cavity, but lets a few percent escape in the laser beam. One or both mirrors have concave curvature to focus the beam within the laser cavity, which is important for good beam quality.

An alternative is to seal the laser cavity with Brewster's angle windows, and mount the mirrors separate from the laser tube. This approach is more expensive, but it avoids losses for the plane-polarized beam and allows selection of the operating wavelength. It is used in some laboratory lasers.

The output power available from helium-neon lasers depends on tube length, gas pressure, and diameter of the discharge bore. Researchers have found that output power for a given tube length is highest when the product of gas pressure (in torr, 760 torr equals one

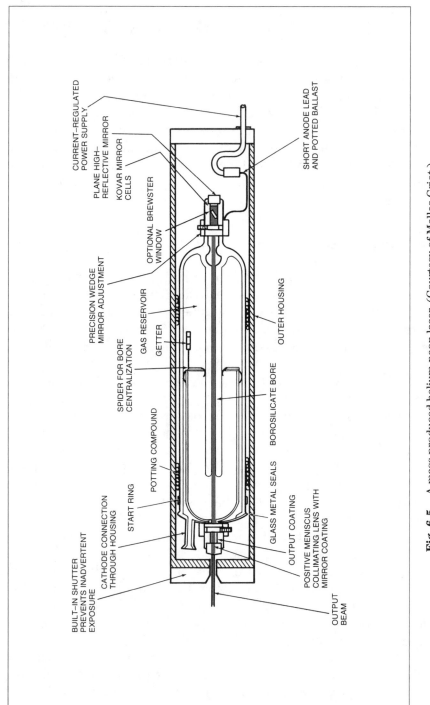

BUILT-IN SHUTTER
PREVENTS INADVERTENT
EXPOSURE

CATHODE CONNECTION
THROUGH HOUSING

START RING

POTTING COMPOUND

SPIDER FOR BORE
CENTRALIZATION

PRECISION WEDGE
MIRROR ADJUSTMENT

OPTIONAL BREWSTER
WINDOW

GAS RESERVOIR

GETTER

CURRENT–REGULATED
POWER SUPPLY

PLANE HIGH–
REFLECTIVE MIRROR

KOVAR MIRROR
CELLS

SHORT ANODE LEAD
AND POTTED BALLAST

OUTER HOUSING

BOROSILICATE BORE

GLASS METAL SEALS

OUTPUT COATING

POSITIVE MENISCUS
COLLIMATING LENS WITH
MIRROR COATING

OUTPUT
BEAM

Fig. 6-5 A mass-produced helium-neon laser. (Courtesy of Melles Griot.)

atmosphere) times bore diameter (in millimeters) is 3.6 to 4. Extending the tube can raise output power somewhat, but the improvements are limited.

One interesting variant on the standard linear helium-neon laser is the ring laser, which can sense rotation. The tube for a ring laser is actually a triangle or square, with mirrors at each corner to reflect the beam from one arm into the next as it travels around the circumference. A ring laser can detect rotation about an axis perpendicular to the ring plane by measuring phase differences in light travelling in different directions around the ring. Ring-laser "gyroscopes" are used in some military and civilian aircraft.

Practical Helium-Neon Lasers

Mass-produced helium-neon lasers can deliver 0.5 to 10 mW of red light. They range in size from some little bigger than fat pens, 10 cm long and 1.6 cm in diameter (4 inches by 5/8 inch) to 30 cm (1 foot) or more long and 2 to 5 cm (1 to 2 inches) in diameter. A few special-purpose helium-neon lasers are considerably larger and can produce up to 60 mW in the red.

Although other visible wavelengths have been available commercially since the mid-1980s, most helium-neon lasers emit the 632.8-nm red line. Output is much weaker, typically no more than a couple of milliwatts, at other visible wavelengths, 543 nm in the green, 594 nm in the yellow-orange, and 612 nm in the red-orange. Powers of 1 to 10 mW are available on infrared lines at 1.153, 1.523, and 3.39 μm. Although a few models can be made to emit at different wavelengths by switching their optics, most helium-neon lasers emit only at a single wavelength.

Mass-produced red helium-neon laser tubes may sell for $20 or less in large quantities, without power supply or accessories, including safety equipment required by federal regulations. Complete general-purpose red helium-neon lasers ready for laboratory or other use start at a couple of hundred dollars each when bought singly; other wavelengths are more expensive. Red helium-neon lasers, tubes, and heads are often available on the surplus market for lower prices.

For most practical purposes, you can consider the light from a red helium-neon laser to be a single wavelength. The typical Doppler broadened bandwidth of a red helium-neon laser is 1.4 GHz, which corresponds to a coherence length of 20 to 30 cm, adequate for holography of small objects. If you need a longer coherence length or a narrower bandwidth, you must buy a special helium-neon laser that is limited to a single longitudinal mode. The 1-MHz bandwidth of a single longitudinal mode corresponds to a coherence length of 200 to 300 m.

Helium-neon lasers typically emit TEM_{00} beams, with diameter about a millimeter and divergence about a milliradian. You can expect a helium-neon laser tube with hard-sealed mirrors to operate for 20,000 hours or more. Because of their low cost, compact size, durability, and ease of operation, helium-neon lasers were long the most widely used visible lasers, but red diode lasers are replacing them in many applications.

RARE-GAS ION LASERS

Argon- and krypton-ion lasers resemble helium-neon lasers enough to be considered cousins. All are driven by electric discharges passing through elements of the rare-gas (Group VIII) column of the periodic table. All normally emit continuous beams of visible light.

However, there are many important differences. Argon- and krypton-ion lasers can generate much more power than helium-neon lasers, and they produce the shorter wavelengths needed for some applications. Typically, their output powers range from a few milliwatts to up to 25 W for argon and somewhat less for krypton. On the other hand, they are more delicate, less efficient, and much more costly. Most of their uses are specialized, but you may have encountered one: they are the preferred sources for laser light shows.

One important note before we look inside rare-gas ion lasers: they are often called simply "ion" lasers. This is not a very good label, because ions emit the light in other gas lasers (such as helium-cadmium), but it is common in the laser world.

Physical Properties

The active medium in rare-gas ion lasers is argon or krypton at a pressure of roughly 0.001 atmosphere. The two gases can be mixed to get emission on both sets of wavelengths, usually for light shows. The gases emit light at many wavelengths in the near-ultraviolet, visible, and near-infrared parts of the spectrum, listed in Table 6-2.

TABLE 6-2 Major wavelengths of ion lasers

Argon	Krypton
275.4 nm	337.4 nm
300.3 nm	350.7 nm
302.4 nm	356.4 nm
305.5 nm	406.7 nm
334.0 nm	413.1 nm
351.1 nm	415.4 nm
363.8 nm	468.0 nm
454.6 nm	476.2 nm
457.9 nm	482.5 nm
465.8 nm	520.8 nm
472.7 nm	530.9 nm
476.5 nm	568.2 nm
488.0 nm (strong)	647.1 nm (strong)
496.5 nm	676.4 nm
501.7 nm	752.5 nm
514.5 nm (strong)	799.3 nm
528.7 nm	
1090.0 nm	

Unlike helium-neon lasers, the emission in argon and krypton lasers comes from atoms that have been ionized by having one or two electrons stripped from their outer shells. Wavelengths shorter than 400 nm come from atoms with two electrons removed (Ar^{+2} or Kr^{+2}). Longer wavelengths come from singly ionized atoms (Ar^+ or Kr^+). Argon is a much more efficient laser gas, but krypton offers a broader choice of visible wavelengths.

The internal kinetics of ion lasers are complicated, but we can give an overview of what happens. As in helium-neon lasers, an initial high-voltage pulse ionizes the gas so that it conducts current. Electrons in the current transfer energy directly to argon or krypton atoms, freeing other electrons and raising the resulting ions to a group of high energy levels, as shown in Fig. 6-6. Three processes can populate the metastable upper laser levels, from which the ions drop to a group of lower laser levels. Argon and krypton lasers emit so many different laser lines because transitions can occur between many pairs of upper and lower levels; there isn't room to show all those levels in the diagram. If the optics permit it, argon and krypton lasers can oscillate simultaneously on several different visible wavelengths, each produced by a transition between a different pair of levels.

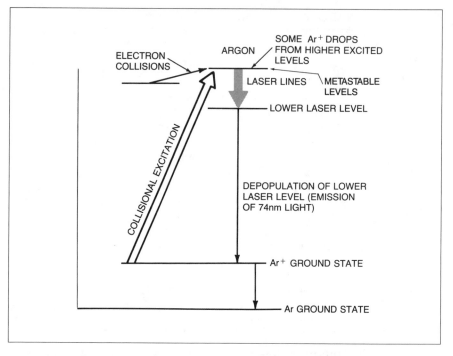

Fig. 6-6   Energy levels that produce visible argon lines.

The lower laser levels all have very short lifetimes. Argon ions quickly drop from the lower laser level (an excited state of the ion) to the ion ground state by emitting extreme-ultraviolet light at 74 nm. The ground-state ion can recapture an electron or again be excited to the upper laser levels. Similar things happen in krypton.

It takes much more energy to excite argon and krypton atoms to their laser levels than it does to excite neon atoms. In the helium-neon laser, it is atoms that emit light; in visible-wavelength argon and krypton lasers, it is ions missing one electron. Discharge currents in argon and krypton lasers are 10 to 70 amperes, more than a thousand times the level in a helium-neon laser, although operating voltages of 90 to 400 V are much lower because the ionized gas has less resistance. This high discharge current heats the laser gas to very high temperatures. (Ultraviolet lines of argon and krypton are emitted by ions missing two electrons, Ar^{+2} and Kr^{+2}. Those states lie much higher above the ground level than the singly ionized states that produce the visible lines, and thus require even stronger currents.)

The spectral linewidth of a single ion-laser line is about 5 GHz, broader than that of a helium-neon laser. This linewidth is large enough to allow modelocking, which generates pulses 90 to 200 picoseconds long. The main use of modelocked pulses is to pump the tunable dye lasers described in Chapter 9.

Argon- and Krypton-Ion Laser Structure

The structures of argon and krypton lasers are quite similar, and manufacturers often use the same basic laser tube for both types. The difference is in the gas fill. A typical argon-ion laser may resemble a helium-neon laser at first glance, although ion lasers tend to be fatter and—especially at higher powers—much larger. (High-power ion lasers are 2 m long.) There are also some more subtle but nonetheless important differences.

Most argon and krypton lasers have Brewster-angle windows on the ends of the tube, with external mirrors defining the laser cavity. As in helium-neon lasers, the gain is low, so care is needed to minimize losses in the laser cavity. That includes avoiding materials that suffer increased losses after being exposed to the extreme-ultraviolet light from decay of the lower laser level in argon. The tube itself is usually a ceramic. Cavity dumpers and modelockers can be inserted in the laser cavity to make the normally continuous-wave laser produce short pulses.

The cavity optics select which wavelengths argon and krypton lasers emit. If the goal is raw output power, the optics can allow oscillation at several wavelengths. Alternatively, wavelength-selective optics can be inserted between the window and rear mirror to limit oscillation to one laser wavelength. Standard cavity optics produce TEM_{00} beams.

As in helium-neon lasers, the discharge is confined to a narrow region in the center of the tube to enhance excitation efficiency. Older argon lasers confined the discharge in a narrow bore like that of the helium-neon laser, but in many newer lasers the discharge is confined by a series of metal disks with central holes that define a "bore." In either case, a return path is needed for the positive ions, which are attracted by the negatively charged cathode. A large gas reservoir is also needed, because laser operation tends to deplete the gas.

With much higher output powers than helium-neon lasers but lower wall-plug efficiency (0.01 to 0.001%), argon and krypton lasers need active cooling. Forced-air cooling is adequate for argon lasers delivering up to a few watts. At higher power levels, or for the less efficient krypton laser and ultraviolet lines, water cooling is needed. Typically, tap water flows through pipes in the laser and then down the drain. (The need for both cooling water and high-voltage service limits where you can use high-power argon- or krypton-ion lasers.)

Practical Argon- and Krypton-Ion Lasers

As we indicated earlier, argon and krypton ions can oscillate on one or many wavelengths, depending on the cavity optics. If bright emission is required on lines throughout the visible spectrum, the two gases can be put into a single tube and excited simultaneously. Such mixed-gas lasers are sometimes used in laser light shows. Most rare-gas ion lasers emit continuous-wave TEM_{00} beams that are vertically polarized, but a few multiwavelength lasers may emit multiple modes. Although argon and krypton lasers are much more powerful than helium-neon lasers, beam diameters and divergences are similar.

The prices of argon lasers have been coming down, but they remain much higher than those of helium-neon lasers. You can expect to pay thousands of dollars for the least expensive argon lasers, which emit several milliwatts. At the high end of the price scale, high-power argon lasers run tens of thousands of dollars. Krypton lasers cost more per watt because they are less efficient than argon.

Operating conditions inside argon and krypton laser tubes are extreme, so lifetimes are shorter than those of helium-neon lasers. Typical rated lifetimes are 1000 to 10,000 hours, with the longer lifetimes for lower-power tubes.

METAL-VAPOR LASERS

Metal vapors emit light in two important families of commercial lasers. In one, typified by the helium-cadmium laser, the light emitter is an ionized metal atom. In the other, typified by copper- and gold-vapor lasers, light is emitted by neutral atoms. The distinction is important because their operating characteristics differ greatly.

Helium-Cadmium and Other Metal Ion Lasers

Helium-cadmium (He-Cd) lasers produce continuous output at slightly higher power levels than helium-neon lasers, from under a milliwatt to tens of milliwatts. He-Cd wavelengths are shorter, 325 and 353.6 nm in the ultraviolet and a stronger 441.6-nm line in the blue. However, despite the similarities in name, power level, and some internal characteristics, the He-Ne and He-Cd lasers are not as closely related as they might at first seem. You can consider the helium-cadmium laser as in between argon and He-Ne lasers in such characteristics as output power and overall efficiency.

Figure 6-7 shows the energy-level structures of helium and cadmium. The laser gets its energy from electric current flowing through a thin capillary bore in the laser tube. The electrons excite helium atoms to high levels. Then the excited helium atoms transfer energy to cadmium atoms, ionizing them and raising them to the upper laser level. (Cadmium normally has only two electrons in an incompletely filled outer shell, and it is much easier to ionize than helium, argon,

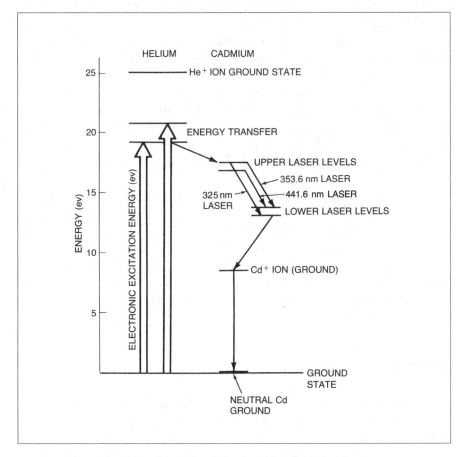

Fig. 6-7   Energy levels in the helium-cadmium laser.

or krypton, all of which normally have full outer electron shells.) Cadmium atoms can be trapped in two metastable states, one of which produces the ultraviolet lines, the other the 441.6-nm line.

Cadmium is a solid at room temperature, so the metal in the laser tube must be heated to about 250 degrees centigrade to produce the several-millitorr pressure of cadmium vapor needed for the laser to operate. Helium pressure is about a thousand times higher, several torr, but still only about 1% of atmospheric pressure.

Like argon and He-Ne lasers, He-Cd laser tubes concentrate the discharge in a narrow bore to excite gas atoms efficiently. Chunks of cadmium metal are put in the tube to replace cadmium that condenses on cool parts of the tube during laser operation. Tubes normally also include a helium reservoir to replace gas that leaks out. Together with zones of the tube designed to collect surplus cadmium so it does not deposit on critical optical surfaces, these reservoirs make He-Cd tubes look different from He-Ne and argon lasers. However, the basic technology is similar, except for the need to heat the metal. Helium-cadmium lasers need discharge voltages around 1500 V and lower currents than argon lasers.

Optical design of helium-cadmium lasers also resembles those of He-Ne and argon lasers. Some He-Cd laser tubes have Brewster-angle windows and external resonator optics; others have cavity mirrors bonded directly to the tube. The output wavelength depends on the optics. Beam diameter and divergence are around 1 mm and one milliradian.

In many practical terms, the He-Cd laser can be seen as an intermediate step between the He-Ne and the argon laser, although the wavelength is shorter than either. Lifetimes of blue He-Cd lasers are several thousand hours, comparable to those of argon lasers. Prices range from a few thousand dollars for a low-power model to over $10,000 for the most powerful types. Other metal-vapor ion lasers have been demonstrated in the laboratory, but none are available commercially.

Copper- and Gold-Vapor Lasers

Neutral metal-vapor atoms are also the basis of a family of gas lasers. The most important members of the family are the copper-vapor laser, which emits at 511 nm in the green and 578 nm in the yellow, and the gold-vapor laser, which operates at 628 nm in the red.

Unlike the other gas lasers we have described so far, the copper- and gold-vapor lasers are limited to pulsed operation. They can generate average powers of tens of watts and thousands of pulses per second, but not a continuous beam.

The restriction to pulsed operation stems from the energy level structure of neutral metal vapors. When an electric discharge passes through copper vapor, collisions with electrons raise copper atoms to one of two excited states. If vapor pressures are low, those states stay excited for only about 10 ns, not long enough to produce laser pulses. However, if the density is increased (to pressures still well below atmospheric pressure), the states' effective lifetimes increase to 10 ms, long enough for stimulated emission and laser operation. That quickly shifts the population to the two lower laser levels, which are metastable, having lifetimes of tens to hundreds of microseconds. The accumulation of atoms in the lower laser levels stops laser action in less than 100 ns. Then the lower-level population decays, and soon the laser is ready to generate another pulse—at repetition rates of many thousand pulses per second. Similar processes occur in gold and several other metals, producing lasers that can generate high average powers in rapid repetitive pulses.

To obtain the 0.1-torr vapor pressure needed for laser action, metallic copper or gold must be heated to 1500 to 1850 degrees centigrade in the laser tube. This takes about half an hour, an exceptionally long warm-up time by laser standards. The metal vapor atoms are excited by collisions with electrons in a pulsed electric discharge passing the length of the tube. Addition of a rare gas (neon, argon, or helium) can improve discharge quality and speed depopulation of the lower laser level. During operation, the laser generates enough waste heat to keep metal vapor pressure high enough for laser operation. (In fact, metal-vapor lasers normally require active cooling with flowing water or forced air.)

Unlike the He-Ne, argon, and He-Cd lasers, copper-vapor lasers have high gain, 10% to 30%/cm. A copper-vapor laser can operate without resonator mirrors, but commercial lasers have a totally reflective rear mirror and an output mirror that transmits about 90% of incident light, reflecting about 10% back into the laser cavity. The laser tube windows are separate from the cavity mirrors, and an uncoated window surface can reflect enough light back to serve as an output mirror.

Two factors control pulsing of gold- and copper-vapor lasers. Pulse length depends on internal kinetics, and on how fast the lower laser fills up to stop the laser pulse, typically tens of nanoseconds. The repetition rate depends on the discharge electronics, because a new electrical pulse must trigger each light pulse. Commercial copper- and gold-vapor lasers have repetition rates of several thousand hertz, which the eye sees as a steady beam. Power can exceed 100 W.

With efficiency of several tenths of a percent, copper- and gold-vapor lasers are the most efficient visible gas lasers and among the highest-powered. Their repetitively pulsed output is not as useful as a continuous beam for many applications, and at tens of thousands of dollars, copper- and gold-vapor lasers are expensive. (Despite what you might think, the cost of gold adds little to the cost of a gold-vapor laser. The laser consumes only about $20 worth of gold in an eight-hour day, and the gold is deposited in the tube, where it can be recovered later).

CARBON-DIOXIDE LASERS

The carbon-dioxide (CO_2) laser is the most versatile gas laser, able to operate either pulsed or continuously, and able to produce the highest continuous power of any laser you can buy. Unlike visible gas lasers, the carbon-dioxide laser operates on a set of vibrational-rotational transitions. This puts its output at much longer wavelengths, 9 to 11 μm in the infrared. There are several important types of carbon-dioxide lasers, but all work on the same transitions.

Basic Physics of CO_2 Lasers

Carbon-dioxide laser lines are emitted during transitions among three vibrational modes of the molecule shown in Fig. 6-8: the symmetric stretching mode v_1, the bending mode v_2, and the asymmetric stretching mode v_3. Each mode is quantized, so the molecule can have 0, 1, 2, or more units of vibrational energy in each mode.

The laser transitions occur when CO_2 molecules drop from the higher-energy asymmetric stretching mode to the lower-energy symmetric stretching or bending modes. The transition to the symmetric stretching mode corresponds to a 10.5-μm photon; dropping to the bending mode (actually, to the second excited level of the bending mode) corresponds to a 9.6-μm photon. However, the laser does not emit those precise wavelengths, because the molecule changes its

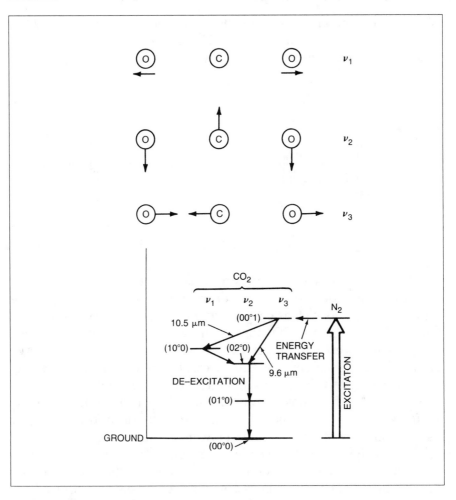

Fig. 6-8   Vibrational modes and transitions of the CO_2 molecule.

rotational state when it changes its vibrational state. The rotational transition energy is smaller than both thermal energy and the vibrational transition energy, so the change can be either up or down. That is, the molecule can speed up or slow down its rotation when moving between vibrational levels. Speeding up its rotation consumes some energy from the vibrational transition, so the emitted wavelength is longer than the nominal transition energy (for example, 10.8 μm rather than 10.5 μm). On the other hand, if the molecule slows its rotation, the energy released adds to the energy from the vibrational transition, resulting in a shorter wavelength, such as 10.3 μm. Because of this phenomenon, carbon-dioxide lasers can emit a family of closely spaced wavelengths, as shown in Fig. 6-9. The average is 10.6 μm, which is often given as the CO_2 laser wavelength.

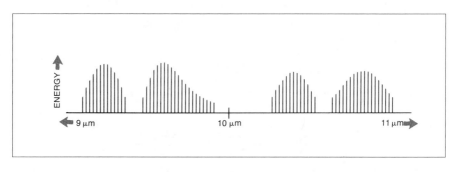

Fig. 6-9   Wavelengths emitted by a carbon-dioxide laser.

An electric discharge passing through the laser gas excites the carbon-dioxide laser. The laser medium contains nitrogen and (usually) helium as well as carbon dioxide. Each gas serves a specific role. Both nitrogen and CO_2 absorb energy from electrons in the discharge. The lowest vibrational level of N_2 has nearly as much energy as the CO_2 molecule's asymmetric stretching mode, so N_2 can readily transfer energy to CO_2. The asymmetric stretching mode is the upper laser level of both groups of transitions. Helium helps to maintain the population inversion by getting CO_2 molecules to drop from the lower laser levels to the ground state or a lower bending level.

The carbon-dioxide laser is an exceptionally versatile system that can operate under a wide range of conditions. (There have even been reports that CO_2 laser lines are emitted by the upper atmosphere of Mars!) Its efficiency is exceptionally high by laser standards and can reach up to 20%. It can produce a steady beam at low gas pressures, or powerful pulses at high pressures. It can operate simultaneously on many lines or be tuned to a single wavelength. Commercial models routinely generate steady beams ranging from milliwatts to many kilowatts. We will discuss the major types of CO_2 lasers next.

Types of CO₂ Lasers

Sealed-Tube Lasers

The simplest CO_2 lasers operate in a sealed tube, like the other gas lasers described earlier. Typically, a high DC voltage causes an electric discharge to pass between positive and negative electrodes at opposite ends of the tube, but the laser can also be excited by a radio-frequency-induced discharge. Because the discharge breaks down CO_2 molecules to form oxygen and carbon monoxide, water or a catalyst must be added to the gas to regenerate CO_2, but this is not a major problem.

The sealed-tube design is convenient and inexpensive; it is widely used for lasers with powers under about 100 W, priced at a couple thousand to a few tens of thousands of dollars. However, other designs are required for higher power. One reason is that the maximum power depends on tube length; for sealed-tube lasers, a rule of thumb has been 50 W/m of gain medium. Mirrors can bend the laser beam inside the cavity so that the laser tube doesn't have to measure a full meter or two end to end; but nonetheless, long tubes do become cumbersome. An added problem at high power levels is the need to remove waste heat.

Note that some gas flow is possible in a sealed-tube laser, by moving gas through the tube and the rest of a sealed system that includes a gas reservoir.

Waveguide CO_2 Lasers

A popular way to make compact carbon-dioxide lasers with powers from under a watt to about 50 W is to shrink the tube's cross section to a millimeter or two across. This is only about 100 times the CO_2 laser's 10-μm wavelength. A dielectric tube of those dimensions functions as an infrared waveguide, guiding the light waves through it (much like hollow pipes guide microwaves). This structure avoids large diffraction losses that otherwise would occur with an output aperture that is so small compared to the wavelength.

Gas must flow through the waveguide for such lasers to operate, but the waveguide can be made part of a sealed system with an internal gas reservoir. Normally, operation is continuous.

Longitudinal-Flow CO_2 Lasers

One way to increase CO_2 laser power is to flow fresh gas along the length of the laser cavity, in the same direction that the discharge current is applied. This consumes gas, but pressures are low, and the gases used in CO_2 lasers are not particularly hazardous or costly. In addition, at least some of the exhaust gas can be recycled by mixing it with fresh gas.

The output power available per unit length is comparable to or somewhat higher than that of sealed-tube CO_2 lasers, but the design is straightforward, and this approach has been widely used in continuous-wave lasers delivering hundreds of watts. As high-power devices, flowing-gas lasers normally cost tens of thousands of dollars and up.

Transverse-Flow CO_2 Lasers

Output power in a continuous CO_2 beam can reach about 10 kW/m of tube length if the laser gas flows perpendicular or transverse to the axis of the laser cavity (the line between the mirrors). This lets gas flow through the laser cavity much faster than if the gas had to pass along the tube axis. The faster gas flow removes waste heat and contaminants. The electric discharge that drives the laser is also applied perpendicular to the tube axis, so it goes through a shorter length of gas, but pressure remains low to maintain a stable discharge.

Like longitudinal-flow lasers, the gas pressure is low and the output beam is continuous. Typically, the gas is recycled, with some fresh gas added. Transverse flow is normally used only in very high power lasers, with outputs in the kilowatt range or above and prices to match. (A diagram of one early 15-kW model labelled part of the flow loop as a "wind tunnel.")

Gas-Dynamic CO_2 Lasers

An electric discharge is not the only way to produce a population inversion in carbon dioxide. Rapid expansion of hot, high-pressure CO_2 (typically mixed with other gases) through nozzles into a near-vacuum can also produce a population inversion, because the expansion reduces gas temperature but does not drop all the molecules to low energy levels. The gas flows transversely through the laser cavity, as in a transverse-flow laser, but does not require electrical excitation.

In the late 1960s, the first "gas-dynamic" carbon-dioxide laser represented a breakthrough in high-power lasers, the first laser to reach the 100-kW range. That breakthrough triggered efforts, beginning around 1970, to develop high-power laser weapons. The Pentagon spent hundreds of millions of dollars learning that gas-dynamic lasers don't make very good weapons.

Transversely Excited Atmospheric (TEA) CO_2 Lasers

All the variations on the carbon-dioxide laser we have described so far operate at pressures well below one atmosphere and normally generate continuous beams (although they can be pulsed). The low pressure is needed because continuous electric discharges are not stable at pressures above about a tenth of an atmosphere.

An alternative is to increase gas pressure to around one atmosphere and pass a pulsed electric discharge through the gas. This works best if the discharge is transverse to the laser axis. Such lasers are called "transversely excited atmospheric-pressure," or "TEA," lasers, although the internal gas pressure is not always one atmosphere.

TEA lasers are compact sources of intense pulses lasting 40 ns to 1 µs. They cover a wide range in power levels, size, and price—the latter ranging from a few thousand dollars to several tens of thousands.

Optics and Wavelength Selection

We mentioned earlier that the many combinations of rotational and vibrational transitions let carbon-dioxide lasers produce a broad range of wavelengths. That wavelength range impacts the design of laser optics.

The carbon-dioxide laser has good but not spectacular gain, so it normally has a totally reflective rear mirror and a partly reflective output mirror. Often the cavity mirrors are made of metal, with output coupling through a hole in the mirror rather than through a partly transmissive coating. Because conventional silica glasses are not transparent at 10 µm, the output windows of a CO_2 laser tube may not look transparent to you, but they are transparent to the laser beam.

Many CO_2 lasers generate the entire range of possible wavelengths, because for applications such as materials working there isn't much difference between 9 and 11 µm. For drilling, welding, cutting, or heat treating, the main concern is delivering the maximum power to a small spot—which requires high power and a good-quality beam. These lasers are made with cavity optics reflective throughout the 9- to 11-µm range, to extract as much energy as possible from the CO_2 laser cavity.

On the other hand, scientific applications often require one specific CO_2 wavelength. Lasers for such applications are made with internal tuning optics to select one line from the entire range of possible transitions. The lines are discrete as long as the laser is operated at low pressure. However, as pressure increases, the individual lines broaden. If TEA lasers are operated at pressures above about 10 atmospheres, the separate rotational lines blend together to form a continuous spectrum throughout the CO_2 laser's operating range, so the laser can be tuned continuously in wavelength.

10-μm Quirks

The 10-μm part of the spectrum is a strange world to those of us used to visible light. As we mentioned earlier, ordinary silica glass is not transparent at 10 μm, but other materials are. Some optical materials are transparent at both visible and 10-μm wavelengths, including salt (sodium chloride) and zinc sulfide. However, many 10-μm window materials do not transmit visible light, so you can't assume that something that looks opaque is not transparent at 10 μm. It may even be emitting a beam.

Infrared viewers can show you carbon-dioxide laser beams, but you need the right kind of infrared viewer. You cannot use the many viewers that operate in the near-infrared at wavelengths near 1 μm, because they do not respond to 10 μm. You need a "thermal" infrared viewer. The name comes from the fact that thermal (heat) radiation from room-temperature objects peaks near 10 μm. The military uses thermal viewers to search for enemy soldiers and vehicles, both of which are hotter than their surroundings, and the military developed much of the infrared technology used with CO_2 lasers.

One other note about carbon-dioxide laser optics: so far, there are no good optical fibers for that wavelength. People have tried to make many materials transparent at 10 μm into optical fibers, but haven't had much success. If you have any bright ideas for 10-μm optical fibers, there are lots of people who would like to talk with you.

CARBON-MONOXIDE LASERS

The carbon-monoxide (CO) laser is something of a less successful brother of the carbon-dioxide laser. Emitting on vibrational-rotational transitions lying mostly between 5 and 6 μm, it can be even more efficient than CO_2. Like CO_2, the CO laser can emit powerful continuous beams. It has been plagued by serious practical problems, including strong absorption of some lines by the atmosphere and the need to cool the gas to below room temperature for efficient operation.

Applications are not widespread, but a few commercial models are available. Developers are testing carbon-monoxide lasers for medical and industrial applications.

EXCIMER LASERS

Excimer lasers are a family in which light is emitted by short-lived molecules made up of one rare-gas atom (e.g., argon, krypton, or xenon) and one halogen (e.g., fluorine, chlorine, or bromine). The most important excimer lasers are listed in Table 6-3. First demonstrated in the mid-1970s, excimer lasers have become important because they are the most powerful practical ultraviolet lasers. They also rely on interesting and unusual physics.

TABLE 6-3 Major excimer lasers

Type	Wavelength
F_2*	157 nm
ArF	193 nm
KrCl	222 nm
KrF	249 nm
XeCl	308 nm
XeF	350 nm

*Not a rare-gas halide, but usually grouped with excimers.

Physical Fundamentals

Rare-gas halides are peculiar molecules that emit laser light on an unusual type of electronic transition. The two atoms are bound only when the molecule is in an excited state, which is the upper laser level. The molecule falls apart when it drops to the ground state, which is the lower laser level. That produces a population inversion in a rather unusual way—there can't be any molecules in the lower laser level because the atoms are not bound together.

Figure 6-10 shows the energy levels of a typical rare-gas halide as a function of the spacing between the two atoms in the molecule, R (the rare gas) and H (the halide). The dip in the excited-state curve shows where the atoms are bound together because they have a minimum energy at that separation and form a metastable molecule.

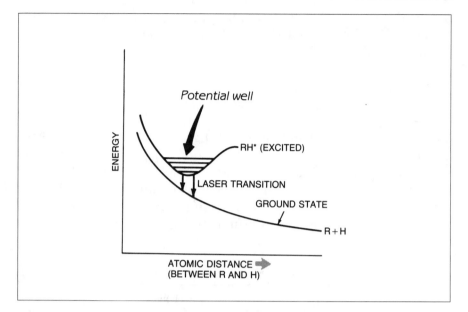

Fig. 6-10   Internal energy of a rare-gas halide molecule in excited and ground states.

The absence of a dip in the ground-state curve shows that the two atoms are not bound together, and that their energy decreases as they drift apart. When the atoms are bound together in the excited state, they can occupy several vibrational levels (shown as horizontal lines) in the low-energy potential well. However, when the molecule drops to the ground state, which has even lower energy, there is no bonding energy to hold the two atoms together, and the molecule falls apart. That reflects something you may remember from chemistry class—rare gases don't like to form compounds, even with elements as highly reactive as halogens.

Excimer lasers are excited by passing a short, intense electrical pulse through a mixture of gases containing the desired rare gas and halogen. Normally, 90% or more of the mixture is a buffer rare gas (typically helium or neon) that does not take part in the reaction. The mixture also contains a few percent of the rare gas (argon, krypton, or xenon) that becomes part of the excimer molecule. It also contains a smaller fraction of molecules that supply the needed halogen atoms, either halogen molecules such as F_2, Cl_2, or Br_2, or molecules that contain halogens such as nitrogen trifluoride (NF_3). Pure halogens can present problems because they are very reactive, and fluorine is so treacherous to handle that the developers of one military high-energy laser that used fluorine spoke of "the fire of the week."

Electrons in the discharge transfer energy to the laser gas, breaking up halogen molecules and causing formation of electronically excited molecules like xenon fluoride (written XeF*, with the * meaning excited). The reactions involved are very complex and depend on the type of gases. The molecules remain excited for about 10 ns and then drop to the ground state and break up. The molecular kinetics (as well as the duration of the driving electrical pulses) normally limit laser operation to pulses lasting tens or hundreds of nanoseconds. The energies involved are large, and output is at ultraviolet wavelengths.

Excimer-laser repetition rates depend more on the power supply than on the gas. The principal limitation is the speed of the high-voltage switches. The highest repetition rates are around 2500 Hz, for a very small laser, but more typical values are tens to a few hundreds of hertz. Pulse energies range from about 10 microjoules to a few joules, and differ somewhat among gases, with KrF and XeCl generally the most energetic. Average power—the product of pulse energy times repetition rate—can reach a couple of hundred watts, although lower values are more common. Note that, in general, the pulse energy tends to decrease with repetition rate.

Structure of Excimer Lasers

Excimer lasers have such high gain that they almost don't need cavity mirrors. In practice, they have fully reflective rear mirrors and uncoated output windows that reflect a few percent of the beam back into the cavity and transmit the rest.

As in other high-gain pulsed lasers, the discharge in an excimer laser is perpendicular to the length of the tube. The usual design is similar to that of a TEA CO_2 laser, except that excimer laser tubes must resist attack by the highly corrosive halogens in the laser gas. Excimer laser tubes are filled with the laser gas mixture, then sealed and operated for a certain number of shots until the gas needs to be replaced. The tube's total volume is much larger—typically 100 to 1000 times—than the volume in which the discharge excites laser action. Often, the gas is passed through a recycling system that helps regenerate the proper gas mixture and extend the life of the gas fill. The laser's pulse energy drops with time, until the spent gas must be pumped out of the laser and replaced. The number of pulses depends on the gas, and can be many millions for longer-lived gases such as xenon chloride.

Although that number of pulses may sound impressive, a little multiplication will show that at high repetition rates it doesn't amount to very much time. A 200-Hz laser generates $200 \times 60 \times 60$ pulses an hour—720,000 pulses! Thus, a gas supply is part of any excimer laser setup.

An alternative design is the waveguide excimer laser, in which the excimer gas flows slowly through a small-bore dielectric waveguide, around a millimeter across. A microwave discharge is applied across the waveguide to excite the gas at repetition rates that can reach thousands of pulses per second—much higher than conventional excimer lasers. With much smaller gas volumes, waveguide excimers produce pulses with much lower energy than standard excimer lasers.

Practical Excimer Lasers

Excimer lasers are the best available pulsed ultraviolet lasers, with wall-plug efficiency as high as a couple of percent. Most are used in research, but they are finding applications in medicine and high-technology industrial systems, including refractive surgery on the eye and the manufacture of semiconductor electronics. These commercial applications have pushed manufacturers to make more reliable excimer lasers, although the use of halogens remains a concern.

Laboratory excimer lasers have long been designed to handle any of several gas mixtures. This reflects the needs of a research laboratory, which one day may be working with the 308-nm xenon-chloride line, and the next day may need the 193-nm argon-fluoride wavelength. The researcher would pump out the old gas mixture, passivate the tube to remove contaminants, then pump out that mixture and replace it with a new laser gas mixture. Industrial excimer lasers are typically made for a specific gas mixture.

Excimer lasers remain complex systems and on the pricy side. Waveguide models cost over $10,000, while typical conventional models cost $40,000 to $100,000.

NITROGEN LASERS

The 337-nm nitrogen laser bears some resemblances to excimer lasers. Like excimer lasers, it has high gain and can produce short ultraviolet pulses with high peak powers. It operates on a combined electronic-vibrational transition of molecular nitrogen, which is excited by a pulsed electric discharge at pressures from about 0.03 to one atmosphere. Excitation is efficient, but other processes in the laser are not. The lower laser level has a 10-µs lifetime, long enough to terminate the population inversion and laser action quickly. As a result, laser efficiency is 0.1% or less, pulse length is limited to a few nanoseconds, and pulse energy is limited to about 10 millijoules.

Because the low pulse energy limits average power, many customers have turned to excimer lasers, which can provide much more power because of their higher pulse energy. However, nitrogen lasers can readily be made small and cheap, and do not require potentially dangerous gases. Compact nitrogen lasers that generate 0.1-millijoule pulses now are available for $1000 to $2000, and are used in measurement and remote sensing systems.

CHEMICAL LASERS

One of the more interesting laser types is the chemical laser. The most common type operates on vibrational transitions of hydrogen fluoride (HF) in the infrared. The standard HF laser emits light on many lines between about 2.6 and 3.0 µm. If normal hydrogen-1 is replaced by the heavier isotope deuterium (hydrogen-2), the wavelength is shifted to 3.6 to 4.0 µm, and the laser is called a deuterium-fluoride (DF) laser.

It is possible to excite chemical lasers by passing an electric discharge through a mixture containing hydrogen and fluorine atoms (often in other molecules, such as sulfur hexafluoride, SF_6). In practice, helium is added as a buffer gas, and other gases may be added to control the production of fluorine. Alternatively, the excitation can be purely chemical, caused by the combustion of hydrogen with fluorine.

The HF laser relies on a chemical chain reaction:

$$H_2 + F \longrightarrow HF^* + H$$
$$H + F_2 \longrightarrow HF^* + F$$
$$H_2 + F \longrightarrow HF^* + H$$
$$H + F_2 \longrightarrow HF^* + F$$

ad infinitum

Each step produces a vibrationally excited HF molecule (HF*), which emits an infrared photon. The chain reaction can continue as long as hydrogen and fluorine atoms remain to keep it going. The chemical laser can operate continuously as long as the gases flow rapidly through the laser and the laser cavity, as shown in Fig. 6-11. Although the population inversion doesn't last very long in any particular group of gas molecules, the gas flows through the laser so fast that it doesn't matter—fresh gas moves into the laser cavity between the resonator optics, maintaining a steady population inversion that can produce a continuous beam.

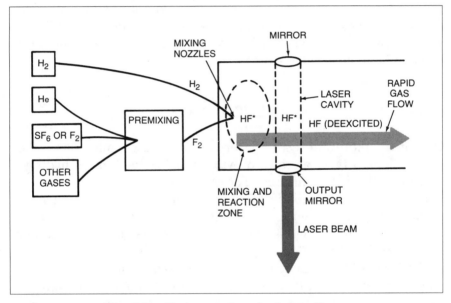

Fig. 6-11   Basic operation of a chemical laser.

The overall effect of a chemical laser is something like a rocket engine. Combustion of fuels containing hydrogen and fluorine generates energy that is released as a laser beam. The reaction consumes the fuels, which must be pumped out of the laser and collected. (Hydrogen fluoride is nasty stuff, too.) The engineering is tricky, but it draws upon earlier work with rocket engines.

Some laboratory-scale hydrogen- and deuterium-fluoride lasers are made that can emit up to tens of watts. However, most interest in chemical lasers has come from military research on high-energy laser weapons. One developmental deuterium-fluoride laser the size of a building is reported to have produced continuous powers of 2,000,000 W at the Army's White Sands Missile Range in New Mexico. However, operation of the laser—called MIRACL, for Mid-InfraRed Advanced Chemical Laser—is possible for only brief intervals and requires many laser specialists. Military researchers are working on another megawatt chemical laser called Alpha, for possible use against missiles in space. However, few people now consider chemical lasers promising weapons, and that budget has been cut.

FAR-INFRARED LASERS

Earlier, we mentioned lasers that can emit infrared wavelengths much longer than the 10-μm output of the carbon-dioxide laser. These are the so-called far-infrared lasers, in which molecular gases emit at wavelengths between 30 and about 1000 μm.

As shown in Fig. 6-12, the energy that powers a far-infrared laser comes from a shorter-wavelength infrared laser beam, typically one of the 10-μm lines of CO_2. A specific narrow wavelength range is needed to excite the molecular gas to an excited vibrational state. The laser transition is between two rotational levels in the excited vibrational state. The CO_2 laser beam enters the far-infrared laser cavity at one end, and excites molecules as it travels along the tube. The far-infrared beam exits at the other end. The end mirrors contain holes, in the back mirror to let the CO_2 beam in, and in the output mirror to let the far-infrared beam out.

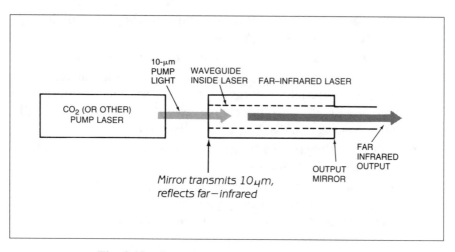

Fig. 6-12   Optical pumping of a far-infrared laser.

Far-infrared lasers are unusual by laser standards because of the strange nature of far-infrared optics. They have few applications, but are an important variant on the gas laser.

WHAT HAVE WE LEARNED

- Gas lasers range from weak to powerful. Their wide variety reflects the ease of developing gas lasers by testing gases in laser tubes.
- Most lasers contain mixtures of gases. For each gas laser, there is an optimum gas mixture and pressure.
- Gas lasers can operate with sealed tubes, but some require flowing gas.
- An electric discharge passing the length of the laser tube excites most low-power lasers. Higher-power lasers are excited with a discharge transverse to the tube. Pulsed lasers can be powered by charging a capacitor and discharging it through the gas.
- In many gas lasers, windows are mounted at Brewster's angle to minimize losses, and mirrors are separate from the tube. Brewster windows polarize the beam.
- Stable resonators are used for low-gain continuous-wave gas lasers. Other lasers with higher gain use different cavity designs.
- Thermal motion of gas atoms causes Doppler broadening, which can dominate a gas laser's bandwidth.
- The red helium-neon laser is the most common gas laser.
- An electric discharge excites helium atoms in a helium-neon laser; they transfer energy to neon to produce a population inversion. Output from a He-Ne laser depends on tube length, gas pressure, and discharge bore diameter.
- Coherence length of a typical helium-neon laser is 20 to 30 cm, adequate for holography.
- Argon- and krypton-ion lasers can emit more visible power and operate at shorter wavelengths than the helium-neon laser. They emit several near-ultraviolet and visible wavelengths.

- Argon and krypton lasers have very similar structures, and somewhat resemble the helium-neon laser.

- Helium-cadmium lasers produce milliwatts continuous-wave at 441.6 nm, and have weaker lines at 325 and 353.6 nm in the ultraviolet. Cadmium is a solid at room temperature, so the tube of a helium-cadmium laser must be heated to produce cadmium vapor.

- Neutral metal atoms emit fast, repetitive pulses in copper- and gold-vapor lasers.

- Copper-vapor lasers have much higher gain than helium-neon, He-Cd, or argon lasers. They are among the highest-powered visible lasers.

- The carbon-dioxide laser is the most versatile and highest-powered gas laser. It operates on vibrational transitions at 9 to 11 μm. CO_2 lasers make transitions among three vibrational modes of the CO_2 molecule. Rotational sublevels create many closely spaced lines.

- Carbon-dioxide lasers in sealed tubes can emit up to 100 W. CO_2 can oscillate in a waveguide structure a millimeter or two across to generate a laser up to about 50 W.

- Gas flows along the length of the tube in longitudinal-flow CO_2 lasers, allowing higher power than sealed tube lasers. Gas flow perpendicular to the laser axis can raise CO_2 output to about 10 kW/m of tube length.

- Rapid expansion of hot, high-pressure CO_2 produces a population inversion in a gas-dynamic laser.

- TEA CO_2 lasers produce high-power pulses when a transverse electric discharge is passed through laser gas at pressure near one atmosphere.

- CO_2 lasers can operate on one line or many lines, depending on the choice of optics.

- Carbon-monoxide lasers emit on vibrational-rotational transitions at 5 to 6 μm.

- Short-lived molecules containing one rare gas atom and one halogen emit intense ultraviolet pulses in excimer lasers. The molecules break up in the ground state, so the lower laser level is always empty.

- Excimer kinetics limit pulse lengths to tens or hundreds of nanoseconds; repetition rate depends on the electronics. Excimer lasers have very high gain, so they need little feedback from the output mirror. One gas fill can generate millions of shots.

- Nitrogen lasers have high gain and emit 337-nm pulses with high peak power, but have low average power and efficiency. They can be made compact and inexpensive.

- Chemical lasers get their energy from the reaction of hydrogen and fluorine to produce vibrationally excited hydrogen fluoride.

- Far-infrared lasers operate on vibrational and rotational transitions at 30 to 1000 μm.

WHAT'S NEXT

In Chapter 7, we will learn about solid-state crystalline and glass lasers, which have different properties than gas lasers. We will cover semiconductor lasers in Chapter 8, and dye and other lasers in Chapter 9.

Quiz for Chapter 6

1. Which of the following general statements is true about gas lasers?
   a. Most gas lasers are excited electrically.
   b. Only materials that are gaseous at room temperature can be used in gas lasers.
   c. The laser tube cannot contain atoms or molecules other than the one species that emits light.
   d. Gas lasers cannot emit continuous beams.
   e. Gas must flow continually through all gas lasers.

2. What type of laser cavity should be used with a low-gain continuous-wave gas laser?
   a. Stable resonator, confocal
   b. Plane-parallel resonator
   c. Unstable resonator
   d. Any of the above
   e. None of the above

3. Which of the following lasers can emit the shortest wavelength in a continuous-wave beam?
   a. Helium-neon
   b. Helium-cadmium
   c. Nitrogen
   d. Argon-fluoride
   e. Carbon-dioxide

4. Which of the following lasers emits on a vibrational transition?
   a. Krypton-fluoride excimer
   b. Helium-neon
   c. Nitrogen
   d. Carbon-dioxide
   e. Krypton-ion

5. What excites the helium-neon laser?
   a. An electric discharge passing through the gas
   b. Transfer of energy from helium to neon
   c. Energy retained by the laser mirrors
   d. A and B
   e. None of the above

6. What kind of helium-neon laser can't you find commercially?
   a. A 2-W continuous-wave laser emitting at 3.39 μm
   b. A 0.5-mW CW laser emitting at 543 nm
   c. A 20-mW CW laser at 632.8 nm

   d. A 0.5-mW CW laser emitting 632.8 nm
   e. You can buy anything if you have enough money.

7. What gas laser has strong lines at 488 and 514.5 nm?
   a. Helium-neon
   b. Copper-vapor
   c. Helium-cadmium
   d. Carbon-dioxide
   e. Argon-ion

8. Which gas laser has the highest overall efficiency?
   a. Nitrogen
   b. Excimer
   c. Carbon-dioxide
   d. Helium-neon
   e. Argon-ion

9. Which of the following lasers cannot emit continuous-wave because its internal kinetics automatically stop pulses?
   a. Excimer
   b. Nitrogen
   c. Copper-vapor
   d. Gold-vapor
   e. All of the above

10. Which of the following is not a type of carbon-dioxide laser?
   a. Gas-dynamic
   b. Sealed-tube
   c. Waveguide
   d. Optically pumped
   e. TEA

CHAPTER 7

Solid-State Lasers

ABOUT THIS CHAPTER

In this chapter, you will learn about solid-state lasers, in which light is emitted by atoms in a crystal or glassy material. After first explaining the basic concepts of solid-state lasers, we will describe the most important types: ruby, neodymium, and vibronic lasers. Then we will mention a few other types that have found some uses.

WHAT IS A SOLID-STATE LASER?

The laser world has come to use the term "solid-state" in a special sense, which we should clarify right away. A solid-state laser is one in which the atoms that emit light are fixed in a crystal or glassy material. It is not the same as a semiconductor diode laser, even though semiconductors are crystalline materials. In the laser world, semiconductor diode lasers belong in a separate category, described in the next chapter.

This may seem confusing if you're familiar with electronics, where "solid-state" is synonymous with semiconductor. That usage dates from the days when transistors started replacing vacuum tubes, and developers wanted to show that the new transistor electronics relied on different principles than the old vacuum tubes. Since semiconductor devices were an outgrowth of solid-state physics, they were called solid-state electronics.

The laser world makes a distinction between solid-state and semiconductor lasers for much the same reason the electronic world differentiates between vacuum-tube and semiconductor electronic devices—they rely on different principles and have different characteristics. Solid-state lasers are electrically nonconducting; they are excited by light from an external source that passes through the crystal. Semiconductor lasers are excited by an electric current passing through a block of material with specific electronic properties, as you will learn in Chapter 8. Semiconductor lasers are often called "diode lasers," because they are electronically a two-terminal device or "diode"—and because diode is a much shorter word than semiconductor. The rest of this chapter describes solid-state—glass or crystalline—lasers.

PRINCIPLES OF SOLID-STATE LASERS

Theodore Maiman's first laser was a solid-state laser, ruby, as were the second and third lasers discovered. The operation of solid-state lasers has been refined greatly since then, but—as with gas lasers—the same basic principles underlie the operation of the entire family of solid-state lasers. Figure 7-1 shows a generic type.

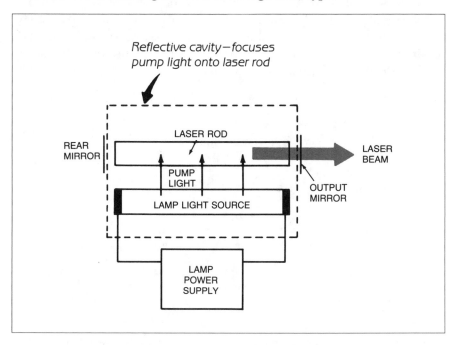

Fig. 7-1   A generic solid-state laser.

The atoms that emit light in solid-state lasers are dispersed in a crystal or glass that contains other elements. The crystal is usually shaped into a rod, with mirrors at each end. Light from an external source—a pulsed flashlamp, a bright continuous arc lamp, or another laser—enters the laser rod and excites the light-emitting atoms. The cavity mirrors form a resonant cavity around the inverted population in the laser rod, providing the feedback needed to generate a laser beam that emerges through the output mirror. If the laser is pumped by a lamp source, like our example in Fig. 7-1, the lamp and the laser rod are enclosed in a reflective cavity that focuses the pump light onto the rod.

Both the characteristics of the laser medium and of the pump light are important in determining how solid-state lasers operate. Let's first look at material characteristics and then at optical pumping techniques and sources.

Solid-State Laser Materials

We saw earlier that, in many gas lasers, only a small fraction of the atoms or molecules in the laser tube belong to the species that actually emit light. The same is true in all solid-state lasers. The atoms that emit light are embedded in a glass or crystalline matrix, as shown in Fig. 7-2. (Actual laser crystals have more complex structures.) From a chemical standpoint, the light-emitting element is

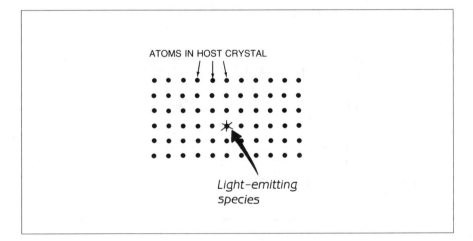

ATOMS IN HOST CRYSTAL

Light–emitting species

Fig. 7-2 Light-emitting atoms in a solid-state laser are embedded in a crystalline or glass host.

usually a dopant added to the compound that serves as the crystalline host. Typically, the light-emitting species accounts for around 1% of the material. (In some developmental laser materials, the light-emitting species is a minor part of the crystalline compound, but these are not in practical use.) Laser characteristics depend on both the light-emitting species and the host.

The light-emitting species is the key ingredient in a solid-state laser. It must have a set of energy levels that let it absorb pump light to populate a metastable upper laser level. (In some developmental solid-state lasers, one species absorbs the light energy and transfers it to a second species, which emits the light.) The best solid-state laser emitters are a handful of related metallic elements: chromium, neodymium, erbium, holmium, cerium, cobalt, and titanium. Chromium is the light-emitting species in ruby lasers and commercial alexandrite lasers. The most common types of solid-state lasers rely on emission from neodymium.

The light-emitting species are often called "ions" because of their nominal chemical valence in host crystals. The most common light emitters have nominal ionic states of $+3$—Cr^{+3}, Nd^{+3}, Er^{+3}, and Ho^{+3}. You should consider those ionization states more nominal than real. Chemically, these elements are not completely ionized in the crystal, and the bonds they form with atoms in the host material are at least somewhat covalent. The "ions" are fixed in the crystal, and the "missing" electrons are nearby, typically bonded with oxygen atoms.

Host Materials

The interaction between the light-emitting species and host material is vital to the operation of solid-state lasers. Candidate host media must meet several requirements. The host must be reasonably transparent to the pump light and absorb very little light at the laser wavelength. Too much absorption at either wavelength could greatly reduce efficiency or make laser action impossible. Excess absorption could also heat the host, which can impair laser action.

Thermal properties of the host also are important. As in other lasers, most pump energy winds up as heat rather than laser light. Because atoms in laser crystals are fixed, they cannot remove heat as efficiently as gas atoms. Other thermal problems can also arise, as we will learn later in this chapter.

The light-emitting species interacts with the host crystal in subtle ways that influence its energy-level structure. Crystalline bonds and effects of adjacent atoms slightly shift energy levels in the light-emitting species. This can change the laser wavelength, usually by a small amount. For example, neodymium emits at 1054 nm when it is doped into phosphate-based glass, and at 1064 nm when in a crystalline host known as "YAG" (for yttrium aluminum garnet—a garnet-like crystal that contains yttrium and aluminum).

The light-emitting atoms in solid-state lasers make electronic transitions as their electrons shift between energy levels. Interactions of the electronic energy levels of the light-emitting atoms with vibrational energy levels in the crystal can produce what are called "vibronic" transitions, in which both electronic and vibrational energy levels change. Because vibrational transitions involve much less energy than electronic transitions, the main factor determining laser wavelength is the electronic energy levels of the light-emitting species. However, if many vibrational transitions are possible, they can smear the electronic transition over a range of wavelengths. As we will see later in this chapter, this gives some solid-state lasers gain over a wide enough bandwidth that their output wavelength can be tuned.

Solid-State Laser Development

There are fewer types of solid-state lasers than there are of gas lasers. The major reason is not that solid-state media are less (or more) suitable for lasers, but the manner in which development must be conducted. To test a gas laser, you need only put the desired gas mixture into an evacuated tube. To test a solid-state laser, you have to grow a crystal containing the desired concentration of light-emitting dopant in a suitable host. Crystal growth is a complex art; it is no easy matter to grow an optically flawless sample suitable for manufacture into a laser rod. Nor is it easy to produce a rod from the crystal. This means that a given investment of time and money is likely to produce more results in developing gas lasers than in developing solid-state lasers.

The problems in making and characterizing solid-state laser materials have led developers to concentrate on a few types that are well-known and satisfy the needs of funding agencies. The neodymium laser has benefitted from years of optimization, but many other promising types have received much less attention. Interest in other solid-state lasers has grown in recent years, leading to some new types described later in this chapter.

Heat Conduction and Optical Distortion

Another vital role of the host material is to dissipate waste heat left over from laser action. Like other lasers, solid-state lasers are inherently inefficient. Only about 1% of the pump energy emerges in the beams of many lasers excited by lamps, although laser excitation can be much more efficient. Some of this energy is lost in the electrical power supply that drives the light source and in the light source itself, but much of it ends up in the solid-state laser material. This makes thermal conductivity of the laser host a very important concern.

Heat accumulation in solid-state laser materials can be bad news in three ways:

1. Excess heat can damage the laser material itself, causing it to warp, crack or soften.

2. Increasing temperature affects population distributions and gain characteristics of the light-emitting species, usually decreasing gain, laser efficiency and output power.

3. Thermal expansion of the laser material can create refractive-index differentials in the laser rod, which can bend light within the rod so that the light does not oscillate properly between the laser cavity mirrors. This increases losses and decreases output power.

The first of these problems is the most dramatic, but when it occurs it may be a consequence of the second and third, which can heat the laser material by decreasing laser efficiency.

In practice, these problems limit the number of materials that can serve as solid-state laser hosts. Certain crystals offer the best heat conductivity, while glasses are used because they are easiest to make into large pieces. Solid-state lasers that dissipate much power usually require active cooling with forced air or flowing water.

Laser Geometry

The archetypal solid-state laser is a rod about the shape of a round pencil, but usually a little shorter and often smaller. The small rod diameter helps ease heat dissipation and, as we will see later, can generate an impressive amount of light. Mirrors are typically mounted at either end of the laser rod, as shown in Fig. 7-1.

The power available from a single-rod laser oscillator is limited, so some high-power lasers use one or more external amplifier stages, as shown in Fig. 7-3. While the first stage of such an oscillator-amplifier has mirrors on both ends (one of which is the partly transparent output mirror), the second and any subsequent stages do not. The beam makes a single pass through their volume, extracting stored light energy accumulated in a population inversion. In some cases, such as in the slab amplifier, the beam may bounce around within the amplifier so that it passes through most of its volume but does not oscillate back and forth over the same path. The mirrors in Fig. 7-3 redirect the light so that the equipment can be made more compact and stay on the page in the diagram or on the optical table in the laboratory.

As you can see from Fig. 7-3, solid-state laser oscillators and amplifiers can take other shapes besides rods. Some amplifiers are flat disks, which are used to generate large-diameter beams. The disk geometry lets the laser material dissipate heat from its surfaces but still generate a large-diameter beam.

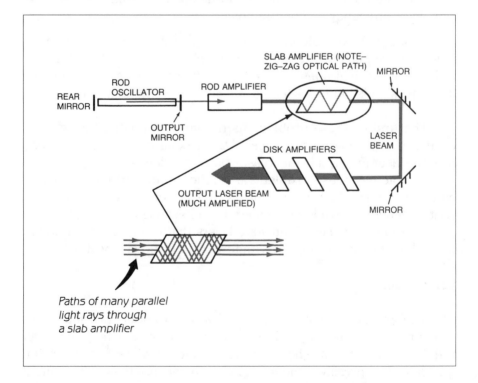

Fig. 7-3 Laser oscillator and amplifiers.

Slabs, which look like parallelograms in cross section, can be used as both oscillators and amplifiers. The upper right part of Fig. 7-3 shows a slab amplifier. The front surface is slanted to refract the incoming light, so it reflects from the parallel side surfaces by total internal reflection until it reaches the end of the slab. If we expanded the input beam to cover the entire input face of the slab laser, we would have light rays following many parallel paths through the slab and generating light energy from most of its volume, as shown at the bottom. Slabs with mirrors can be used as oscillators.

The amount of energy that a laser material can store increases with its volume. Thus, the larger the volume, the more energy can be extracted in a pulse. Waste heat produced also increases with volume, but heat dissipation increases with surface area. Disk and slab geometries are attractive for solid-state lasers because they have more surface areas than large-diameter rods, so they can better dissipate waste heat and can thus operate at higher overall power levels.

Reduce the diameter of a solid-state laser rod enough, and you wind up with a fiber. In fact, optical-fiber lasers and amplifiers are used in optical communications. The light-emitting atoms are doped into the core of the optical fiber, which confines the light they emit so it passes along the length of the fiber. Optical fibers with mirrors on each end can serve as oscillators, but more typically they serve as amplifiers to increase the strength of signals in a communication system.

Material Quality and Size

A final factor in considering solid-state laser materials is the ease of making rods of the required size and optical quality. The material must have a uniform refractive index, so light can oscillate smoothly back and forth in the laser cavity. The rod must be free of flaws that scatter or absorb light. (Light absorption at such flaws would lead to localized heating that could crack or shatter the laser rod.) Some crystals with otherwise attractive characteristics are too difficult to grow or too prone to flaws to be of any practical use.

The growth of laser crystals is at best a slow and agonizing process, which requires carefully controlled conditions. The crystals are grown in blocks called "boules," from which rods must be drilled. The cutting of the rod itself is a difficult process and may reveal flaws that were not detected in the boules. These problems must be traded off against the good thermal characteristics of crystalline materials.

Glass is much easier than crystals to produce in large slabs of uniform optical quality. Offsetting that attraction is its poor heat dissipation, which limits the repetition rate and prevents continuous-wave operation. Those limitations are acceptable tradeoffs in some cases, where the advantages of the higher pulse powers available from larger-size glass amplifiers offset the much lower repetition rates of glass lasers.

OPTICAL PUMPING AND SOURCES

Because solid-state laser materials are nonconducting, the only practical way to produce a population inversion is to illuminate the laser material with a bright light. Photons raise the light-emitting species to an excited state that leads to the laser transition. The choice of the source depends on the nature of the laser material.

Pumping Bands and Absorption

We mentioned earlier that all materials have characteristic energy levels and transitions. Each species emits light on certain transitions when they drop from excited states. Likewise, they absorb light at characteristic wavelengths when they are in the ground state or other low levels. This absorption is an essential part of optical pumping.

Absorption can be at a narrow or broad range of wavelengths, depending upon the transitions involved. Conceptually, a narrow range of wavelengths may seem simpler, because they involve a transition between a simple pair of energy levels. However, that is often a poor arrangement for optical pumping, because lasers are the only light sources that emit light only at a narrow band of wavelengths. The problem is that most lasers, except for semiconductor types, only convert a small fraction of the input energy into light. Flashlamps and arc lamps convert more of the input energy into light than most lasers, but semiconductor lasers can be even more efficient.

Efficient pumping requires finding a light source that matches the absorption bands of the laser material. Figure 7-4 shows the absorption spectrum of neodymium-YAG (Nd-YAG), the most common solid-state laser medium. This version has been smoothed out; actual measurements show many sharp spikes. Nd-YAG has long been

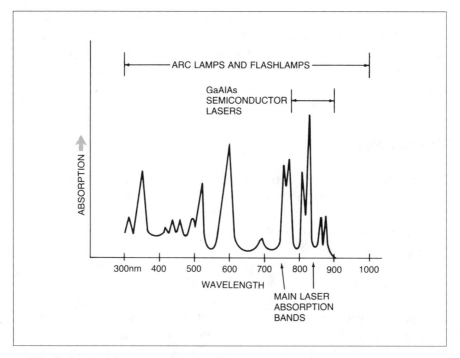

Fig. 7-4 Broadband absorption of neodymium, compared with output of typical pump sources.

pumped with flashlamps and arc lamps, which emit visible, near-infrared, and near-ultraviolet light that overlaps the Nd-YAG absorption spectrum. These lamps are widely used because of their high intensity, reasonable light-producing efficiency, and ready availability. Alternatively, Nd-YAG can be pumped with semiconductor lasers made of gallium aluminum arsenide (GaAlAs), which emits near-infrared light at wavelengths that include some of the strongest neodymium absorption lines. Each approach has its advantages, as we will learn later.

Absorption spikes, including some which are too narrow to show in this version of the neodymium spectrum, come from transitions between well-defined energy levels. The broadband absorption comes from interaction of vibrational energy levels in the crystal with electronic energy levels of individual atoms. The vibrational transitions involve small enough energy changes that they can combine with electronic transitions to absorb light over a continuous range of wavelengths, such as near 800 nm in the near-infrared. This broadband absorption makes it practical to pump solid-state lasers with conventional light sources such as flashlamps or arc lamps, which emit a wide range of wavelengths, and makes it easier to match the wavelengths of the few lasers that can make efficient pump sources. The absorption bands are normally not as wide as the range of wavelengths available from pump lamps, but they do allow absorption of much of the pump light.

Flashlamp Pumping

The first laser was pumped with the bright flash of light from a xenon flashlamp, and flashlamp pumping remains common today. The extremely intense pulse of light from a flashlamp can excite most available atoms, and thus produce a population inversion even in a three-level laser material like ruby.

Flashlamps come in various shapes. Maiman used a helical lamp, shaped like a spring, but the most common type used to pump lasers today is a long linear tube. The lamp is filled with a gas such as xenon, and contains electrodes at each end. Applying a brief high-voltage electrical pulse between the ends of the tube makes the gas break down electrically and conduct current, emitting a bright flash of light. The whole process requires about a thousandth of a second. Typically, the pulse from a solid-state laser will be slightly shorter than the flashlamp pulse, because the flashlamp does not instantaneously generate enough power to raise the laser above threshold.

Flashlamp pumping is common for pulsed solid-state lasers. It is limited in repetition rate by the switching electronics and the lamp itself. Q-switching and other techniques can produce shorter pulses, as described in Chapter 5.

Arc Lamp Pumping

Continuous-wave solid-state lasers can be pumped with electric arc lamps, in which a steady electric current flows through a gas-filled tube, producing intense light. This light is bright enough to sustain a continuous population inversion in some solid-state laser materials that are capable of continuous laser operation. (Not all solid-state laser materials can operate continuous-wave.) In practice, lasers pumped with arc lamps can be pulsed with Q switches, mode-lockers, or cavity dumpers. This allows pulsing at higher repetition rates than are possible with flashlamps.

Laser Pumping

Pumping a solid-state laser with another laser can be an attractive alternative to lamps, if the pump laser line matches an absorption peak of the solid-state laser material. The high absorption at the peak means that pump photons are converted into laser output more efficiently than when pumping with the broad spectrum of light from a lamp. However, the overall efficiency is high only if the pump laser is itself efficient. In practice, this makes semiconductor lasers the most attractive pump lasers, because high-power commercial types can convert well over 10% of the input energy into laser output. (Over 50% efficiency has been demonstrated in the laboratory.)

The highest-power semiconductor lasers are GaAlAs, which emit at 750 to 900 nm in the near-infrared. As shown in Fig. 7-4, that band includes a strong absorption peak near 800 nm in neodymium, so arrays of GaAlAs lasers are widely used to pump neodymium ions in YAG and other hosts. Semiconductor lasers emitting other wavelengths can pump other solid-state lasers.

Diode laser pumping offers important practical advantages. The diodes themselves are compact, and their high efficiency reduces the need for bulky equipment to dissipate waste heat. While the power from individual semiconductor lasers is limited, as we will see in Chapter 8, they can be fabricated in arrays that together produce impressive output power. However, commercial diode-pumped solid-state lasers still can't match the powers of lamp-pumped lasers. In addition, high-power semiconductor lasers remain considerably more expensive than pump lamps, limiting their practical applications.

As we will see later in this chapter, other types of lasers also can pump some solid-state lasers. For example, argon lasers often pump titanium-sapphire lasers, which cannot be pumped efficiently by lamps.

Pumping Geometries

Laser designers no longer use flashlamps of the helical or coiled-spring design like those used in early solid-state lasers. The most common types are linear lamps. Figure 7-5 shows three common ways to transfer light from a linear flashlamp (or arc lamp) to a laser rod, along with a laser rod slipped inside a helical lamp.

Figure 7-5B shows how linear lamps are placed close to each other in a reflective cylinder in a side view; Fig. 7-5C gives an end view. An alternative, shown in Fig. 7-5D, is to put the lamp and rod at

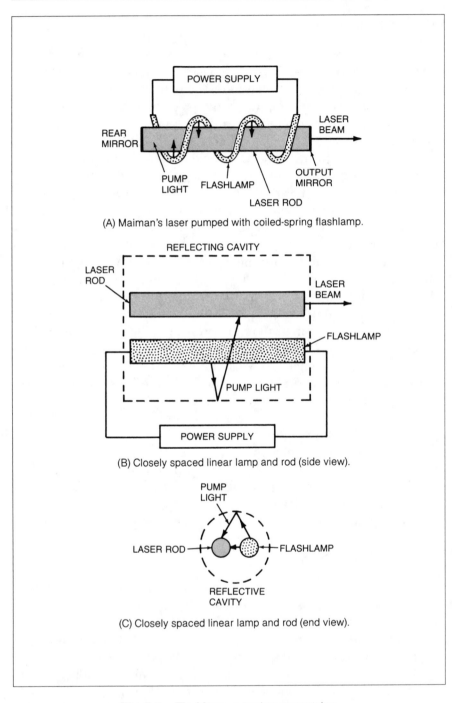

(A) Maiman's laser pumped with coiled-spring flashlamp.

(B) Closely spaced linear lamp and rod (side view).

(C) Closely spaced linear lamp and rod (end view).

Fig. 7-5 Flashlamp pumping geometries.

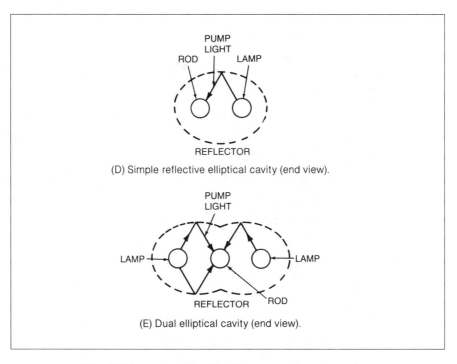

(D) Simple reflective elliptical cavity (end view).

(E) Dual elliptical cavity (end view).

Fig. 7-5 (continued) Flashlamp pumping geometries.

the two foci of a reflective elliptical cavity. Light radiating from one focus of the cavity is reflected to the other focus from any point on the ellipse. If the pump lamp is at one focus, the reflective elliptical cavity focuses all its light onto the laser rod at the other focus.

The same principle can be used in a dual-elliptical cavity, shown in Fig. 7-5E. The laser rod is put at the common focus of two overlapping ellipses. A lamp is at the other focus of each ellipse, so the cavity focuses light from both lamps onto the laser rod. (Although the figure implies that the cavity is filled with air, some pump cavities are filled with flowing water that removes waste heat from the laser.)

Different geometries are used for laser pumping. Semiconductor lasers may be arranged along the side of a laser rod or slab, so they illuminate its entire length, or they may be focused along the length of a rod or fiber laser. Figure 7-6 shows some geometries used in diode-pumped solid-state lasers. Titanium-sapphire lasers are pumped with external laser beams (usually argon-ion), which typically strike the laser crystal at an angle from the side.

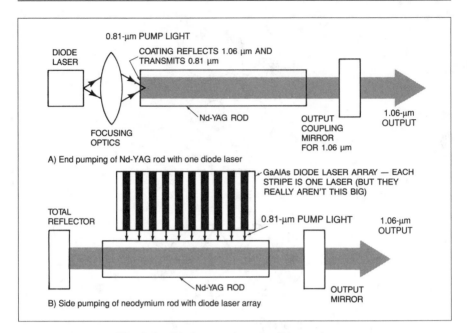

Fig. 7-6   Diode-pumping of solid-state lasers.

RUBY LASERS

Ruby was the first laser material, but it is far from ideal. The three-level ruby laser system is inherently less efficient than the four-level neodymium laser, and is limited to pulsed operation at low repetition rates. However, ruby remains important for a few applications that require its high-power red pulses.

Laser Medium and Physics

Natural ruby is a gemstone, but ruby lasers use a synthetic ruby made by doping aluminum oxide (which crystallizes to form sapphire) with 0.01 to 0.5% chromium. Sapphire is naturally clear, but the chromium atoms give it a pink or reddish color.

Ruby is a three-level laser system, with its energy-level structure shown in Fig. 7-7. Ground-state chromium atoms absorb light in two pump bands, one centered near 550 nm and the other at about 400 nm. These are well-matched to the output of xenon flash-lamps. After about 100 ns, excited chromium atoms release energy as

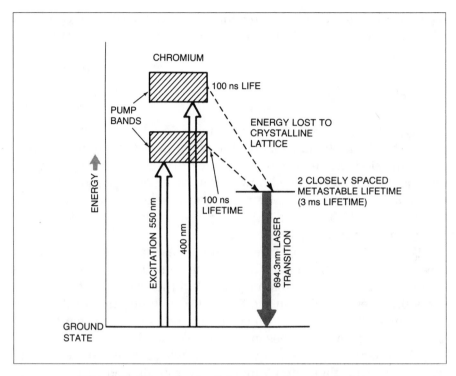

Fig. 7-7   Energy levels of chromium atoms in ruby laser.

vibrations of the crystalline lattice and drop to a pair of closely spaced metastable levels. (They are too closely spaced to separate on the illustration.) These metastable states have a 3-ms lifetime at room temperature, long enough to serve as the upper level of a 694.3-nm laser transition to the ground state.

The three-level energy structure makes ruby relatively inefficient, with typical wall-plug efficiency of 0.1 to 1%. Because the ground state is the lower laser level, chromium atoms that have not been excited can absorb the 694.3-nm laser line. This makes it essential to pump all of a ruby rod to avoid excess absorption. However, a ruby rod can store more energy than neodymium, and can readily be Q-switched to produce short pulses with energies of a few joules. It also can be used in an oscillator-amplifier configuration to raise total laser energy.

Ruby conducts heat well and resists damage from excess optical energy as long as its surface is kept clean of dirt that could absorb excess light and heat the material. However, because it is a three-level laser, its laser properties degrade rapidly with increasing temperature, so it must be operated at low repetition rates (typically no more than a few pulses per second except for very small rods) to prevent heat build-up.

Practical Ruby Lasers

Typically, ruby rods are 3 to 25 mm (1/8 to 1 inch) in diameter, and up to about 20 cm (8 inches) long. Without Q-switching, energy in a multiple-transverse mode pulse can reach 100 joules in 1 ms, although it is typically much lower. Pulse energies are much lower if oscillation is limited to TEM_{00} mode. Q-switching can compress pulse duration to 10 to 35 ns, which limits pulse energy to a few joules but gives peak power in the 100-MW range.

Ruby lasers often operate in a dual-pulse mode for holographic measurements. In this mode, a Q switch generates two short pulses during a single long flashlamp pulse. In other words, the Q switch turns the laser on twice during the millisecond duration of the flash-lamp pulse. The two pulses record holograms on the same film, so they can show small changes in the shape of an object.

Ruby lasers require active cooling with forced air or flowing water, but even then their repetition rate is limited to low levels, except for the smallest rods.

NEODYMIUM LASERS

The most common solid-state lasers now are types in which light-emitting neodymium atoms are embedded in a glass or crystalline matrix. Neodymium lasers share a common energy-level structure, but they differ in some ways because of the properties of the different hosts. The most common host is yttrium-aluminum garnet, a hard, brittle crystal known by the acronym "YAG." Silicate and phosphate glasses are also useful hosts for neodymium. Dozens of other neodymium hosts have been tested, but only a few have been developed extensively. Three of the most important alternatives are yttrium lithium fluoride (YLF), yttrium vanadate (YVO_4), and yttrium aluminate (YALO, from its chemical formula $YAlO_3$). The main wavelength of neodymium lasers is 1064 nm in Nd-YAG and slightly different wavelengths in other hosts. However, for most practical purposes there is no real difference among these wavelengths.

The main energy levels in the four-level neodymium laser system are shown in Fig. 7-8. In this general diagram, we round the wavelength to 1.06 µm (equivalent to 1060 nm) to indicate that we are talking about the entire family of neodymium lasers, not a specific type. Interactions between neodymium atoms and the host material can change wavelength by about 1%. The two primary pump bands shown are in the 700- to 850-nm range, which raise neodymium atoms to one of two broad high-energy bands. However, as indicated in Fig. 7-4, there also are other important absorption lines. Traditionally, neodymium lasers have been pumped by flashlamps or arc lamps. However, 808-nm GaAlAs semiconductor lasers have become important pump sources in recent years.

The optically excited neodymium atoms quickly decay to the metastable upper laser level, releasing their excess energy to the crystalline lattice. In this state, they are stimulated to emit on the main 1.06-µm laser transition, dropping to a lower laser level that they quickly leave, again by transferring energy to the crystal.

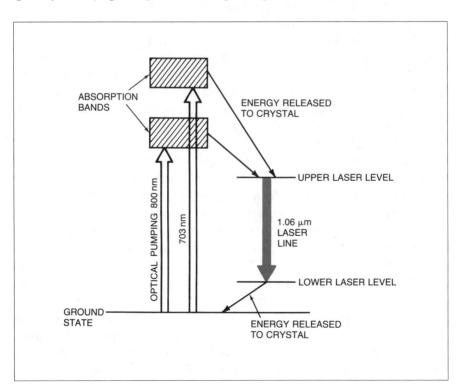

Fig. 7-8   Major energy levels in 1.06-µm neodymium laser.

Our energy-level diagram in Fig. 7-8 simplifies the situation by omitting the splitting of the upper and lower laser levels in neodymium. This splitting creates several other laser transitions in the near-infrared, but all of them are weaker than the main 1.06-μm line. The strongest, at 1318 nm in YAG, can produce about 20% as much power as the 1064-nm YAG line, and is important because that wavelength is useful in fiber-optic systems.

Neodymium-YAG

In neodymium-YAG, neodymium is an impurity that takes the place of some yttrium atoms in the YAG crystal. YAG's chemical formula is $Y_3Al_5O_{12}$; its crystalline structure is similar to that of garnet. The crystal has good thermal, optical, and mechanical properties, but it is hard to grow. The crystal is grown in blocks called "boules" from which rods are drilled. Typical YAG rods are 6 to 9 mm (0.24 to 0.35 inches) in diameter and up to 10 cm (4 inches) long. YAG can also be formed into slabs with flat surfaces.

The small size of Nd-YAG rods and the optical properties of neodymium atoms limit the amount of energy that can be stored in a typical rod to about half a joule, far less than in a ruby rod. However, most of that energy can be removed from the rod in a Q-switched pulse, and the energy can be replenished quickly—in well under the 1-ms duration of a flashlamp pulse—so a repetitively pulsed Nd-YAG rod can generate high average powers, as well as high peak power in Q-switched pulses.

The thermal and optical properties of Nd-YAG let it be pumped continuously with an arc lamp or by a series of flashlamp pulses. Maximum average power from an Nd-YAG laser can exceed 1000 W, although most operate at much lower powers. The peak power can reach tens or hundreds of kilowatts in a millisecond-long pulse generated by a flashlamp, or over 100 MW in a Q-switched pulse of 10 to 20 ns.

When doped in YAG, neodymium has a high gain and can be operated as an oscillator with a stable or unstable resonator. Although energy storage in an oscillator rod is limited, oscillator-amplifier configurations can reach higher powers. The oscillator and amplifier need not use the same host material—often, oscillators are Nd-YAG or another crystal, and the amplifiers are Nd-glass.

Neodymium-YLF Lasers

Yttrium lithium fluoride (YLiF$_4$) does not conduct heat as well as YAG and is softer, but its refractive index changes less with temperature than YAG, so it suffers fewer heat-related problems. As in YAG, the neodymium atoms are impurities that replace some yttrium atoms in the crystal. Nd-YLF can also store more energy than Nd-YAG, so it can generate higher-energy Q-switched pulses. The crystal is birefringent, so it generates two wavelengths, 1047 and 1053 nm, each with its own polarization orientation. Commercial versions operate pulsed or continuously.

Neodymium-Glass Lasers

The principal attraction of glass as a solid-state laser host is the well-developed technology for making large pieces of laser glass with good optical quality. Also, because neodymium-glass has lower gain than Nd-YAG, a glass laser rod can store more energy than an equal-sized YAG rod. Thus, Nd-glass lasers can generate higher-energy pulses than YAG lasers. The principal tradeoff is that glass has poorer thermal characteristics, so it needs more time to cool between pulses. Thus, glass laser oscillators cannot be pumped continuously and normally operate at much lower repetition rates than Nd-YAG lasers.

Laser glass can be made in a much wider variety of shapes than Nd-YAG. For example, the Nova laser used in fusion research at the Lawrence Livermore National Laboratory includes both glass rod amplifiers up to 5 cm in diameter and disk amplifiers up to 46 by 85 cm (18 by 33.5 inches) across. As shown in Fig. 7-3, the disk amplifiers are thin to assist in heat removal, and tilted at an angle to avoid potentially harmful reflections. Laser glass can also be cast in slabs, which can be used as amplifiers, as shown in Fig. 7-3, or can themselves serve as oscillators.

Nd-glass lasers emit broader bandwidth light than Nd-YAG. This is significant in modelocking, because it lets you generate shorter pulses with Nd-glass than with Nd-YAG. The wavelength emitted depends on the glass composition; it is 1062 nm for silicate glass, 1054 nm for phosphate glass, and 1080 nm for fused silica.

Neodymium Laser Configurations

Neodymium lasers are pumped in the same way as other solid-state lasers, using flashlamps, arc lamps, or semiconductor lasers. The pumping arrangements used with flashlamps and arc lamps were shown in Fig. 7-5; diode pumping was shown in Fig. 7-6.

The gain in neodymium lasers is high enough that they can use either stable or unstable resonators. An unstable resonator has the advantage of extracting laser energy from more of the laser medium, which is usually a rod but can also be a slab. Close to the laser, some unstable resonators produce beams with a bright ring around a central point of minimum intensity, which looks like a ring or doughnut in cross section. However, far from the laser the hole vanishes to produce a bright central spot. A stable resonator can produce the standard Gaussian TEM_{00} beam with a bright central spot, but it does not extract laser energy from as much of the laser volume. Unstable resonators have been growing in popularity because, for most applications, output power and energy are more important than near-field beam quality.

As we mentioned earlier, an external amplifier can boost the output power and energy from a neodymium oscillator. Oscillator and amplifier stages need not be made of materials based on the same host, but the host wavelengths must be close enough that the oscillator wavelength falls within the gain bandwidth of the laser amplifier.

Practical Neodymium Lasers

The neodymium laser is extremely versatile and can take many forms. Some neodymium lasers are made for general laboratory use, while others are made for specific jobs, including measuring the ranges to military targets, treating eye disease, or drilling holes. This wide range of uses means that neodymium lasers can look quite different from each other. A compact diode-pumped laboratory laser is very different from a massive water-cooled high-power drilling laser built to stand on a factory floor. Yet Nd-YAG rods lie at the core of both.

The wide range of applications leads to a range of design choices. General-purpose laboratory lasers are built to give the user as many options as possible. Typically, room is left in the laser cavity or on the outside of the laser to add accessories to change the wavelength and/or pulse length. On the other hand, lasers designed for specific applications, such as drilling holes, may allow few modifications not essential to performing their major job.

The overall efficiency of lamp-pumped neodymium lasers is in the same 0.1 to 1% range as ruby lasers. The smallest models can operate without active cooling, but larger types require either forced-air cooling (i.e., a fan) or flowing-water cooling. Remember that a 100-W laser that is 1% efficient generates 10 kW of waste heat!

Overall efficiency can exceed 10% when pumping with a semiconductor laser, generating much less waste heat. Combined with the inherent small size of semiconductor lasers, this means that many diode-pumped lasers are small enough to hold in your hand. In practice, most diode-pumped neodymium lasers have limited power levels, largely because diode power levels remain limited, and the pump diodes themselves are expensive. However, average powers to 1 kW have been demonstrated in the laboratory from a diode-pumped laser head about the size of a grapefruit that was actively cooled by flowing liquid to remove the excess heat generated in the semiconductor lasers. (Diode lasers are sensitive to excess heat, so designers must pay more attention to removing waste heat from the semiconductor lasers than from the Nd-YAG rod.) That technology is being investigated for industrial applications.

Changing Wavelength and Pulse Length

We saw in Chapter 5 that a variety of accessories can change a laser's wavelength and pulse duration. These accessories are often used with neodymium lasers.

Harmonic Generation

The near-infrared wavelength of neodymium lasers is fine for some purposes, but visible or ultraviolet light is better for many others. Fortunately, neodymium lasers generate high enough powers that nonlinear harmonic generation can readily produce shorter wavelengths. Pulses produce higher harmonic powers, but harmonics can also be produced from continuous-wave beams.

In Chapter 5, we showed how nonlinear interactions between light waves and certain materials could generate light at different wavelengths. The simplest such interaction is frequency doubling. Twice the frequency corresponds to half the wavelength, so frequency-doubling the 1064-nm output of an Nd-YAG laser produces green light at a wavelength of 532 nm. A slightly more complex optical arrangement, such as shown in Fig. 7-9, can generate the third harmonic at 355 nm and the fourth harmonic at 266 nm, both in the ultraviolet. Those wavelengths also have many applications. Note that some light always remains at the input wavelengths and must be blocked by a filter or removed by a wavelength-selective beamsplitter. Nd-YLF and Nd-YVO$_4$ lasers are also used for harmonic generation.

Harmonic generation from neodymium lasers has become so common that some lasers are packaged with a harmonic generator inside. This is why you may hear people talking about "green" neodymium lasers, even though the neodymium atom does not emit green light. You should always remember that the neodymium laser's primary wavelength is the infrared.

Changing Pulse Length

Without external pulse control, a flashlamp-pumped Nd-YAG or Nd-glass laser will produce pulses about a millisecond long. Those are usable for some purposes, including some welding and drilling. However, many other applications require pulses that are much shorter or have much higher peak power. Those can be generated by Q-switching, cavity-dumping, or modelocking, as described in Chapter 5.

With a flashlamp-pumped Nd-YAG laser, a Q switch can generate pulses lasting 3 to 30 ns, with the repetition rate depending on that of the flashlamp. A Q switch can also generate short pulses from a continuously pumped Nd-YAG laser, but those pulses will be longer, often hundreds of nanoseconds, because of differences in how fast energy accumulates within the laser. Cavity-dumped pulses are also shorter with flashlamp-pumped lasers than with continuous-wave models. However, modelocking of either type produces pulses of the same duration, 30 to 200 ps, because it depends on the range of wavelengths emitted by the neodymium atoms.

Q switches, in particular, are so often built into neodymium lasers that they are sold as an Nd-YAG laser emitting short pulses with high peak power. Again, it pays to remember that the generation of these short pulses depends on the internal Q switch.

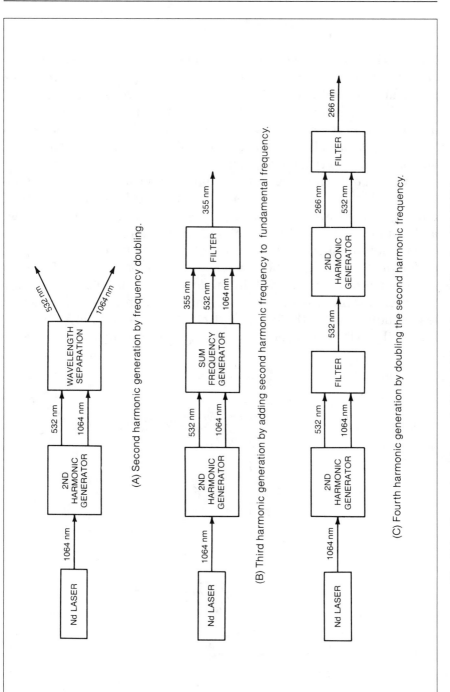

Fig. 7-9 Generation of second, third and fourth harmonics of an Nd-YAG laser.

VIBRONIC SOLID-STATE LASERS

Vibronic lasers are closely related to other solid-state lasers but have subtly different energy-level structures. The upper and lower laser levels in ruby and neodymium lasers are single states, as shown in Figs. 7-7 and 7-8. However, the lower level in vibronic lasers is actually a band of energy levels, as shown in Fig. 7-10. This band represents vibrational sublevels of a single electronic energy level, arising from vibrations of the crystalline lattice.

When an atom drops from the upper laser level (which may be a single electronic state or—as shown in Fig. 7-10—the bottom of a band) to the lower band of laser levels, it changes both electronic and vibrational energy. This compound transition is called *vibronic*, a contraction of vibrational-electronic. A vibronic transition can occur over a range of energies, because the excited atom can drop from the upper level to anywhere within the lower vibronic band. Normally, the laser transition is to the upper part of the lower band, which is less populated than the lower part closer to the ground state. This means that vibronic lasers can emit light at a comparatively broad

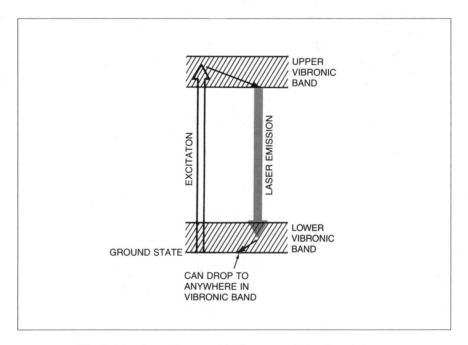

Fig. 7-10   General energy level structure of a vibronic laser.

range of wavelengths, which can vary up to ± 20% from a central wavelength. Table 7-1 lists emission ranges and compositions of some vibronic lasers.

The ranges in Table 7-1 are the spectral ranges over which vibronic lasers can show laser gain—the gain bandwidths we discussed earlier. Without special optics in the laser cavity, vibronic lasers (like other types) emit light at the wavelength where gain is highest, typically toward the middle of the range. However, if the cavity optics are adjusted so only certain wavelengths can oscillate in the laser cavity, as described in Chapter 4, a vibronic laser can oscillate at essentially any wavelength within its gain bandwidth. The combination of tunable output and the ability to generate wavelengths not available from other solid-state lasers has pushed development of vibronic lasers for military, research, and civilian applications.

Tunability is what sets vibronic lasers apart from other solid-state types. Otherwise, they work in much the same way. Vibronic lasers are pumped optically by a flashlamp, an arc lamp, or another laser (including diode lasers). Alexandrite rods can even lase when inserted into laser cavities designed for neodymium rods, although they work better in cavities designed especially for alexandrite. This similarity should not be surprising, especially for alexandrite and the other chromium-doped materials, because they use energy levels closely related to those in the ruby laser. (Ruby rods can also work in cavities designed for neodymium rods.)

TABLE 7-1 Some vibronic lasers

Name	Composition	Output Range (nm)*
Alexandrite	Chromium-doped $BeAl_2O_4$	701–858**
Co-MgF_2	Cobalt-doped MgF_2	1750–2500***
Cr-GSGG	Chromium-doped $Gd_3Sc_2Ga_3O_{12}$	740–850
Cr-Emerald	Chromium-doped $Be_3Al_2(SiO_3)_6$	729–842
Cr-Forsterite	Chromium-doped Mg_2SiO_4	1167–1345
LiCAF	Cr-doped $LiCaAlF_6$	720–840
LiSAF	Cr-doped $LiSrAlF_6$	780–920
Thulium-YAG	Thulium-doped YAG	1870–2160
Ti-sapphire	Titanium-doped Al_2O_3	660–1180

*Approximate ranges; the exact tuning range depends on operating conditions and laser design

**To 858 nm at elevated temperatures; only to 826 nm at room temperature

***At room temperature; as short as 1500 nm at cryogenic temperatures

Alexandrite Lasers

Alexandrite was the first vibronic laser developed commercially, and it remains an important type. The light-emitting species is chromium, added to $BeAl_2O_4$ (a mineral known as alexandrite) in concentrations of about 0.01 to 0.4%. Its energy levels are similar to those of the ruby laser, as shown in Fig. 7-11, except that the ground state has a vibronic band. Alexandrite can emit on a 680.4-nm fixed-

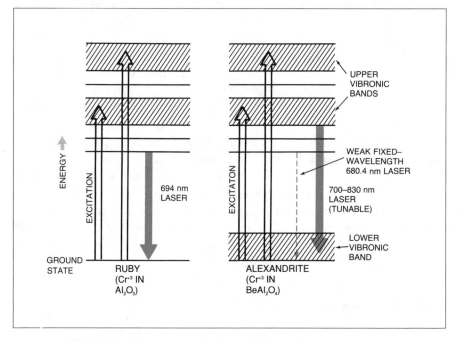

Fig. 7-11   Comparison of ruby and alexandrite energy levels.

wavelength transition to the ground state that is equivalent to the 694.3-nm ruby transition, but it is not as efficient as ruby. Like ruby, alexandrite has pump bands at 380 to 630 nm, which allow optical pumping with a flashlamp or arc lamp. It can also be pumped with red diode lasers, but not with near-infrared GaAlAs diode lasers. Its strongest room-temperature laser emission is at 700 to 830 nm.

Alexandrite has peculiar kinetics because two electronically excited states function together as the upper laser level. One is the bottom of a vibronic band of energy levels, and the other is a fixed state with a longer lifetime and only slightly less energy. This combination makes gain of alexandrite increase with temperature, unlike most other lasers in which gain drops as temperature rises. At high temperatures, alexandrite lasers can operate at wavelengths to 858 nm.

The lower laser level is a band of vibrationally excited states of the ground electronic level. When a chromium atom emits a laser photon, it drops to one of the vibrationally excited states and then releases the extra energy (remaining between the lower laser level and the ground state) as a vibration in the crystal lattice. Rising temperature increases the steady-state population of the vibrational sublevels slightly above the ground state. This makes it increasingly difficult to produce a population inversion with respect to these low-lying levels, which correspond to the higher-energy end of alexandrite's tuning range. Thus, raising the temperature of alexandrite and other vibronic lasers makes it impossible to reach the shortest wavelengths in their tuning ranges.

Alexandrite has lower gain than neodymium lasers, so more care must be taken in cavity design. On the other hand, this lets an alexandrite rod store more energy than an equal-sized Nd-YAG rod. (Alexandrite rods are similar in size to Nd-YAG rods.) Like Nd-YAG, alexandrite can operate pulsed or continuous-wave. In pulsed operation, average powers can reach 100 W, lower than the most powerful Nd-YAG lasers, but enough to make it among the more powerful lasers available.

Alexandrite has its advocates, but at this writing it remains primarily a research laser, with no widespread applications outside of the laboratory. However, it has potential uses ranging from target designation and range finding on the battlefield to medical treatment.

Titanium-Sapphire

Titanium-doped sapphire has two major attractions: wide tunability and good material characteristics. Laser emission is tunable between 660 nm in the red and 1180 nm in the near-infrared. Sapphire is easy to grow, with good thermal characteristics. While 30-cm Ti-sapphire crystals can be grown, the gain is high enough that typical continuous-wave lasers use crystals under 1 cm long. Ti-sapphire lasers can operate pulsed or continuous-wave. They have rapidly gained favor as laboratory sources of light with tunable wavelength.

The biggest practical problem is pumping the material. The 3.2-μs lifetime of the upper laser level is too short for flashlamp pumping, and the pump bands are near 500 nm, wavelengths too short for diode lasers. The most common pump sources are argon-ion lasers, frequency-doubled neodymium lasers, and copper-vapor lasers. Diode lasers can be used as pumps only indirectly, to pump neodymium lasers that are then doubled in frequency to excite the Ti-sapphire laser.

The laser crystal contains about 0.1% titanium, added to sapphire (Al_2O_3) to replace aluminum in the crystal lattice. In that sense it is similar to the ruby laser, in which chromium atoms replace aluminum in the sapphire lattice. (Note that chromium is used in other vibronic solid-state lasers.) The titanium atom interacts strongly with the host crystal, and this combines with the structure of the titanium energy levels to make the range of transition energies exceptionally broad. This gives the Ti-sapphire laser the broadest wavelength range of any solid-state laser. Power is highest at 700 to 900 nm, but the tuning range is much broader, from 660 to 1180 nm.

Their exceptionally broad bandwidth lets Ti-sapphire lasers directly generate extremely short pulses. At this writing, the shortest modelocked pulses generated directly from a Ti-sapphire laser are 11 femtoseconds (11×10^{-15} s), produced at Washington State University. Those are the shortest pulses ever produced directly from any laser, although compression of pulses from a tunable dye laser (see Chapter 9) has generated pulses lasting only 6 femtoseconds.

Harmonic generation is possible with pulsed Ti-sapphire lasers, generating the second harmonic at 350 to 470 nm, the third harmonic at 235 to 300 nm, and the fourth harmonic near 210 nm.

Other Vibronic Lasers

Table 7-1 lists several other types of vibronic lasers in advanced development or small-scale commercial use. While they are considered promising, they have yet to find wide applications.

Crystals in the garnet family (other than YAG) have good thermal and optical properties that make them attractive hosts for solid-state vibronic lasers. So far the best-investigated type is gadolinium scandium gallium garnet, known as GSGG, a crystal with the chemical formula $Gd_3Sc_2Ga_3O_{12}$. This material has been doped with chromium as a vibronic laser, as well as with neodymium. Laser action also has been demonstrated in double-doped crystals, Cr-Nd-GSGG. However, important practical problems remain, including the difficulty in pumping with a flashlamp and the high cost of scandium.

Cobalt-doped magnesium fluoride (Co-MgF$_2$) emits pulses that are broadly tunable between 1750 and 2500 nm at room temperature. It can operate continuously at cryogenic temperatures at wavelengths as short as 1500 nm. However, the power is limited, and these wavelengths are not in wide use.

OTHER SOLID-STATE LASERS

Many other solid-state lasers have been demonstrated in the laboratory, but only a handful are practical enough to be available commercially. There are four principal types: erbium, holmium, and color-center lasers, and optical parametric oscillators (OPOs). Erbium lasers come in two distinct types, fiber lasers and amplifiers used for fiber-optic communications, and bulk lasers that emit at other infrared wavelengths. Erbium and holmium lasers are analogous to neodymium lasers; color-center lasers are a distinct type and operate in a quite different manner. OPOs are not exactly lasers, but they look and act so much like solid-state lasers that we will cover them here.

Erbium-Doped Fiber Lasers and Amplifiers

Erbium in a glass host forms a three-level laser with gain over a wavelength range centered at 1550 nm. This is an important wavelength for fiber-optic communications, because it is the range where silica glass optical fibers absorb the least light. Thus, optical signals can travel the greatest distance with the least loss at 1550 nm.

Early fiber-optic systems amplified weak optical signals by sensing them with a detector to produce electronic signals that could be amplified and used to drive semiconductor laser transmitters. However, a conceptually simpler approach is to find a way to amplify the optical signal by increasing the intensity of the light. This can be done by passing the weak signal through a length of optical fiber with a core doped with erbium. If the erbium atoms are excited by an external light source at a different wavelength, the weak signal can stimulate them to emit light at the signal wavelength, 1550 nm, increasing the signal strength. Such optical fiber amplifiers promise to find important applications in long-distance fiber-optic systems.

Figure 7-12 shows the basic concept of an optical fiber amplifier. Light from a semiconductor laser excites the erbium atoms (absorption peaks are at 980 and 1480 nm, so either wavelength can be used). They, in turn, amplify the 1550-nm signal that is entering the fiber amplifier and coupled into the next section of the fiber-optic system. While a little pump light may enter the output fiber, it is eventually lost in the transmission system.

Erbium-doped fibers can also be used as oscillators if mirrors are put at both ends. However, the main interest is in amplifiers to boost signal strength and extend transmission distance.

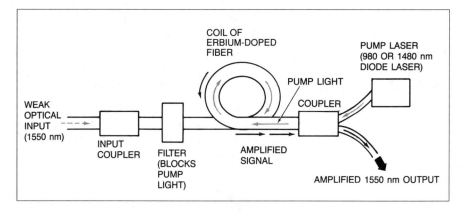

Fig. 7-12   Erbium-doped fiber amplifier.

Bulk Erbium Lasers

Bulk erbium lasers emitting at 1540 nm have been developed because that wavelength poses very little hazard to the human eye. That makes it attractive for use in military range finders and target designators used in war games and training exercises involving friendly troops, because the standard neodymium laser range finders and designators pose serious eye hazards.

Bulk erbium lasers are also made for operation at wavelengths near 2900 nm in YAG and some other hosts. That wavelength is being investigated for medical applications because it is absorbed very strongly by tissue, and unlike other laser wavelengths it can cut bone and hard tissue.

Holmium Lasers

Holmium is another rare-earth element that can lase when doped in YLF or YAG. Its best output wavelength is near 2100 nm, which is also being investigated for medical applications.

Color-Center Lasers

Color-center lasers have, as their active media, crystals that are intentionally doped with impurities that introduce flaws in the crystalline lattice. The flaws themselves absorb and emit light as the atoms at the flaw, or "color center," change position. The transitions are not sharp because some energy is converted into strain and vibration in the crystal. This lets color center lasers both absorb and emit light over a broad range of wavelengths. In practice, color-center lasers must be pumped with laser light and cooled to low temperatures, but they have gain over a broad range and thus can be tuned in wavelength. Commercial color-center lasers emit near-infrared light, either near 1550 nm or between about 2300 and 3300 nm.

The need for a costly pump laser and for cryogenic cooling to produce the color centers limits the practical applications of color-center lasers. However, their ability to produce narrow-line, tunable output not otherwise available at those wavelengths has led to some applications.

Optical Parametric Oscillators

The optical parametric oscillator (OPO) is not strictly speaking a laser, but it behaves so much like a solid-state laser that it deserves mention. As described in Chapter 5, the OPO is a nonlinear device, which generates two longer wavelengths when pumped by a shorter wavelength. The frequencies of the two longer-wavelength beams add together to equal that of the shorter wavelength beam, which supplies the energy. The nonlinear crystal is placed in a cavity which is resonant (like a laser cavity) at one or both of the longer wavelengths to be generated. The output wavelength can be tuned by adjusting the crystal or the cavity.

As nonlinear devices, OPOs work best when pumped by pulses with high peak power, but specially designed OPOs can operate continuous-wave. They tend to work best when the two generated wavelengths are closely spaced, at roughly twice the pump wavelength, but differences of a factor of three or four are possible.

Research on OPOs goes back many years, but early versions were impractical because they were too vulnerable to optical damage. That has changed with the availability of new nonlinear materials, and over the last few years many new types have been demonstrated in the laboratory, with output in the visible and infrared. A few of these have begun to reach the commercial laser market.

One major application is a source at 1.54 μm. Diode lasers and erbium-doped glass lasers and fibers can also generate this wavelength, but OPOs promise considerably higher powers and repetition rates. Those capabilities are attractive because 1.54 μm is a particularly eyesafe wavelength where the human eye is quite insensitive to light. That makes high-power 1.54 μm sources important for applications which require sending beams where people may be, including laser radars, and range-finders and target designators used in military training.

Another application is wavelength-tunable light sources between about 1.5 and 5 μm. That is a region where many molecules have characteristic absorption bands, useful for spectroscopy, but few tunable lasers are available. Synchronous pumping with modelocked lasers, described in Chapter 5, can also generate femtosecond pulses from OPOs, important for a variety of research applications.

WHAT HAVE WE LEARNED

- Light-emitting atoms in solid-state lasers are fixed in a crystal or glass and excited by light from an external source. Semiconductor lasers are not considered solid-state types.

- The element that emits light in a solid-state laser accounts for only a small fraction of the crystal or glass substance of the laser rod. The most common light emitters are the ions Cr^{+3}, Nd^{+3}, Er^{+3}, Ti^{+3} and Ho^{+3}.

- Transparency, thermal conductivity, and interaction with the light-emitting species are important features of host materials.

- New crystals are harder to develop than new gas lasers, so solid-state laser development has concentrated on fewer materials.

- Excess heat can damage laser media, decrease laser gain, or cause optical distortion in the rod. Optical distortion can impair laser action.

- Solid-state laser oscillators and amplifiers can be rods, slabs, disks, or fibers.

- Crystal growth is a major practical concern in solid-state lasers. Glass is the easiest material to produce in large sizes.

- Flashlamps of various shapes remain common pumps for solid-state lasers because the intense flash can readily produce a population inversion. Linear pump lamps are used in reflective cavities that focus light onto the laser rod. Arc lamps pump some continuous solid-state lasers.

- Semiconductor diode lasers are efficient pumps for solid-state lasers, but other laser pumping is inefficient.

- Ruby is a three-level laser system, but it remains in use despite its inherent inefficiency because it can produce high-power red pulses.

- The synthetic ruby used in ruby lasers is made by doping aluminum oxide with chromium; it emits red light at 694 nm. Energy in a ruby-laser pulse can reach 100 joules. Q-switched pulses of 10 to 35 ns can have peak power to 100 MW.

- The most important solid-state lasers are 1.06-μm neodymium types, with the host either a glass or a crystal.

- Nd-YAG has good thermal, optical, and mechanical properties, but it is hard to grow. Nd-YAG can operate continuous-wave or pulsed. The main attraction of glass as an Nd host is the ability to produce large blocks of good optical quality.

- Semiconductor laser pumping of neodymium lasers is more efficient than flashlamp pumping.

- Harmonic generation can convert 1064-nm Nd-YAG output to 532, 355, or 266 nm.

- Nd-YAG lasers are often Q-switched to generate nanosecond pulses with high peak power.

- The lower laser levels of vibronic lasers are bands broad enough that such lasers can be tuned over a range of wavelengths. Tunability sets vibronic lasers apart from other solid-state lasers.

- Alexandrite, like ruby, emits on lines of chromium. Its output wavelengths are 700 to 830 nm. Its tunability comes from the vibrational band that is its lower laser level.
- Titanium-sapphire can be tuned from 660 to 1180 nm and is easy to grow.
- Erbium-doped fiber amplifiers are used to increase the strength of signals at 1550 nm in fiber-optic systems.

WHAT'S NEXT

In Chapter 8, we will learn about one of the fastest-moving areas of laser technology, semiconductor lasers.

Quiz for Chapter 7

1. A host material for a solid-state laser must meet which of the following criteria?
   a. Must be transparent at the pump wavelength
   b. Must be transparent at the laser wavelength
   c. Must be able to conduct away waste heat
   d. A and B only
   e. A, B, and C
2. Which of the following is the best pump source for a solid-state laser?
   a. A flashlight
   b. A flashlamp
   c. A helium-neon laser
   d. An electrical discharge
   e. A fluorescent tube
3. What type of laser is the most efficient pump for neodymium lasers?
   a. GaAlAs semiconductor
   b. Argon-ion
   c. Helium-neon at 632.8 nm
   d. Ruby
   e. InGaAsP semiconductor
4. Which of the following optical pumping arrangements is no longer used with solid-state lasers?
   a. Semiconductor-laser pumping from the sides of the rod
   b. Pumping with a close-coupled linear flashlamp
   c. Pumping with a linear flashlamp in an elliptical cavity
   d. Pumping with a helical (spring-like) flashlamp surrounding the rod
   e. All of the above are widely used

5. What are the pump bands of ruby lasers?
  a. 750 to 900 nm
  b. One at 400 nm, another at 550 nm
  c. 694.3 nm
  d. 1064 nm
  e. None of the above

6. What makes it feasible to pump neodymium laser with semiconductor lasers?
  a. High efficiency of semiconductor lasers
  b. High power available from semiconductor lasers
  c. Pump bands coincide with semiconductor laser output
  d. A and B
  e. A, B, and C

7. In which of the following characteristics is neodymium-doped glass better than Nd-YAG?
  a. Thermal characteristics
  b. Higher repetition rate
  c. Ease of producing large blocks
  d. Higher laser gain
  e. Much shorter output wavelength

8. Which of the following wavelengths cannot be readily generated from an Nd-YAG laser?
  a. 266 nm
  b. 355 nm
  c. 477 nm
  d. 532 nm
  e. 1064 nm

9. What differentiates vibronic lasers from other solid-state lasers?
  a. Much broader gain bandwidths and tunable output
  b. Higher gain
  c. Shorter wavelengths
  d. Can be pumped efficiently with semiconductor lasers
  e. All of the above

10. Which of the following lists includes only vibronic lasers?
  a. Nd-YAG, Ho-YLF, alexandrite, ruby
  b. Ruby, Alexandrite
  c. Titanium-sapphire, ruby, color-center lasers
  d. Titanium-sapphire, alexandrite
  e. Nd-glass, Erbium-glass, gallium-aluminum-arsenide

CHAPTER 8

Semiconductor Lasers

ABOUT THIS CHAPTER

Semiconductor laser technology has been growing explosively for several years. It is based on a combination of optical and semiconductor electronic technology that gives semiconductor lasers their own distinct properties. Power levels have steadily increased, both for individual lasers and for monolithic arrays that can generate continuous powers of 10 W or more. In this chapter, we will explore the principles of semiconductor lasers and learn about the diverse types now available.

EVOLUTION AND BASIC CONCEPTS

The roots of semiconductor laser technology go back to the 1950s, when semiconductor physics was new. As far back as 1953, noted physicist John von Neumann considered the possibility of light amplification by stimulated emission in semiconductors, but he never formally proposed the idea. In 1957, Yasushi Watanabe and Jun-ichi Nishizawa applied for a Japanese patent on a "semiconductor maser." The most detailed proposals for semiconductor lasers emerged in 1961 from Nikolai Basov's group at the Lebedev Physics Institute in Moscow. Basov and another Russian maser-laser pioneer, Aleksander Prokhorov, later shared the 1964 Nobel Prize in Physics with Charles Townes for other research.

Those early ideas were based on phenomena that occur when a current flows through a junction between two regions of a semiconductor doped with different elements. If a current flows in the right direction through certain materials, it raises electrons into excited states that can eventually emit light. We'll explain that later in more detail.

In 1962, four independent groups in the United States succeeded in making semiconductor lasers within weeks of each other. (The winner, in a photo finish, was Robert N. Hall of General Electric Research and Development Laboratories in Schenectady, New York.) Those lasers stimulated tremendous interest, but they only worked when high-current pulses passed through them, and they required cooling to the 77 K temperature of liquid nitrogen. It was not until 1970 that independent groups in the Soviet Union and at AT&T Bell Laboratories made the first semiconductor lasers able to produce a continuous-wave beam at room temperature. More years passed before semiconductor lasers could operate for long periods at room temperature without self-destructing.

The state of the art in semiconductor lasers has continued to advance at an amazing rate. Arrays of many semiconductor laser stripes on a single chip can produce continuous powers of tens of watts in the laboratory. Tens of millions of inexpensive semiconductor lasers are mass-produced for use in compact disc audio players, CD-ROM drives, and other equipment. Sophisticated semiconductor lasers can transmit billions of bits per second through optical fibers. Red semiconductor lasers are readily available, and orange, green and blue types have been demonstrated in the laboratory.

Some tough technological problems do remain. The new technology for green and blue semiconductor lasers has yet to be tamed for practical use. High-power semiconductor lasers and some other types remain expensive. We will learn why as we explore the fundamentals of semiconductor lasers.

PROPERTIES OF SEMICONDUCTORS

A semiconductor is a material with electrical properties partway between those of a conductor and an insulator. The outermost electrons in a conductor (e.g., a metal) are free to move in the material as an electrical current. The outer electrons in an insulator cannot move in the same way, so they cannot carry current. In a semiconductor, only a few of the electrons are free to move in the material.

The major difference among conductors, semiconductors, and insulators is in how the electrons are bound to atoms in the material. (In all cases, we are talking about electrons in the outermost shell of the atoms.) Insulators are materials that hold onto their electrons tightly, because the outer electrons form bonds between atoms. For example, the outer electrons in silicon dioxide (SiO_2 or silica, the primary component of glass) are bonded to both silicon and oxygen atoms. Conductors are materials in which the outer electrons are only very loosely bound to atoms; metallic aluminum, with three electrons in its outer shell, is an example. In semiconductors such as pure silicon, most electrons are bound to the crystalline lattice, but a few can escape to travel within the material.

The electrons in semiconductors can fall into the two energy bands shown in Fig. 8-1. The lower energy level is the *valence* band, for electrons that form bonds with adjacent atoms. A pure semiconductor crystal such as silicon or germanium has exactly enough electrons to fill all the niches in the valence band. At room temperature, a few electrons have enough energy to reach the higher-energy "conduction" band, where they can freely move about. The number of electrons in each band depends on the size of the energy difference or "band gap" between conduction and valence bands, and on the temperature, according to the Boltzmann law:

$$N_2/N_1 = \exp[-(E_2 - E_1)/kT]$$

where N_2 and N_1 are the numbers of electrons in the conduction and

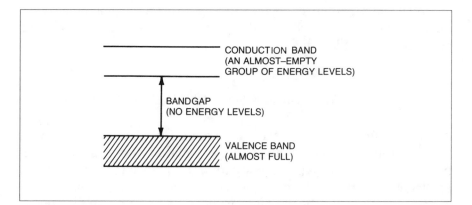

Fig. 8-1 Valence band, conduction band, and band gap in a pure semiconductor.

valence bands, respectively, $E_2 - E_1$ is the energy difference between the two states (the band gap), k is the Boltzmann constant, and T is the temperature (in Kelvin or absolute). We used the same formula in Chapter 3 to describe populations of energy levels in atoms or molecules.

If you plug in numbers, you find that there aren't many electrons in the conduction band at room temperature. For silicon, in which the band gap is 1.1 eV, only 3×10^{-19} of the electrons from the valence band are in the conduction band at room temperature. That allows silicon to conduct some current, but leaves it with a high resistance.

Doping Semiconductors

Silicon and germanium are pure semiconductors, each having four electrons in their outer shell. When they form crystals, each of the four electrons forms a bond with an adjacent atom, as shown in Fig. 8-2A. As we saw above, very few of those electrons—less than one in a million trillion—can escape from those bonds in a pure silicon crystal at room temperature.

Something different happens if an impurity atom is added that has a different number of electrons in its outer shell. If there are only a few impurity atoms, they fit into the crystal in the same position as would atoms of the semiconductor. Figure 8-2C shows what happens when an atom with five outer electrons is added to silicon. Four of the electrons form bonds with the surrounding silicon atoms. The fifth can easily be elevated to the conduction band, where it remains free in the crystal lattice. A semiconductor doped with such electron donors is called *n-type*, because the doping produces negative current carriers. Elements such as phosphorous or arsenic are used as electron donors in n-type silicon.

Something more complicated happens if the impurity atom has only three electrons. All three electrons form bonds with surrounding atoms, but a "hole" remains in the crystal where the impurity atom should contribute a fourth electron, as shown in Fig. 8-2B. An electron from elsewhere in the crystal can move to fill that hole. If it comes from another atom in the crystal, that electron leaves behind another "hole" where it used to be. That makes it possible for the hole to effectively move and to carry current in the semiconductor. Because the hole is the absence of a negatively charged electron, it is said to have a positive charge, and materials doped with elements with three electrons in their outer shell (electron acceptors) are called *p-type* semiconductors. Elements such as aluminum or gallium are electron acceptors in silicon.

To complete the picture, we should note that pure semiconductors are called *intrinsic* materials, because their natural conductivity is intrinsic to the semiconductor. Intrinsic semiconductors are sometimes called *i-type* materials, but they aren't used often.

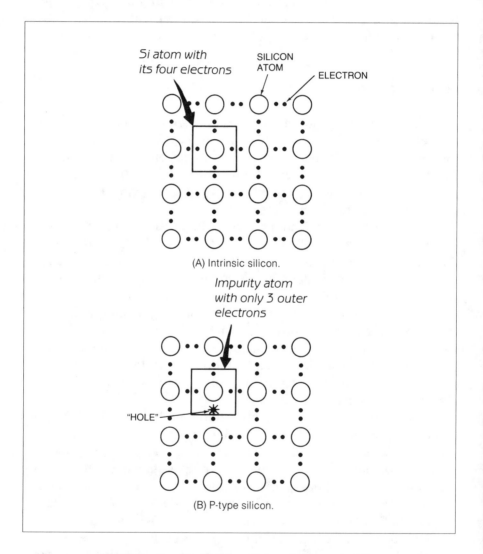

Fig. 8-2   Bonding in pure and doped silicon crystals.

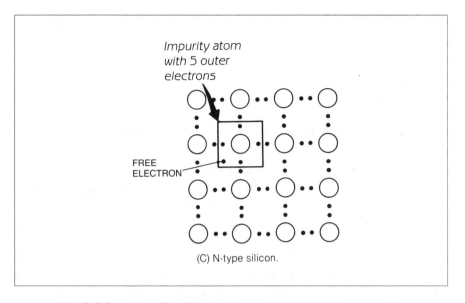

Fig. 8-2 (continued)   Bonding in pure and doped silicon crystals.

In practice, n- or p-type semiconductors are much more conductive than intrinsic materials, because the impurities provide many more current carriers. The degree of conductivity depends on the amount of doping. Impurity levels and conductivity can differ between different n- or p-type materials, even in different layers of the same device.

Compound Semiconductors

Silicon and germanium were the first semiconductors used, but there are many others. Most are compounds of two or more elements that have similar crystal structures and function as semiconductors. Most electronic components are made of silicon, because it is easy to make into complex devices. However, silicon has its limits. The most important one in the laser world is that it is a very poor light emitter. Compound semiconductors are much better, and they are used for all practical semiconductor lasers.

The most important compound semiconductors for laser applications are the so-called III-V semiconductors. They get their name because they contain equal amounts of elements from group IIIa and group Va of the periodic table. We list the important elements next:

GROUP IIIA   GROUP VA
Aluminum (Al)   Nitrogen (N)
Gallium (Ga)   Phosphorous (P)
Indium (In)   Arsenic (As)
                                    Antimony (Sb)

The simplest of these materials are "binary" compounds containing two elements, such as gallium arsenide (GaAs). More complex compounds are also possible, and are often desirable because they allow adjustment of material properties. For example, replacing some of the gallium in gallium arsenide with aluminum changes the energy band gap of the semiconductor (something that, as we will see later, is of crucial importance in semiconductor lasers). The aluminum atoms must replace gallium atoms because the number of atoms with three outer electrons (gallium and aluminum) must equal the number with five outer electrons (arsenic). Such compounds containing three elements are called "ternary" and are written in the form:

$$Ga_{1-x}Al_xAs$$

where x is a number between 0 and 1. This format indicates what we said above, that the number of gallium atoms plus the number of aluminum atoms must equal the number of arsenic atoms.

You can get even more flexibility and control over material properties by adding a fourth element to make a "quaternary" compound. This is something like

$$In_{1-x}Ga_xAs_{1-y}P_y$$

where both x and y are numbers between 0 and 1. In this case, the total number of indium and gallium atoms must equal the number of arsenic and phosphorous atoms. Another possibility is

$$In_{1-x-y}Ga_xAl_yP$$

where both x and y are numbers between 0 and 1, and their sum is also less than 1. In this case, the total number of indium, gallium, and aluminum atoms must equal the number of phosphorous atoms to form a semiconductor crystal.

The more elements in the mixture, the harder it gets to grow good crystals. Other complications also arise because the elements don't always mix well together. However, ternary and quaternary compounds have important attractions. By continuously varying the concentrations of indium and gallium and phosphorous and arsenic, for example, you can vary the characteristics of InGaAsP in the range between those of GaAs, InAs, InP, and GaP. Thus, you can make a variety of compounds with different properties.

There is a further practical complication, which we will mention here but which does not become important until later in our story. Changing the blend of elements in a ternary or quaternary semiconductor also changes the spacing of atoms in the crystal. For example, adding indium to gallium arsenide increases the lattice spacing. This is important because, normally, ternary and quaternary compounds are made only as thin layers on substrates of binary semiconductors, like gallium arsenide or indium phosphide. (The reason is that they are difficult to grow.) To grow good layers, the lattice spacing of the compound has to be reasonably close to that of the substrate.

Other families of semiconducting solids exist besides the III-V compounds. The family of II-VI compounds is also important for semiconductor lasers. These include elements that have two and six outer electrons; they come mostly—but not entirely—from columns IIB and VI of the periodic table. The most important constituents of these lasers are:

COLUMN IIB	COLUMN IVB	COLUMN VI
Zinc	Tin	Sulfur
Cadmium	Lead	Selenium
Mercury		Tellurium

The II-VI family falls into two broad groups. Compounds of zinc and cadmium with group VI elements have large band gaps and can emit visible light. Compounds of lead and tin with group VI elements have small band gaps and emit in the infrared. We will describe both families in more detail later in this chapter.

Semiconductor Band Gaps

We can break semiconductor properties into two broad categories: electronic and optical. Electronic properties include concentration of current carriers, conductivity, and mobility of electrons. These are important in the operation of all semiconductor electronics, including lasers.

In semiconductor lasers (and in other optical devices), optical properties are also important. The key parameter behind most of them is the difference in energy between the bottom of the conduction band and the top of the valence band, called the band gap. The band gap gets its name because semiconductors have no energy levels between the valence band (the energy levels involved in bonding within the crystal) and the conduction band (the energy levels of the free electrons that carry current). An electron that drops from the conduction band must lose enough energy to go to the top of the valence band. Likewise, an electron at the top of the valence band must jump the entire band gap to reach the conduction band. There is no middle ground. An electron must gain or lose at least the band-gap energy for it to make a transition between the conduction and valence bands.

The implications of this for optical properties of semiconductors are shown in Fig. 8-3. At a particular wavelength λ_{bandgap}, the photon energy $h\nu_{\text{bandgap}}$ equals the band-gap energy, E_{bandgap}. Photons with longer wavelengths have less than the band-gap energy. That means that light with wavelength greater than λ_{bandgap} lack enough energy to raise valence electrons to the conduction level, so the valence electrons can't absorb them. Thus, the semiconductor is transparent to wavelengths longer than λ_{bandgap}.

On the other hand, the semiconductor readily absorbs wavelengths shorter than λ_{bandgap} because that light has more than enough energy to raise valence electrons to the conduction band. That makes the semiconductor opaque at those shorter wavelengths. Note, however, that this picture is somewhat oversimplified; in practice, semiconductors do not switch from complete absorption to complete transparency at λ_{bandgap}, although the change can be fairly sharp.

If the band-gap energy E_{bandgap} is measured in electronvolts, its relationship to band-gap wavelength λ_{bandgap} (in nanometers) is:

$$E_{\text{bandgap}} = 1240/\lambda_{\text{bandgap}}$$

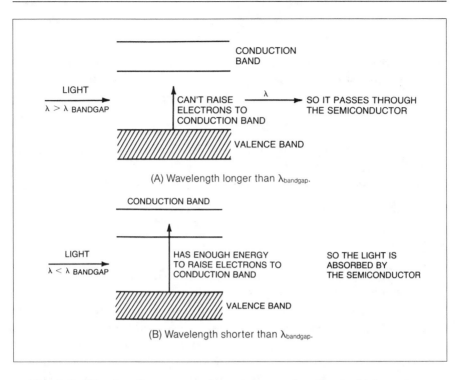

Fig. 8-3 How band-gap energy determines wavelengths at which semiconductors absorb, transmit and emit light.

The light-emitting properties of semiconductors likewise depend on band-gap energy, which is very important for lasers. Electrons must release at least the band-gap energy when they drop from the conduction band down into the valence band. In some semiconductors, like silicon, most of the energy is lost as heat. In others, the energy can be released as a photon of light—with photon energy dependent on band-gap energy. While there are real-world complications to this picture, the basic point is that the wavelength of light released by electrons in a semiconductor is proportional to the band-gap energy.

The notion of band gap does not apply only to semiconductors. We should note here that insulators also have band gaps—but band gaps too large for electrons to cross under normal circumstances. In conductors, in contrast, the band gap is essentially zero, because there is no gap between valence and conduction electrons.

LIGHT EMISSION AT JUNCTIONS

While the bulk properties of semiconductors are important in determining how they will function in electronic devices, the "action" in a semiconductor usually takes place at the boundary, or *junction* between p- and n-type regions. This includes light emission, which occurs at the junction itself. Thus, it is very important to see what occurs at semiconductor junctions. Because this is a book about lasers, not about semiconductors, we will take a rather qualitative look at junctions, considering only the properties important to semiconductor lasers.

Nature of Junctions

Doping creates semiconductor junctions. A junction is the boundary zone between two parts of a semiconductor crystal with different doping. In practice, the semiconductor begins as a substrate with a modest level of p- or n-type doping. For light-emitting devices, the usual substrate is a wafer of a binary semiconductor, such as GaAs or InP. Thin layers of different composition (such as GaAlAs or InGaAsP) are deposited on one side of the substrate. Dopants are also diffused into the material from that side of the crystal, forming one (or in practice, usually more) successive layers of different doping. Patterns may be etched into the surface of the semiconductor, or new layers may be deposited in ways that form patterns on the top surface. The combination of surface patterns and layer structure control the function of a semiconductor device.

Suppose that the initial substrate contains a modest doping of p-type impurities. A simple junction can be formed by diffusing a higher concentration of n-type impurities into the crystal from the top. This converts the top layer of the semiconductor to n-type material, while the bottom remains p-type material. The concentration of n-type impurities (roughly the electron concentration) drops off with distance from the top of the crystal, until at some distance from the surface, the concentration drops to the same level as the p-type impurities (roughly the hole concentration), as shown in Fig. 8-4. That level is considered the junction between the n- and p-type materials. In practice, that level is thin by human standards (0.1 to 1 μm thick), but it represents many atomic layers.

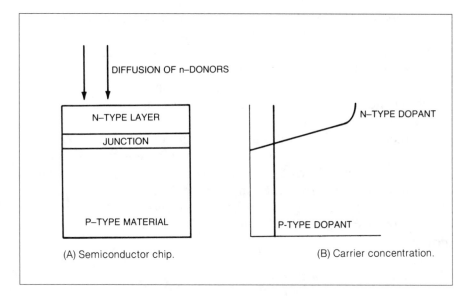

Fig. 8-4   Doping of a semiconductor crystal to form a junction.

Alternatively, the n-type layer can be formed by depositing additional material on the surface of the semiconductor. Techniques such as molecular beam epitaxy make it possible to deposit thin layers, essentially atom by atom, with well-controlled composition that differs from the substrate composition. These techniques also produce semiconductor junctions, but give the fine control over layer structure needed for high-performance semiconductor lasers.

The electrical performance of a semiconductor junction depends on whether or not a voltage (or bias) is applied across the junction and—if so—the direction in which the voltage is applied. We will describe each of the three possible cases in qualitative terms.

Unbiased Junctions

If there is no bias across the junction, charge carriers are distributed through the crystal in roughly the same way as impurities. Electron carriers are common throughout the n region, while holes are common in the p region. Near the junction, the two types of carriers are present in roughly equal concentrations and can cancel each other out by a process called *recombination*. The way this works is important.

Remember that a hole really is the absence of an electron in the valence band. A hole moves about the crystal when valence electrons shift their positions, creating a new hole while filling an existing one. However, a hole can also be filled by capturing a "loose" electron from the conduction band, which releases energy as it drops to the valence band. It is this energy that produces light in light-emitting diodes and semiconductor lasers.

If there is no electrical bias across the junction, recombination on average is balanced by the creation of new electron-hole pairs, and there is no net flow of current carriers in the crystal. Individual electrons and holes form, move, and recombine, but no current flows. Much the same thing happens in still air, where the gas as a whole stays in the same place, although individual atoms and molecules move.

Reverse-Biased Junctions

Now suppose that a positive voltage is applied to the n side of a junction and a negative voltage applied to the p side. As shown in Fig. 8-5, the positive electrode attracts electrons in the n-type material, while the negative electrode attracts holes in the p-type material. This pulls all the current carriers away from the junction so that, essentially, no current flows through the device. (There is, however, a

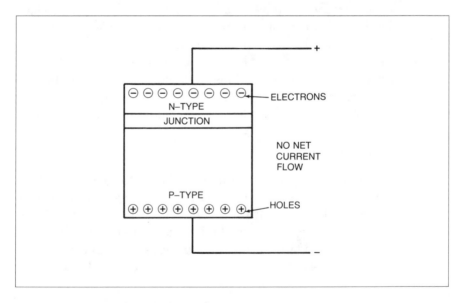

Fig. 8-5   A reverse-biased semiconductor junction.

small leakage current, because the semiconductor does not have infinite resistance to electrical current.)

You will recognize this *reverse biasing* if you are familiar with semiconductor electronics. This is the state in which a diode does not conduct current, as long as the voltage is kept below a breakdown threshold. It also does not emit light.

Forward-Biased Junctions

A semiconductor diode is said to be *forward-biased* if a positive voltage is applied to the p side and a negative voltage to the n side. As shown in Fig. 8-6, this attracts the p carriers to the n side of the device and vice versa, making them cross the junction. A strong current starts to flow once the voltage passes the band-gap energy measured in electronvolts, typically 0.5 to 2 eV.

In an electronic diode, the important thing is that current starts flowing when it is forward-biased. In optical devices, something else important also happens at the junction. Electrons from the n-type material recombine with holes from the p-type material, dropping into the valence band and releasing energy at the junction. The amount of energy lost equals the band gap, and it accounts for the

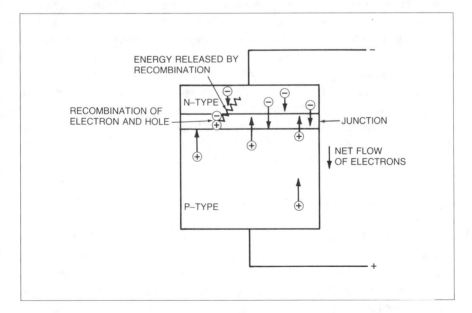

Fig. 8-6   Forward-biasing triggers flow of electrons in a semiconductor diode.

voltage drop across a forward-biased junction. (The larger the band gap, the larger the voltage drop.)

In some semiconductors, such as silicon, the recombination energy is released as heat, warming the device. In others, such as gallium arsenide, much of the recombination energy is released as light. (The difference depends on details of semiconductor physics that you don't want to worry about.) Light emission from recombining electrons and holes is the basis of light-emitting diodes (LEDs) and semiconductor lasers. The process that produces recombination is also called "current injection," so semiconductor lasers are sometimes called "injection lasers" as well as "diode lasers" or "laser diodes."

Optical Excitation at Junctions

Before moving on to the workings of LEDs and diode lasers, we should mention another process that occurs in semiconductor diodes: optical excitation. If a photon with more than the band-gap energy strikes a semiconductor, it can raise an electron from the valence band to the conduction band, creating an electron and a hole—a pair of current carriers. This effect can be used to detect light, by generating an electric current proportional to the amount of illumination or by changing the semiconductor's conductivity, again by an amount proportional to the light intensity. It can also convert light energy into electricity in solar cells.

Light-sensitive semiconductors contain internal junctions and are called "photodiodes." They are the most common type of detector used with lasers, especially with semiconductor lasers. We mentioned them briefly in Chapter 5. We won't go into their operation here, but you should realize that semiconductor devices play a vital role in converting optical signals back into electronic form.

Light-Emitting Diodes (LEDs)

An LED is a forward-biased semiconductor diode in which recombination at the junction produces light. As we mentioned earlier, only certain materials are suitable for LEDs. A sampling of these are listed in Table 8-1. Gallium arsenide doped with silicon is by far the most efficient, but its output is in the infrared.

LEDs are not lasers. The light they produce is spontaneous emission, like that from a light bulb. Like a light bulb, an LED junction

TABLE 8-1 Some important LED materials

Material	Dopant	Peak Wavelength or Range (nm)	Status
ZnTe/ZnSe	Antimony	459 (blue)	Developmental
SiC	—	470 (blue)	Commercial
GaN/GaAlN	Magnesium	480 (blue)	Prototype
GaP	Nitrogen	550–590 (green–yellow)	Commercial
$GaAs_{0.15}P_{0.85}$	Nitrogen	589 (yellow)	Commercial
$GaAs_{0.35}P_{0.65}$	Nitrogen	632 (orange)	Commercial
$GaAs_{0.6}P_{0.4}$	—	650 (red)	Commercial
GaP	Zn, O	700 (red)	Commercial
GaAs	Zn	900 (infrared)	Commercial
GaAs	Si	910–1020 (infrared)	Commercial

radiates light in every direction, as shown in Fig. 8-7. To get the most efficient output, the junction should be close to the surface of the device. This reduces the chance that the emitted light will be absorbed by other parts of the device. (That absorption is possible because a semiconductor material can absorb light at the same transition that it emits on.)

The structure shown in Fig. 8-7 is typical in LEDs for use as small lamps, indicators, or displays. The LED may be packaged with a lens

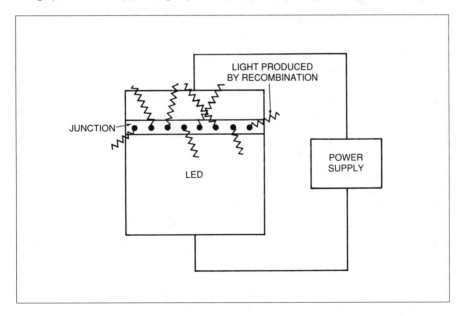

Fig. 8-7 An LED emits light in all directions from the junction.

that concentrates output emerging from the top surface. Other structures can concentrate the output into smaller areas, for applications such as coupling the light into an optical fiber. However, LEDs do not generate well-focused beams like lasers, and their output is spontaneous rather than stimulated emission.

Nature of Semiconductor Lasers

Semiconductor lasers are closely related to LEDs. Both generate light from the recombination of electron-hole pairs at a forward-biased junction. In both, the light output is proportional to the drive current, and the output wavelength depends on the material's band gap. Semiconductor lasers can produce low levels of incoherent emission—functioning like LEDs—when the current passing through them is below the threshold for laser action. However, there are also many important differences.

The first important difference is in current level. A modest current flows through an LED, causing some recombination that generates light. Much higher current passes through a diode laser, sufficient to produce a large concentration of recombining electron-hole pairs at the junction. The current concentration must be high enough for emission from the electron-hole pairs to dominate over absorption at the junction. That is, there must be a population inversion at the junction.

Spontaneous emission can go in any direction. However, as in other lasers, a pair of reflective surfaces confines stimulated emission to a certain direction in a semiconductor laser. In most semiconductor lasers, this direction is in the plane of the junction layer. The lasers are made by cleaving a wafer into many small chips. The cleaved edges, perpendicular to the junction, form the reflective surfaces. Typically, structures within the laser confine stimulated emission to a stripe in the junction plane, as shown in Fig. 8-8. As we will see later, this diagram is very much simpler than actual semiconductor lasers. (We will see later that mirrors can also be formed at the top and bottom of the wafer, producing a "vertical cavity laser," in contrast to the horizontal cavity lasers that emit in the junction plane.)

The feedback that concentrates stimulated emission in the junction plane of horizontal cavity lasers comes from the ends of the semiconductor crystal, which are smooth "facets" formed by cleaving the chip. Semiconductors have high refractive indexes, so typically the facets reflect about 30% of the light back into the material, providing enough feedback for laser action in the high-gain semiconductor laser material. Some semiconductor lasers emit from both facets, but they are normally packaged so only one output emerges from the case. Others have reflective coatings on one facet, and only emit from the other facet. Figure 8-8 shows the beam emerging from only one end of the laser.

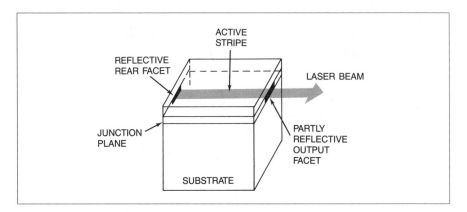

Fig. 8-8   Reflective facets produce laser action in the junction plane of a semiconductor laser.

At low current levels, diode lasers generate some spontaneous emission by the same processes that drive LEDs. However, as the current level increases, diode lasers pass a threshold where the population becomes inverted and laser action begins. This laser threshold is most visible in plots such as that shown in Fig. 8-9, which shows output power as a function of drive current. Below threshold, light is emitted with low efficiency. Once the current has passed the threshold (which is a characteristic of the individual laser), output rises steeply, showing the presence of stimulated emission.

Threshold current is a very important parameter in semiconductor lasers. Below threshold, light output is very inefficient, and most energy in the drive current is lost as heat. Light emission is much more efficient above threshold, so more of the input electrical energy is converted to light. The higher the laser threshold, the more electrical power is lost as heat that must be dissipated before you get any laser output. Because this heat tends to shorten the lifetime of a semiconductor laser, lifetime declines as threshold current increases.

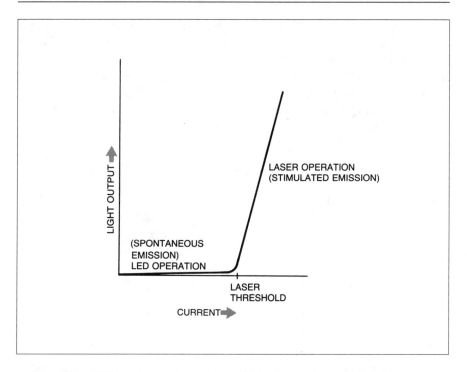

Fig. 8-9 Light output of a semiconductor laser above and below laser threshold current.

We should note that the figure of merit for semiconductor laser lifetime is often measured as threshold current density rather than the absolute value of threshold current. Threshold current density equals the current divided by the active area of the laser stripe, and is given in amperes per square centimeter flowing through the laser junction. It measures the stress on the laser material.

Direct vs. Indirect Band-Gap Materials

All semiconductor laser materials can operate as LEDs, but many LEDs cannot operate as semiconductor lasers. The reason lies in the complexities of semiconductor physics that determine the nature of the energy levels. In what are called *direct band-gap* materials, a transition from the conduction band to the valence band can release essentially all its energy in the form of light. However, in what are called *indirect band-gap* materials, some of the transition energy must go into vibrations of the crystal lattice. That makes it hard for indirect band-gap materials to generate light by recombination of current carriers.

Silicon is an example of an indirect band-gap material, which does not emit light efficiently. Gallium arsenide is a direct band-gap material that makes good LEDs and lasers, although some other compound semiconductors have indirect band gaps. It is possible for some indirect band-gap materials to work as LEDs if they are heavily doped with impurities, but they do not work as lasers. This has presented a problem in developing short-wavelength diode lasers, because many semiconductors with the large band gaps needed to generate short wavelengths have indirect band gaps.

STRUCTURES OF SEMICONDUCTOR LASERS

As we warned earlier, Fig. 8-8 gave a very oversimplified view of semiconductor lasers. Such simple structures were used in the first diode lasers, but developers who wanted to produce continuous-wave output at room temperature had to turn to much more complex structures to reduce threshold current and increase efficiency and lifetime.

A diode laser must contain at least three layers—a p-type layer, an n-type layer, and a separate active or junction layer (either n- or p-type) where recombination occurs to produce the light. High-performance diode lasers usually contain more layers. All this complexity comes in a tiny package. Typical diode laser chips are 300 to 500 µm long and a fraction of that distance wide and thick. Bare laser chips are sometimes used in research laboratories, but for most applications they are packaged in small cases, many of which look like the metal cases used with old discrete transistors.

Homostructure and Heterostructure Lasers

The boundaries between the active layer and the surrounding layers are very important in determining the efficiency of a diode laser. In the first semiconductor lasers, these layers were the same material. Because they used the same material for both layers, they were called *homostructure* or *homojunction* lasers. Later developers turned to what are called *heterostructure* or *heterojunction* lasers, in which adjacent layers are made of different materials. (Note that homojunctions and heterojunctions are not exactly the same as the simple p-n junctions we have been describing throughout this chapter, but represent boundaries between different layers in a multilayer semiconductor structure.)

The advantage of using different materials is that they confine light better in the active layer, making stimulated emission more efficient. Figure 8-10 shows the difference between homostructure and heterostructure lasers. In a homostructure laser (Fig. 8-10A), the semiconductor has essentially uniform refractive index throughout, so light can diffuse from the active layer into surrounding layers. However, the layers in a heterostructure laser have different compositions and thus different refractive indexes. In a double-heterostructure laser (Fig. 8-10C), the active layer has a higher refractive index than either of the two adjacent layers. This creates a waveguide effect (much like that in an optical fiber) which confines light to the active layer. Virtually all commercial diode lasers use variations on this structure. Single-heterostructure lasers (Fig. 8-10B) also exist, in which the layer on one side of the active layer is made of the same material as the active layer, while that on the other side is made of a different material. This gives better confinement than a homojunction laser, but not as good confinement as a double-heterostructure laser.

The differences in light confinement make big differences in functional characteristics. Homostructure lasers let so much light leak out of the active layer that they are too inefficient for practical use. Single-heterostructure lasers are more efficient, but they do not confine light well enough to emit continuously at room temperature. Their main uses are to generate pulses lasting under 1 μs with peak powers of 1 W to a few tens of watts at low repetition rates.

Double-heterostructure lasers are much more efficient and can generate a continuous-wave beam at room temperature. That approach is used in most commercial diode lasers, and has been refined in many ways. Some refinements are intended to confine the light to a narrow stripe within the junction layer. Other refinements involve making thinner and thinner layers that also control the flow of current and light within the semiconductor.

Stripe-Geometry Lasers

We mentioned earlier that most diode lasers emit light from a narrow stripe in the junction plane rather than its entire width. Lasers with comparatively wide emitting regions can offer higher power because of the larger emitting volume. However, narrow stripe-geometry lasers offer better beam quality. Typically, the stripes are only a few micrometers wide. The limits of the stripes are set in two

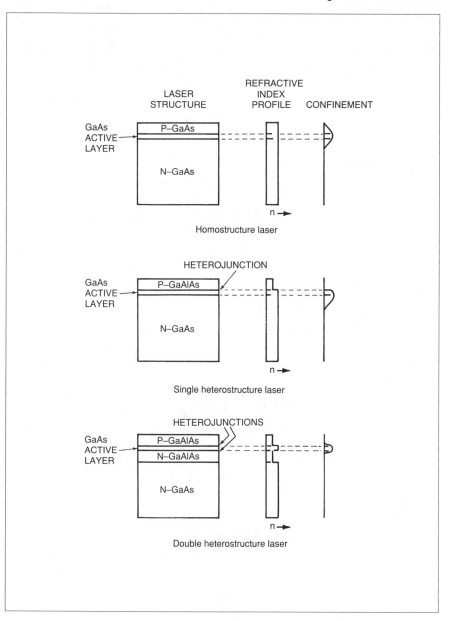

Fig. 8-10 Homostructure and heterostructure lasers, with refractive index profile and light confinement.

ways: by boundaries where laser gain drops off, or by changes in the refractive index of the material itself.

The basic concept of a *gain-guided* laser is shown in Fig. 8-11A. An insulating layer at the top of the laser chip blocks current flow at the sides, and confines it to a narrow stripe the length of the chip. Only in that narrow stripe does enough current flow to produce a population inversion and the right conditions for laser gain. There is no gain at the sides, so those regions do not emit light, even though no physical boundary separates the stripe from the rest of the active layer.

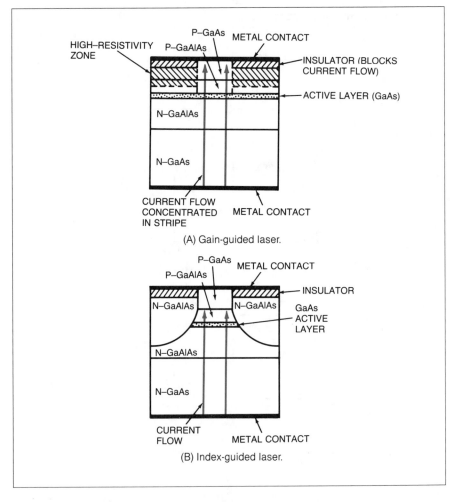

(A) Gain-guided laser.

(B) Index-guided laser.

Fig. 8-11   Gain-guided and index-guided lasers.

Index-guided lasers extend the concept behind double-heterostructure lasers to provide horizontal as well as vertical confinement. The stripe in an index-guided laser is defined by a change in semiconductor composition and refractive index. In Fig. 8-11B, the current flows only through the central "mesa" and the stripe of the active layer buried below it. The layers that border the sides of the active layer have a different composition and different refractive index, trapping the light in the active stripe as the adjacent layers confine light in the active layer in a double-heterostructure laser. Lasers in which the light emitting stripe is entirely "buried" by other layers—as in Fig. 8-11B—are sometimes called *buried heterostructure* lasers.

Index-guided lasers are used for most diode laser applications, and are available in many variations. Gain-guided lasers are simpler to make, but they do not confine light as well, limiting beam quality. However, their poorer confinement of light is an advantage in generating high powers because it spreads light out over a larger area to reduce the chance of optical damage to the emitting surface.

Laser Arrays

In our examples so far, each laser chip contains only one laser stripe. However, many parallel laser stripes can be fabricated on the same semiconductor substrate, as shown in Fig. 8-12. These monolithic laser arrays can produce much higher power than single laser stripes, although their beam quality tends to be poorer.

Single-stripe lasers can generate continuous powers between a fraction of a milliwatt and hundreds of milliwatts. Arrays with tens or even hundreds of parallel laser stripes can generate continuous-wave powers above 10 W. In repetitively pulsed or "quasi-continuous-wave" mode, linear arrays can generate tens of watts. If several such repetitively pulsed linear arrays are mounted together in a stack, the peak power can reach hundreds of watts.

At high powers, diode lasers become more efficient in converting input electrical energy into optical energy. This is because the marginal efficiency above laser threshold is high. However, high-power semiconductor lasers still generate considerable waste heat, which must be dissipated to avoid damaging the lasers. Because semiconductor lasers are small, this requires active cooling, with considerable attention to heat removal at high powers.

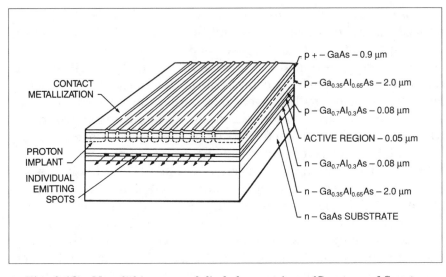

Fig. 8-12   Monolithic array of diode-laser stripes. (Courtesy of Spectra Diode Laboratories.)

So far, most high-power arrays have been of gallium aluminum arsenide, emitting near 800 nm for such applications as pumping solid-state neodymium lasers, as described in Chapter 7. Power levels may continue to increase as the technology develops, opening up new applications for diode lasers. The technology is also spreading to other semiconductor materials.

Quantum Wells, Wires and Dots

Improvements in semiconductor fabrication have made it possible to deposit thinner and thinner layers. Some are only a few atomic layers thick. This permits a new type of confinement based on the quantum properties of matter.

Standard semiconductor physics assumes that the material is a continuum rather than made up of individual atoms. That approximation works fine for layers more than about 20 nm—roughly 40 atoms —thick. However, the quantum mechanical properties of atoms and electrons become important at smaller scales, changing energy levels and the properties of matter. That opens up new possibilities for the development of semiconductor lasers.

Layers only a few nanometers thick can serve as "quantum wells." The basic idea of a quantum well is to sandwich a thin layer with a small band gap between two thicker layers with larger band gaps, as shown in Fig. 8-13. Electrons passing through the semi-conductor can be captured in the quantum well. They have enough energy to remain free in the small-band-gap quantum well layer, but not enough to enter the higher-band-gap confinement layer. (The confinement layer has to be thicker than the quantum well to keep the electrons from leaking out.) This makes quantum wells good traps for electrons, which are confined very tightly in the thin layer.

That tight confinement is a big advantage if the quantum well is put at the junction of a semiconductor laser. By concentrating the electrons as they recombine with holes, it improves the laser efficiency and decreases its threshold current. The confinement layers also have a different refractive index than the quantum well, so they can confine light as well.

The volume of a single quantum well is small, which can limit the amount of light it can generate. Fortunately, quantum wells can be stacked in alternating layers with confinement layers, producing a multiple quantum well structure that contains a larger volume of light-emitting material. This allows the production of higher powers.

Quantum wells confine electrons in only one dimension—in the plane of the quantum well layer. Developers are trying to extend the

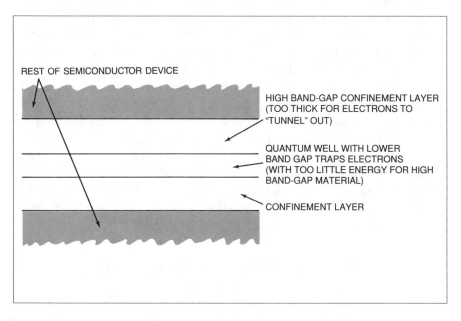

Fig. 8-13   A quantum well traps electrons.

idea to providing confinement in two dimensions (a narrow stripe called a *quantum wire*) and in three dimensions (a *quantum dot*). Those devices remain in development, but quantum well lasers are now standard products.

Strained Layers

We saw earlier the importance of matching the interatomic spacing of semiconductor compounds being grown on the same wafer. If the match is not good enough, flaws will develop where the crystalline lattice does not match. These defects can prove fatal to semiconductor lasers as well as to other devices.

In bulk semiconductors, the match between layers must be very good—within about 0.1%. That poses a big practical problem, because the lattice spacing of semiconductor compounds can differ by considerably more than that amount. Lattice spacing does not change much when aluminum replaces gallium in a compound like gallium arsenide; both gallium arsenide and aluminum arsenide have lattice spacing of about 0.566 nm. That match allowed the development of GaAlAs lasers, which could be lattice-matched to GaAs substrates. However, other compounds don't work that way. The lattice spacing of indium phosphide is 0.587 nm, while that of indium arsenide is 0.606 nm, so if you add too much arsenic to InP, you can't match the lattice spacing of InP. That effectively limits the range of wavelengths available from bulk semiconductor lasers, even when a fourth element is added to help maintain a constant lattice spacing.

Fortunately, the lattice-matching requirement becomes less stringent as layers get thinner. "Strained layers" only a few nanometers thick can tolerate much more strain than thicker layers, allowing lattice mismatches to around 1%. That makes it possible to fabricate devices with layers that would otherwise be incompatible and to generate new wavelengths. (We will discuss these lasers later in this chapter in a section on materials.) In addition, strained layers seem to improve laser characteristics in other ways, so strained-layer lasers can have lower thresholds and longer lifetimes than bulk lasers.

BEAM CHARACTERISTICS AND STRUCTURE

As with other lasers, the characteristics of diode-laser beams depend heavily on their structure and lasing characteristics. These differ markedly from those of other types of lasers.

Diode laser materials have very high gain, so laser oscillation requires only a small amount of laser material in the resonant cavity. Typical cavities in the plane of the laser junction are only 300 to 500 μm long. Semiconductor laser action is also possible perpendicular to the laser junction if highly reflective mirrors are placed above and below the active layer, as in the developmental vertical cavity lasers we will describe later. Here we will concentrate on the most common "horizontal cavity" diode lasers, in which the flat cleaved edges or "facets" of the semiconductor chip form the resonator mirrors.

These standard semiconductor lasers have the basic resonator structure shown in Fig. 8-8. The materials used in semiconductor lasers have high refractive indexes; for example, the refractive index of gallium arsenide is 3.34 at 780 nm. This causes high reflection at the semiconductor-air interface, so about 30% of the light is reflected back into the semiconductor. This is adequate feedback for laser oscillation. The rear facet may (or may not) have a reflective coating; the front facet generally does not. Without a totally reflective coating, the rear facet also emits a beam, which is normally trapped in the laser package or directed to a detector in the package that monitors laser power. As in other high-gain lasers, the cavity does not limit oscillation to a TEM_{00} mode.

The short resonators of normal horizontal cavity diode lasers are several hundred wavelengths long, compared to hundreds of thousands of wavelengths long for a typical visible-wavelength gas laser. This means that the cavity modes—the wavelengths that fit into the cavity according to the equation $2L = n\lambda$—are much more widely spaced than in longer gas lasers. As a result, most diode lasers normally oscillate at only one wavelength (i.e., on one longitudinal mode) at a time. However, the gain curve of semiconductors is broad, and diode lasers can drift (or "mode hop") from one wavelength to another during operation.

Beam Size and Divergence

We saw earlier that beam divergence depends on the nature of the resonant cavity and the size of a laser's emitting area. In diode lasers, the small cavity combines with an unusually shaped emitting area to produce a noncircular beam that spreads unusually rapidly.

The shape of a diode-laser beam depends mainly on its emitting area. We have assumed that gas and solid-state lasers emit round beams, and for most purposes this is a good assumption. Most diode laser beams are not round because the light-emitting ends of the active layers are not circular. (Strictly speaking, this is true only for horizontal cavity lasers, which oscillate in the plane of the active layer.) The active regions are typically a fraction of a micrometer thick and several micrometers across. (These dimensions are difficult to show to scale, so we haven't tried in our figures.) At the facet ends, they look like long, thin rectangles, but the beams they produce are smoother around the edges and are long, thin ovals.

Earlier, we saw that beam divergence roughly equals the wavelength divided by the output-beam diameter. Beams that are not circular diverge at different rates in the two perpendicular directions. As we saw earlier, the divergence is larger in the direction of the smaller dimension (active-region thickness) and smaller in the direction of greater active-region width.

For a gas laser, the beam diameter at the output mirror is typically over a thousand times the wavelength, but for a semiconductor laser the ratio is much smaller. For a typical 800-nm GaAlAs/GaAs laser, the light-emitting stripe is about 10 wavelengths wide and only about a quarter of a wavelength high. These values are small enough that the simple approximations used earlier to calculate divergence break down. It is easier to give typical measured values for beam divergence: 10 degrees (0.17 radian) in the direction parallel to the active layer and 40 degrees (0.70 radian) in the direction perpendicular to the active layer. Compare those to the typical 0.001 radian divergence of a helium-neon laser, and you realize you have a rather different beast. The beam from a diode laser spreads out more rapidly than one from a good flashlight!

Fortunately, it is possible to add external optics to correct for this broad beam divergence. A cylindrical lens, which refracts light in one direction but not in the perpendicular direction, can make the beam circular in shape. Collimating lenses can focus the rapidly diverging beam to a much narrower divergence, like a flashlight lens focuses light from a flashlight bulb into a beam. Semiconductor laser pointers made with such optics generate beams that seem as tightly focused as gas laser beams.

Modes and Coherence

Narrow-stripe diode lasers can oscillate in a single transverse mode, and are often called "single-mode" lasers. However, as with gas lasers, that single transverse mode can include more than one longitudinal mode. As long as a diode laser emits multiple longitudinal modes, its bandwidth remains large enough to severely limit coherence length. Limiting oscillation to a single longitudinal mode can give coherence lengths of tens of meters, but it requires special structures, which are described later.

The poor coherence length of standard semiconductor lasers sets them apart from gas lasers. Unlike their broad beam divergence, it cannot be corrected just by adding simple external optics. This is a problem for holography, interferometry, and a few other applications that require coherent light. However, coherence is unnecessary for most applications, and in some cases can introduce extraneous noise in the form of speckle, as described in Chapter 4. Thus, the short coherence length of semiconductor lasers is not always bad news.

Modulation

The output of a diode laser depends on the drive current flowing through it. Once the drive current has passed laser threshold, light output increases roughly linearly with current, as shown in Fig. 8-9. (The light-current curves of some diode lasers have "kinks" that mark sudden changes in light emission or current level, but for many purposes lasers with such kinks are considered defective.)

This simple dependence of light output on drive current makes direct modulation merely a matter of changing the drive current. The laser emission responds quickly, especially when the light-emitting stripe is narrow and current flow is confined to a small area. The best commercial diode lasers have bandwidths in the gigahertz range (billions of hertz), although not all diode lasers can respond that rapidly. The highest bandwidths of laboratory lasers are about 28 GHz.

Some problems can occur at very high speeds. One that may limit the speed of direct modulation for fiber-optic communications is a change in wavelength, or "chirp," that accompanies direct modulation. The change in electron density during the current pulse changes the semiconductor's refractive index n. This enters into the equation for the laser oscillation wavelength λ (measured in air) given by the requirement that the round-trip length of the cavity $2L$ be equal to an integral number of wavelengths in the laser medium.

$$\lambda = 2Ln/N$$

The change is not large, but because the refractive index of optical fibers varies with wavelength, pulses containing a range of wavelengths spread out as they travel through the fiber. If the spreading is too large, the pulses will interfere with each other, limiting transmission speed. To avoid this problem, developers of ultra-high-speed fiber-optic systems are working on external modulators that modulate the beams from continuous-wave lasers without changing their wavelengths.

LASER RESONATOR STRUCTURES

Earlier in this chapter, we saw that standard semiconductor diode lasers have a simple resonator structure. Cleaving a wafer into diode-laser chips forms facets perpendicular to the active layer that reflect some light back into the laser cavity and transmit other light which becomes the laser beam. Uncoated laser chips emit from both ends but are usually packaged to produce only a single output beam. One facet can be coated so that it reflects all incident laser light back into the semiconductor, which then emits only from the other facet. Some diode lasers are packaged so that light emerging from the rear facet is monitored by a detector attached to a feedback circuit, which controls drive current to maintain output power at a constant level.

Such lasers follow the standard rules of laser resonators, but as we saw earlier, their output differs in important ways from that of other lasers. In addition, semiconductor lasers may be made with other resonator structures. Some are intended to provide output over a very narrow range of wavelengths, others to emit from the surface rather than the edge of the chip. Some are coated on both ends to suppress reflections so they can function as amplifiers but not as oscillators. We describe four of the most important variants below; many more have been studied in the laboratory.

Semiconductor Laser Amplifiers

Semiconductor lasers oscillate and generate their own signal because light is reflected back and forth between the facets on the edges of the chip. Semiconductor laser materials can also be made to amplify the light from an external source. That light can stimulate emission from the active layer of the laser at the same wavelength, increasing the strength of the signal.

The basic process is similar to laser oscillation, but there are some important differences. The light that stimulates emission comes from an external source, not from inside the laser. The goal is not to build up a strong beam, but to accurately amplify a weak signal with wavelength in the laser's gain bandwidth. This means that the light should make a single pass in one direction (from input to output) rather than oscillate between a pair of cavity mirrors. In semiconductor laser amplifiers, it is essential to apply an antireflective coating to suppress reflection from the laser facets in order to avoid feedback that might start laser oscillation.

Semiconductor laser amplifiers have been developed for use in fiber-optic communications. Unfortunately, they don't work as well as the optical-fiber amplifiers described in Chapter 7, and thus have found few applications so far. However, development is continuing, particularly at the 1.3-µm wavelength, where good fiber amplifiers are not readily available.

Distributed-Feedback Lasers

Some applications require semiconductor lasers to emit only a very narrow range of wavelengths. This is difficult with conventional diode lasers. The short cavity length of diode lasers does limit the number of longitudinal modes possible, because there are few values of N that solve the equation

$$2L = N\lambda$$

If cavity length L is only 500 µm for a typical gallium-arsenide laser emitting near 900 nm, then the longitudinal cavity modes are about 1 nm apart.

Semiconductor lasers have gain bandwidth broad enough to span several longitudinal modes. The processes that amplify laser emission normally suppress some—but not all—side modes, and often one mode dominates. Unfortunately, changes in temperature and other operating conditions can make a diode laser "hop" from one dominant mode to an adjacent one. This causes undesirable instability in laser power and output wavelength, which can lead to problems in certain fiber-optic systems and other applications.

The cure is a diode-laser design that includes a mechanism to stabilize the laser's output wavelength. One leading approach is the distributed-feedback laser, shown in Fig. 8-14, which contains a diffraction grating that scatters light back into the active layer. Strictly speaking, a distributed-feedback laser is one in which the grating is on the part of the active layer where there is laser gain. If the grating is in a part of the active layer where there is no gain, it stabilizes the wavelength by a process called "distributed Bragg reflection." Both cases are shown in Fig. 8-14; the difference is more important to specialists than to laser users. Feedback from the grating causes interference effects that allow oscillation only at certain wavelengths, which the interference reinforces. The overall effect is to limit laser oscillation to a stable single longitudinal mode, giving the laser very narrow bandwidth.

The grating spacing D needed to select a particular wavelength λ in a distributed-feedback laser depends on the refractive index n in the laser media and a positive integer m that gives the order of distributed-feedback coupling,

$$D = m\lambda/2n$$

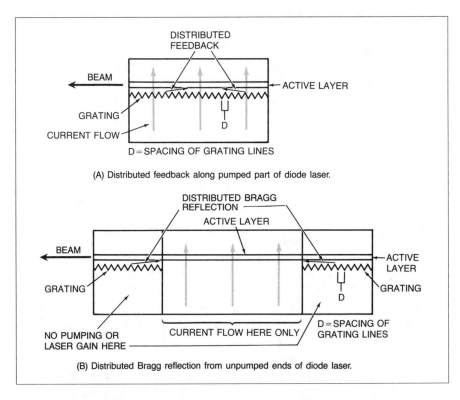

(A) Distributed feedback along pumped part of diode laser.

(B) Distributed Bragg reflection from unpumped ends of diode laser.

Fig. 8-14   Two types of distributed-feedback lasers.

Typical values of m are 1 or 2. If we plug in the appropriate numbers for a 1550-nm distributed-feedback laser made of InGaAsP (n = 3.4), we find that grating spacing D is 228 nm for m = 1 and 456 nm for m = 2. The oscillation wavelength shifts slightly with temperature because the refractive index n is a function of temperature, but this change is much smaller than the wavelength shifts in conventional diode lasers. This means that the structure of a distributed-feedback laser sets its wavelength at a fixed value, although minor variations in that range are possible.

Distributed-feedback lasers are so far the most successful type of "single-frequency" diode laser. They are made commercially and used in fiber-optic telecommunication systems.

External Cavity and Tunable Lasers

Another way to produce narrow-line output from a semiconductor laser is to place a semiconductor laser in an external cavity with suitable tuning optics. Typically, as shown in Fig. 8-15, the rear facet is coated so that it transmits nearly all light circulating in the active layer. This lets light travel to focusing and tuning optics, which form the rear of the cavity. Adjusting the tuning optics (a diffraction grating in the figure) selects the wavelength resonant in the laser cavity.

While semiconductor laser chips are less than a millimeter long, external-cavity lasers have cavity lengths up to a meter long. This can generate a narrow-line output that is tunable over the gain width of the semiconductor, with a total range that can reach 200 nm.

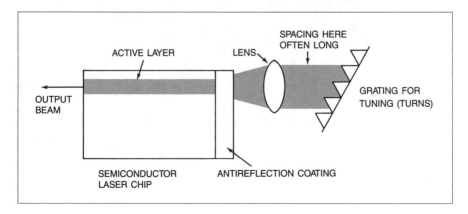

Fig. 8-15   External-cavity semiconductor laser.

This tunability is an important feature that makes external-cavity lasers attractive for certain applications. It gives semiconductor lasers the same broad tunability that has been critical to the use of titanium-sapphire and dye lasers. While individual lasers generate reasonable powers over only limited ranges, a series of semiconductor lasers can be used with a common external cavity to span a wider range of wavelengths, with reasonable power levels. Such lasers have only recently reached the market.

Vertical Cavity Lasers

The newest approach to semiconductor lasers is the vertical cavity laser, with mirrors above and below the active layer. Oscillation in these lasers is vertical, with the cavity perpendicular to the plane of the laser cavity, as shown in Fig. 8-16. The beam emerges from the surface of the wafer. At this writing, the technology is in development, not production, but it offers interesting opportunities and differs in important ways from standard semiconductor lasers.

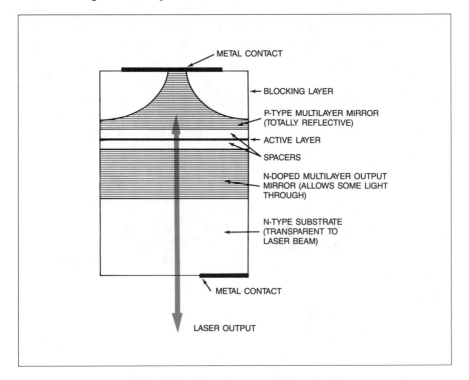

Fig. 8-16   Vertical cavity semiconductor laser.

One difference is in mirror reflectivity. Because the light passes through only a very short length of gain medium, overall laser gain is low. Thus, the mirrors must be highly reflective. They are made by depositing reflective layers on the top and bottom of the wafer, or by depositing multiple layers of semiconductor material to form a reflective structure.

Each laser occupies only a very small area on the chip, as small as a few micrometers across, so many vertical cavity lasers can be tightly packed in an array. One group has packed over a million such lasers on a single gallium-arsenide chip, with a density greater than two million lasers/cm². The minute size of the lasers means that they have low threshold currents—researchers have reported values below one milliampere, and have projected "ultimate" threshold currents as low as one microampere. Developers are trying many variations in making arrays, such as giving each laser element a different wavelength and modulating each laser in the array independently, or driving all the elements in the array in unison to produce a single coherent beam.

Vertical cavity lasers typically emit light from a round area on the surface, so their beams are more symmetric than those from side-emitting lasers. Surface emission is particularly sought for optical interconnection between integrated circuits (in which signals would be routed optically between circuits), and for optical computing and signal processing. Developers are investigating practical applications and improving device fabrication.

WAVELENGTHS AND MATERIALS

Earlier, we indicated that a diode laser's output wavelength depended on the size of the material's band gap. The photon energy does not exactly equal the band-gap energy, but it is close. If band-gap energy E is given in electronvolts (as is common), you can quickly calculate the approximate emission wavelength λ (in nanometers) from the formula:

$$\lambda = 1240/E$$

The band-gap energy in electronvolts is also close to the voltage drop across a forward-biased diode.

The band-gap energy, and thus the emission wavelength, depend on lattice spacing and semiconductor composition. Both lattice spacing and band-gap energy are fixed for a pure binary semiconductor such as gallium arsenide, indium phosphide, or lead sulfide. However, you can get other wavelengths by forming semiconductor crystals from mixtures of three or four elements. You can think of the process as blending different materials to get intermediate characteristics. For example, a mixture of 20% AlAs and 80% GaAs has band-gap energy and lattice spacing partway between pure AlAs and pure GaAs. The formula for such a compound can be written $Ga_{0.8}Al_{0.2}As$ or simply GaAlAs. (The order of the two elements that replace each other—aluminum and gallium—is arbitrary; some people write AlGaAs.)

Things get more complex if the compound semiconductor contains four elements. Depending on the relative concentrations of the elements, such "quaternary" semiconductors can have properties somewhere in a broad range. For example, as shown in Fig. 8-17, InGaAsP (more precisely $In_xGa_{1-x}As_{1-y}P_y$) can have lattice spacing and band gap within an area defined by four possible binary

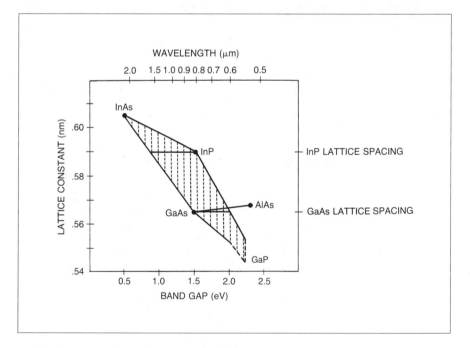

Fig. 8-17 Lattice constant vs. band gap for InGaAsP can fall within the shaded area defined by properties of InAs, InP, GaAs, and GaP.

compounds: InAs, InP, GaAs, and GaP. The material's properties can vary within the entire space because the In-Ga and As-P ratios can be adjusted independently. (The odd shape of the area in the figure arises because pure GaP is an indirect band-gap material, unsuitable for lasers and somewhat different from the other binary compounds. The dashed line indicates the indirect band-gap region.)

As we mentioned earlier, lattice spacing is also a crucial variable for semiconductor laser materials. To avoid flaws that degrade device performance, the entire structure of bulk lasers (with layers thicker than about 20 nm) must be made of compounds with nearly identical lattice spacing. This means that all compositions in a bulk InGaAsP structure must fall roughly on a horizontal line in Fig. 8-17. The use of thin strained layers relaxes this constraint somewhat, as mentioned earlier.

In practice, the lattice spacing is set by the choice of substrate material. Binary compounds are normally used as substrates because they are much easier to produce, and the substrate accounts for most of the device volume. However, the choice of a binary compound restricts the choice of other materials. If the substrate is GaAs, you can only use mixtures with a similar lattice spacing. Look closely at the figure, and you will note that most compounds with spacing close to GaAs spacing are those with larger band gaps and output wavelengths shorter than the nominal 900 nm of GaAs. If you want longer wavelengths from a bulk InGaAsP laser, you must use InP as a substrate.

The substrate is important enough that developers often indicate it in describing the structure of semiconductor lasers. The composition of the active layer is first, then that of the substrate. For example, an InGaAsP/InP laser has an active layer of InGaAsP and a substrate of InP. Other deposited layers may have other compositions, generally ternary (three-element) compounds that are lattice-matched to the substrate and active layer. For quantum-well lasers, developers sometimes specify the composition of the confinement layer as well as that of the quantum well itself.

Developers long concentrated primarily on three families of semiconductor laser materials: GaAlAs, InGaAsP, and lead-salt materials. Recent laboratory demonstrations of blue and green laser diodes and LEDs have stimulated interest in other compounds. Table 8-2 lists the wavelength ranges of important semiconductor laser materials in commercial production and in development.

TABLE 8-2 Major semiconductor laser materials

Material	Wavelength Range	Comments
GaN/AlGaN	Blue–green	Candidate, but not yet demonstrated
ZnSSe	447–480 nm	Laboratory
ZnCdSe	490–525 nm	Laboratory
AlGaInP/GaAs*	620–680 nm	Commercial
$Ga_{0.5}In_{0.5}P$/GaAs*	670–680 nm	Commercial
GaAlAs/GaAs*	750–870 nm	Commercial; works at shorter wavelengths, but short-lived
GaAs/GaAs (pure)	904 nm	Commercial
$In_{0.2}Ga_{0.8}As$/GaAs*	980–1050 nm	Strained-layer; commercial
InGaAsP/InP*	1100–1650 nm	Commercial
PbCdS	2.7–4.2 µm	Requires cooling
PbSSe	4.2–8 µm	Requires cooling
PbSnTe	6.5–30 µm	Requires cooling
PbSnSe	8–30 µm	Requires cooling

*Capable of continuous-wave operation at room temperature over at least part of wavelength range

Blue and Green Diode Lasers

Intense laboratory development during 1991 and 1992 led to demonstrations of semiconductor lasers emitting between 447 and 525 nm, in the previously unreachable blue and green parts of the spectrum. At this writing, the technology is very young, and it remains to be seen if practical devices can be made. Nonetheless, the new lasers deserve at least brief mention.

The most extensive work so far has been with the zinc sulfide family of II-VI materials (which include compounds of cadmium, manganese, selenium, and tellurium). These materials have large band gaps, and like other semiconductors can be doped to form n- and p-type materials. However, the p-type materials suffer from very high resistance, and it has proven difficult to make good electrical contacts. As a result, the devices have high resistance, and the best ones require drive voltages in the 10-V range—well above the few volts needed to provide the band-gap energy.

The shortest wavelength at this writing is 447 nm, from a double-heterostructure ZnSe laser that emitted a continuous beam at 77 K, the temperature of liquid nitrogen. That device has not been operated at room temperature. Raising the temperature would shift output to longer wavelengths—about 470 nm. ZnCdSe lasers have operated at room temperature (although not for long) at 520 nm; when cooled to 77 K they emit at 495 nm.

Developers are also studying prospects for blue gallium-nitride semiconductor lasers. The compound has a direct band gap, and blue LEDs have been made from it, but at this writing no GaN or AlGaN semiconductor lasers have been reported.

We should note that the harmonic generation techniques described in Chapter 5 can be used with high-power semiconductor lasers to produce blue light. For example, frequency-doubling the 820-nm output of a high-power GaAlAs laser produces violet light at 410 nm. Likewise, 860-nm GaAlAs lasers can be frequency-doubled to 430 nm, and 980-nm InGaAs lasers can be doubled to 490 nm. The availability of high-power semiconductor lasers encourages harmonic generation, but the process is inefficient unless the beams are concentrated to high intensities in the nonlinear material.

Frequency-doubled diode lasers are beginning to reach the market and are being tested for some applications. They have a developmental head start on blue-green II-VI diode lasers in the zinc selenide family and may not suffer the same lifetime limitations that have discouraged some II-VI developers. However, harmonic generation is optically more complex.

You should be aware that the label "blue semiconductor laser" is sometimes used sloppily. Strictly speaking, it should be used only to describe semiconductor lasers that emit blue light directly. However, some companies have used it to describe diode lasers with output doubled into the blue.

Red Diode Lasers

The first diode lasers to generate a beam easily visible to the human eye, at 670 nm, reached the market in the late 1980s. Since then, red lasers have been developed at even shorter wavelengths, reaching about 630 nm in commercial products and considerably shorter wavelengths in the laboratory. They are part of a family of devices in which the active layer is an unusual family of quaternary compounds, AlGaInP, in which the number of aluminum, gallium, and indium atoms together equal the number of phosphorous atoms. (You could write this $Al_yGa_xIn_{1-x-y}P$.) They are grown on GaAs substrates.

First to be developed were lasers emitting at 670 nm with active layers of $Ga_{0.5}In_{0.5}P$. Surrounding those active layers were layers in which aluminum replaced some of the gallium to raise the band gap and thus improve confinement. Later, lasers were developed with active layers of AlGaInP, which emit at shorter wavelengths. Shorter wavelengths are important because at wavelengths longer than the 633-nm helium-neon line, the human eye's sensitivity drops by a factor of 10 for every increase in wavelength of 25 to 30 nm. Other materials also respond more strongly to light with shorter wavelengths.

Experimental AlGaInP lasers have operated at wavelengths shorter than the 633-nm helium-neon line. However, at this writing the shortest red wavelengths available from semiconductor lasers are about 630 to 635 nm. Somewhat shorter wavelengths may become available within the next few years.

In general, power levels are lower at shorter wavelengths than at longer ones. Powers of only a few milliwatts are available at the shortest wavelengths in the 630-nm range. Hundreds of milliwatts are available from monolithic AlGaInP laser arrays at 670 to 680 nm, but those powers cannot yet match the highest levels from GaAlAs lasers at near-infrared wavelengths.

Gallium-Arsenide Lasers

The best-developed family of diode lasers are those made from GaAlAs and GaAs. They usually are called simply "gallium-arsenide" lasers, because GaAs is the substrate material and the aluminum concentration is normally small.

The nominal wavelength of pure GaAs lasers is 904 nm, but adding aluminum gives shorter wavelengths. However, too much aluminum increases the likelihood of flaws that can quickly cause laser failure. The shortest wavelengths readily available from commercial GaAlAs lasers are about 750 nm. Shorter-wavelength GaAlAs lasers have been made in the laboratory, but they are short-lived.

The near-infrared wavelengths of GaAlAs lasers are adequate for many purposes but ideal for few. They can be used in applications including fiber-optic communications, pumping of solid-state lasers, reading optical disks, and writing in laser printers. However, other wavelengths would be better for many applications. Optical fibers have lower attenuation and broader bandwidth at longer wavelengths, which led to the development of InGaAsP lasers. Shorter wavelengths would allow smaller focal spots, and would work better with many light-sensitive materials. In addition, shorter wavelengths would be visible, letting diode lasers replace helium-neon lasers for jobs that require visible light.

The strength of GaAlAs laser technology is that the material is well-developed and comparatively easy to use. These are far from trivial advantages, because they permit production of both high-power and low-cost devices. All commercial high-power monolithic laser arrays are made of GaAlAs. The 780- to 870-nm output may not be ideal for all purposes, but it matches the absorption bands of neodymium lasers, and can be used for other applications where high power from a small laser is advantageous. Power levels are reaching levels where frequency-doubling may become attractive, leading to the generation of second-harmonic wavelengths of 400 to 435 nm.

Low manufacturing costs have been crucial to the development of consumer products using diode lasers. The most successful of these is the compact disc audio player, which uses a GaAlAs laser to play back digital sound from discs 12 cm (4-3/4 inches) in diameter. The least expensive players sell for under $100. Production of the 780-nm GaAlAs lasers used in CD players now exceeds 20 million a year, and they sell for as little as a few dollars each in quantity.

InGaAsP Lasers

The development of InGaAsP lasers was stimulated by the realization that the lowest loss of optical fibers was beyond the range of GaAlAs lasers at wavelengths longer than 1 μm. GaAlAs/GaAs is an unusually simple system because the addition of modest quantities of aluminum does not appreciably change the lattice constant, so GaAlAs lasers can be lattice-matched to GaAs substrates. In most other cases, addition of a third element to a binary semiconductor changes the lattice constant too much to make bulk lasers. This problem can be avoided by adding a fourth element, forming a quaternary semiconductor. InGaAsP was chosen because it offered the right wavelengths.

The characteristics of $In_xGa_{1-x}As_yP_{1-y}$ can be varied continuously by varying x and y independently. When lattice-matched to an InP substrate, InGaAsP lasers can emit 1100 to 1650 nm. Standard bulk laser designs use InGaAsP for the active layer, but other layers deposited on the InP substrate are often not the quaternary compound.

Most work has concentrated on two specific wavelengths—the

1310-nm wavelength where conventional single-mode optical fibers have their highest transmission bandwidth, and the 1550-nm wavelength where they have their lowest attenuation. Lasers emitting at 1310 nm were developed quickly, but the commercial development of 1550-nm lasers took more time because materials problems caused low production yields.

The need to precisely control concentrations of four elements makes InGaAsP a difficult material to produce, making costs—and prices—much higher than for GaAlAs lasers. Prices have dropped as manufacturing techniques have improved, but InGaAsP lasers remain high-performance devices for applications that demand them. That is particularly true for 1550-nm lasers, which are more costly than 1310-nm lasers.

Practical production limitations make the 1310- and 1550-nm wavelengths nominal values. The actual values of production lasers are typically specified only to within ± 20 or 30 nm, because it is hard to control composition precisely enough to make lasers with output at a specific wavelength.

Strained-Layer Compound Lasers

The development of strained-layer technology makes it possible to produce semiconductor lasers from some compounds that could not otherwise be grown on binary semiconductor substrates like GaAs or InP. The most important of these so far is InGaAs, which can generate wavelengths between 900 and 1100 nm that are not available from lattice-matched InGaAsP or GaAlAs lasers. Most commercial models emit the 980-nm wavelength used to pump erbium-doped fiber amplifiers.

Lead-Salt Diode Lasers

As we mentioned earlier, lead-salt diode lasers are not members of the III-V compound semiconductor family that includes GaAlAs and InGaAsP lasers. They emit light at longer infrared wavelengths than III-V lasers, as shown in Table 8-2, require cooling to low temperatures (generally under 100 K or –170° C), and are used in specialized scientific applications.

Lead-salt lasers have a different crystalline structure than III-V semiconductors. They also have smaller band gaps, under half an electronvolt. That small band gap leads to emission at much longer wavelengths, beyond 2.7 µm. The size of the band gap depends on the semiconductor composition and, like that of III-V compounds, can be varied continuously by adjusting the relative concentrations of two elements in a ternary compound. For example, changing the composition of $PbS_{1-x}Se_x$ from PbS ($x = 0$) to PbSe ($x = 1$) changes the emission wavelength from 4.2 to 8 µm. The longer the wavelength, the lower the temperature to which the laser must be cooled.

The only practical use of lead-salt diode lasers is in infrared spectroscopy, the measurement of how materials behave when illuminated with infrared light. Infrared measurements can be made with other light sources, but the diode laser emits a narrower range of wavelengths. The nominal wavelength of an individual laser is set when it is made, but the emitted wavelength can be adjusted somewhat around this value by changing the pressure (which affects refractive index), magnetic field, and temperature.

SPECIALIZATION OF DIODE LASERS

An unpackaged diode laser is a tiny chip of semiconductor barely visible to the human eye. The bare laser must be packaged for practical applications. Some are packaged for general-purpose use, often in metal housings similar to standard transistor cans but with windows for optical output. However, most are assembled in packages designed for specific applications.

Optical Disc Lasers

The vast majority of GaAlAs semiconductor lasers are packaged to play compact disc audio recordings. The laser is mounted directly in a playback head that includes focusing optics and a light detector.

Because of the economies of scale in semiconductor production, identical diode-laser chips may be packaged in other ways for other applications where low cost is more important than matching laser characteristics specifically to the application. For example, compact disc lasers are packaged in light pens and used in some inexpensive fiber-optic transmitters. Semiconductor lasers in compact disc packages may also be used for other applications because their low cost offsets any performance limitations.

Fiber-Optic Lasers

Both GaAlAs and InGaAsP diode lasers are used in fiber-optic systems. The short-wavelength GaAlAs lasers generally cost less, but fiber losses are higher at their operating wavelength, so they are normally used only in short-distance transmission. Longer-wavelength but more expensive InGaAsP lasers are used when transmission distance is more important. For less demanding applications where data rates are modest (typically 10 Mb/s or less) and distances are short, LEDs of either GaAlAs or InGaAsP can be used in transmitters at significant cost savings.

Laser packaging is a prime consideration in, and a major contributor to the cost of transmitters for, single-mode fiber systems. The laser's light-emitting area must be coupled to the light-carrying core of a single-mode fiber. The problem is that the fiber cores are only 8 to 10 μm in diameter and the laser's light-emitting areas are even smaller, imposing tight tolerances on alignment. The laser can be packaged in a housing that mates with a fiber-optic connector, or in housing that includes a fiber-optic "pigtail" that can be spliced into the fiber-optic system.

Developers are working on ways to reduce costs of fiber-optic lasers by automating packaging. This may reduce performance as measured on some scales, such as output power emerging from the connector or fiber pigtail. However, developers hope that reducing the cost can increase the market greatly, and make it possible for fiber-optic connections to reach all the way to homes.

High-Power Lasers

Another fast-developing area of diode-laser technology is high-power monolithic arrays. As we mentioned earlier, the diode array is a large-area chip containing many parallel laser stripes. These lasers are all driven by the same source. Although individual stripes have been modulated separately in some laboratory versions, the prime emphasis now is on generating power, without worrying too much about modulation speed or beam quality. The major application of these laser arrays so far has been in pumping neodymium-YAG lasers, as mentioned in Chapter 7, for which neither beam quality nor modulation speed are critical.

Heat dissipation remains a concern for diode-laser arrays, despite their high efficiency. Thus, such lasers are packaged with heat sinks to remove excess heat before it destroys the laser. This cooling requirement could impose a limit on power levels available from diode lasers.

WHAT HAVE WE LEARNED

- Semiconductor lasers emit light when current flows through a junction of differently doped materials.
- Most outer-shell electrons in semiconductors are in low-energy valence bands that form bonds between adjacent atoms. Some outer-shell electrons lie in higher-energy conduction bands and are free to conduct current in the crystal.
- N-type semiconductors are doped with elements that release electrons to carry current. P-type semiconductors are doped with elements that form holes as current carriers.
- III-V compounds such as gallium arsenide are called compound semiconductors. Laser properties such as wavelength depend on their composition.
- Ternary semiconductors contain three elements, such as $Ga_{1-x}Al_xAs$. Quaternary semiconductors contain four elements, such as $In_{1-x}Ga_xAs_{1-y}P_y$.
- Semiconductors absorb photons with more than the band-gap energy, and transmit photons with less energy. Emission wavelengths of semiconductors depend on band-gap energy.
- Light emission in a semiconductor laser takes place at p-n junctions, where free electrons fall into "holes," eliminating both current carriers and emitting light energy by a process called "recombination."
- If a positive bias is applied to the n material and a negative bias to the p material, no current flows across the junction.
- A semiconductor junction carries current if a positive voltage is applied to the p material and a negative voltage to the n material.
- A photon with energy greater than the band gap can create an electron-hole carrier pair, an effect useful in detecting light.

- LEDs are forward-biased semiconductor diodes that generate light by recombination. Only some semiconductor materials work as LEDs.
- Semiconductor lasers produce stimulated emission. They operate at higher currents than LEDs and concentrate emission in the junction plane.
- Stimulated emission is concentrated in the junction plane of most semiconductor lasers by feedback from cleaved ends of the chip.
- Only direct-band-gap semiconductors can operate as lasers.
- Complex internal structures are needed for continuous-wave room-temperature diode lasers.
- Normally, laser action is concentrated in a narrow stripe on the active layer.
- Quantum wells trap current carriers in a thin layer with a smaller band gap than the surrounding layers, enhancing laser operation.
- Very thin layers can accommodate more strain than thicker layers, reducing the need to precisely match the atomic spacing of different layers.
- Arrays of many laser stripes in parallel on one chip can produce high power output.
- Distributed-feedback lasers have very narrow emission bandwidth.
- Most diode laser beams are oval in cross section and diverge very rapidly.
- The emitting stripe of an 800-nm GaAlAs laser is about 10 wavelengths wide and a quarter-wavelength thick.
- Horizontal cavity lasers emit from the edge of the semiconductor chip. Vertical cavity lasers emit from the surface.
- External-cavity diode lasers can be tuned in output wavelength and can have a narrow linewidth.
- Diode lasers have short coherence lengths unless they operate in a single longitudinal mode.
- Diode lasers are simple to modulate by changing their drive current.
- ZnSe lasers and related types emit green and blue light. Frequency-doubling of GaAlAs lasers also produces blue light.

- AlGaInP lasers emit visible red light.
- GaAlAs is the best-developed and highest-power diode-laser material. Wavelengths are 750 to 870 nm.
- InGaAsP lasers emit at 1100 to 1650 nm and are used mostly in fiber-optic communications.
- Lead-salt lasers emit at wavelengths longer than 2.7 μm and require cryogenic cooling. Their major uses are in research.
- Packaging is critical to the applications of semiconductor lasers.

WHAT'S NEXT

In Chapter 9 we will talk about types of lasers that didn't fall into our major categories: tunable organic-dye lasers, free-electron lasers, and X-ray lasers.

Quiz for Chapter 8

1. What type of semiconductor is doped with impurities that create holes as current carriers?
   a. Intrinsic
   b. n-type
   c. p-type
   d. Insulating
   e. None of the above

2. Which of the following is a quaternary III-V semiconductor?
   a. InGaAsP
   b. PbSnSSe
   c. GaAlAs
   d. GaAs
   e. NSbAsP

3. A semiconductor has a band gap of 1.5 eV. At about what wavelength will it emit light if it can operate as a laser?
   a. 1500 nm
   b. 1000 nm
   c. 827 nm
   d. 678 nm
   e. 667 nm

4. What type of semiconductor junction can function as a laser?
   a. Unbiased junction
   b. Forward-biased junction
   c. Reverse-biased junction
   d. All of the above
   e. None of the above

5. How does a semiconductor laser operate when the drive current is below laser threshold?
   a. As a reverse-biased diode

b. As a photodetector
c. As an LED
d. As a perfect insulator
e. None of the above

6. What gives a double-heterostructure laser better efficiency than a homostructure laser?
   a. Reverse biasing
   b. Better confinement of stimulated emission in the active layer
   c. Restriction of current flow to the active layer
   d. Lower levels of spontaneous emission
   e. A homostructure laser is more efficient

7. Which type of semiconductor lasers emit a single longitudinal mode?
   a. Homojunction
   b. Buried-heterostructure
   c. Monolithic arrays
   d. Distributed-feedback
   e. Those built for compact disc players

8. If you want a distributed-feedback laser to emit light at 1000 nm, and the semiconductor has a refractive index of 3.2 at that wavelength, what should the spacing be for a first-order grating ($m = 1$)?
   a. 100 nm
   b. 156 nm
   c. 250 nm
   d. 500 nm
   e. 617 nm

9. Which family of semiconductor lasers has produced the highest power levels?
   a. GaAlAs
   b. InGaAs
   c. InGaAsP
   d. Lead-salt
   e. GaInP

10. Which family of semiconductor lasers has produced the shortest wavelength (in a device able to operate continuously at room temperature)?
    a. GaAlAs
    b. InGaAs
    c. InGaAsP
    d. Lead-salt
    e. AlGaInP

CHAPTER 9

Other Lasers

ABOUT THIS CHAPTER

Not all lasers fit neatly into the three categories of gas, solid-state, and semiconductor. In this chapter, we will learn about three important types that are categories in themselves. One, the dye laser, is an important tool in scientific research and is finding growing uses in medicine. The other two, the free-electron laser and the X-ray laser, remain in development. At the end we will briefly mention a few other lasers.

TUNABLE DYE LASERS

Most lasers we have learned about emit light at fixed wavelengths, except for the vibronic solid-state lasers covered in Chapter 7 and a few tunable diode lasers described in Chapter 8. However, for many applications it would be helpful to be able to change or tune the laser wavelength. In some cases, the goal is to produce a precise wavelength not available from other lasers. In others, the goal is to scan a range of wavelengths. This premium on tunability led to the development of the organic dye laser, which is widely used in research and has found some medical applications, although it is being challenged in some areas by tunable solid-state lasers.

The active media in dye lasers are organic dyes, a family of large and complex molecules. They are called dyes because they are brightly colored, a property that comes from their complex sets of electronic and vibrational energy levels. The vibrational energy levels create many sublevels of the electronic states, as in solid-state vibronic lasers. Thus, transitions between those energy levels can occur at an unusually broad range of wavelengths, allowing broadband laser oscillation. Different dye molecules can function as laser media over different ranges of wavelengths, as we will explain in more detail below.

Laser dyes are normally dissolved in liquid solvents, which serve as hosts for the active medium just as glasses or crystals are hosts for the active media in solid-state lasers. Dyes have produced laser light in both the vapor phase and when doped in transparent solids, but liquid dye lasers have been far more successful. Dye lasers operate throughout the visible, near-infrared, and near-ultraviolet, but no single dye has that broad a tuning range. To cover the entire visible spectrum, you must switch among several dyes, each with a limited tuning range.

Applications Influence Dye-Laser Design

With comparatively high gain, the dye laser is a versatile and flexible tool. Its performance can be adjusted in many ways, creating variations on the basic dye-laser theme. However, this versatility comes at the expense of cost, complexity, and durability.

Dye lasers can generate pulsed or continuous beams at wavelengths in the visible, near-infrared, and near-ultraviolet that are not available from other lasers. However, if another laser will do the job, you should probably use it. Dye lasers are fine for their special purposes, but it makes no more sense to use one to replace a helium-neon laser than to buy an expensive all-band short-wave radio to listen to a local AM radio station.

Because of its versatility, the dye laser has been adapted to meet specific needs. It is fair to say that the technology has been shaped by its applications. Before we examine dye lasers in more detail, let's look at their major uses.

Tunable Wavelength Output

The major application of dye lasers has long been to generate light that is tunable across the visible spectrum. Initially, virtually all dye lasers were used for spectroscopy, the measurement of the spectral lines characteristic of various materials. Now other applications have emerged, especially in medicine, where specific wavelengths are needed for diagnostics and therapy. However, more dye lasers continue to be used in research and development.

Most of these applications require laser emission that can be adjusted in wavelength, often across much of the visible spectrum or into the near-infrared or near-ultraviolet. This means that the laser cavity must include wavelength-changing optics. The desired tuning ranges usually span more than the range of a single dye, so the design must allow for the switching of laser dyes.

Research lasers have some added requirements. They must let users adjust many characteristics, including power levels and spectral bandwidth, to match experimental requirements that can easily change. Many researchers are especially concerned with narrow bandwidth and stability. Some prefer continuous-wave emission; others prefer pulsed operation.

Wavelength adjustment is also important in medicine, but medical systems have different requirements. The laser wavelength must be matched to the requirements of specific procedures, such as bleaching birthmarks or shattering kidney stones. This may be done by factory adjustment of the laser so that it emits only the specific wavelengths sought for the application. Physicians don't need the same flexibility as research scientists, but they do need instruments that are reliable and easy to use.

Short-Pulse Research

Another major use of dye lasers is in research on ultra-short light pulses. The attraction of dye lasers for this application is that their unusually broad gain bandwidth lets them generate exceptionally short pulses. (Titanium-sapphire lasers have broader gain bandwidth than individual dyes, and are rapidly becoming important in short-pulse research, but their output is in the near-infrared.) As we learned earlier, there is a fundamental relationship between the minimum duration of a light pulse and its maximum spectral bandwidth:

$$\text{Pulse Length} = 0.441/\text{Bandwidth}$$

where pulse length is measured in seconds and bandwidth in hertz. Thus, if you want 1-ns (10^{-9} s) pulses, the spectral bandwidth must be at least 441 MHz. If you want picosecond (10^{-12} s) pulses, the bandwidth must be at least 441 GHz. This limit applies to all pulses, whether directly generated by a laser or modified outside the laser.

This transform limit pushes short-pulse researchers to use the laser with the broadest possible bandwidth. Dye lasers have long been used, and some commercial models are designed expressly to produce such short pulses. The same limitation causes researchers who want extremely narrow spectral bandwidth to use continuous-wave lasers.

Basic Structure of Dye Lasers

Figure 9-1 shows the internal structure of a generalized dye laser. Short-pulse lasers usually lack tuning optics, which, as we saw in Chapter 4, is a prism or diffraction grating that deflects light of different wavelengths at different angles. The tuning optics and cavity mirrors are arranged so that only one wavelength can oscillate in the laser cavity. Other wavelengths are deflected from the cavity axis and are not amplified, even if they fall within the laser's gain bandwidth.

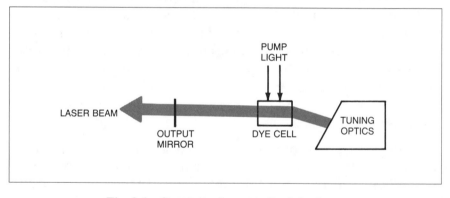

Fig. 9-1   Structure of a generalized dye laser.

Laser dyes get their energy by absorbing light at wavelengths shorter than they emit. The bright light needed to produce a population inversion can come from various pump sources, including flashlamps and pulsed or continuous-wave lasers. As we will see later in this chapter, these light sources differ considerably, so this illustration is very general.

The solution containing the laser dye is placed in the laser cavity. Our generalized picture shows a dye cell, without indicating whether the cell is sealed or the liquid flows through it. In pulsed lasers with modest average power, the dye cell may be a sealed container called a "cuvette." At higher powers, the dye solution may be pumped rapidly through the cell to prevent heating and degradation. In continuous-wave lasers, the dye may pass through the tightly focused pump-laser beam in an unconfined jet. (This avoids optical damage to the windows of the dye cell.) Laser dyes have high gain, so you do not need a long length of dye in the laser cavity, but they do require high pump intensity, so the beam must be tightly focused or pulsed to provide high peak power.

Most dye lasers have room for accessories inside and outside the cavity, including modelockers to reduce pulse duration and optics to narrow the spectral linewidth. We don't show them here, but such accessories can play a vital role in operation of dye lasers.

In talking about dye lasers, we may make some functional distinctions that are not apparent in some commercial dye lasers. For example, the pump source in flashlamp-pumped dye lasers is built into the laser itself. Also, many dye lasers are sold with sets of accessories, and might be packaged as picosecond dye lasers or other specialized types. You should remember that these are packaging choices, not physical requirements.

Following our description of the functional elements of dye lasers, we will look in more detail at arrangements for tuning laser wavelength, the dyes themselves, and the various types of optical pumping.

Tuning and Cavity Design

Most dye-laser cavities are designed to allow wavelength tuning, except those in short-pulse or fixed-wavelength lasers. Figure 9-2 shows three basic approaches to tuning dye-laser wavelength, with a prism, a diffraction grating, or other wavelength-selective elements such as tunable etalons or filters. Actual laser pump cavities often have extra mirrors to direct pump and laser beams between components.

Prisms and diffraction gratings disperse light at different angles as a function of wavelength, so wavelength is tuned by turning either these elements or some other optical components. In Fig. 9-2A, the grating is turned to change the wavelength, but in Fig. 9-2B, the prism stays fixed while the rear cavity mirror moves to select a particular wavelength refracted by the prism.

In the third example, Fig. 9-2C, the main tuning element is a wedge-shaped etalon. As we saw in Chapter 5, an etalon functions like a miniature optical cavity to limit oscillation to a narrow range of

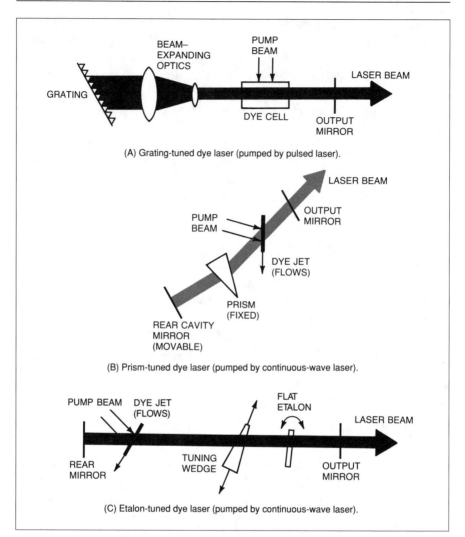

Fig. 9-2 Three examples of wavelength-tunable laser cavities used with dye lasers.

wavelengths within the laser's gain bandwidth. In a tunable etalon, the distance between the two reflective surfaces is changed to select the oscillation wavelength. Normally, the etalon is a flat or wedge-shaped chunk of glass, and tuning is done by moving it so that the beam passes through a different part of it in the laser cavity. A wedge-shaped etalon can be moved up and down in the laser beam so the light travels through different thicknesses—and thus sees different spacings of the two reflective surfaces. An alternative is the fine-tuning flat etalon (at the right in Fig. 9-2C), which is rotated to change the distance that the laser beam travels between the surfaces.

Another approach to tuning, not shown, is the use of a bire-fringent filter. It is placed in the laser cavity and turned to change the wavelength that can oscillate.

Most laser dyes have gain bandwidths from 10 to 70 nm. Even without a tuning element, the natural line-narrowing inherent in laser oscillation limits linewidth to a few nanometers. Simple tuning prisms and gratings limit linewidth to a much smaller range, several thou-sandths of a nanometer. Addition of an etalon can limit oscillation to an even narrower range of wavelengths. With active control systems and sophisticated additional optics, the linewidth of a continuous-wave dye laser can be reduced to well below 1 MHz—less than one part in 10^9 at visible wavelengths, or under 5×10^{-7} nm.

The designs of dye-laser cavities are heavily influenced by the need to tune their output wavelength. This takes dye lasers beyond the standard straight-line cavity, where two mirrors are on opposite ends of a laser rod or tube. We simplified the cavities in Fig. 9-2 to make it clear how they worked, but most dye-laser cavities include extra mirrors for focusing and beam direction because the cavity components are not in a straight line. Even more complex arrange-ments are sometimes used to optimize performance, such as the ring cavity shown in Fig. 9-3. At first glance, you might think light could pass in either direction around the ring, but this design contains com-ponents that prevent oscillation in one direction. Again, this drawing has been simplified by removing some components used in commer-cial lasers. The ring cavity limits laser oscillation to a narrower range of wavelengths than other cavities; it can also be used in short-pulse lasers.

As we indicated earlier, organic dyes have high gain under proper pumping conditions, so the cavity contains only a short length of active medium. The high gain of dyes makes them suitable for use in oscillator-amplifier configurations when high power is required.

We should note that the tuning techniques and cavity designs we have discussed here can be used with any type of tunable laser. The cavities used with tunable solid-state lasers are quite similar, and some dye lasers can operate with a titanium-sapphire replacing the dye cell.

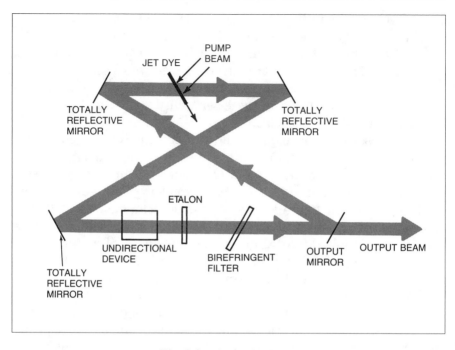

Fig. 9-3 A ring dye laser.

Properties of Organic Dyes

The organic dyes used in lasers are large and complex molecules with multiple rings and complex spectra. The dye molecules are fluorescent, meaning that they can absorb a short-wavelength photon and almost immediately "fluoresce," or emit light at a longer wavelength. This fluorescence occurs without a population inversion. Light from an external source excites the molecule, which then drops back down the energy-level ladder, emitting light to release energy. This is the process by which minerals and other materials fluoresce—emitting visible light when illuminated by shorter-wavelength ultraviolet light.

In a dye laser, the molecules are excited by a light bright enough to produce a population inversion on the laser transition. Lifetimes of the upper laser levels are typically a few nanoseconds, giving them high gain but low energy storage. Laser emission is on electronic transitions with vibrational sublevels, as shown in Fig. 9-4. We have seen similar effects in other lasers, like the carbon-dioxide gas lasers, where rotational sublevels of vibrational transitions create many distinct lines. In organic dyes, the levels are so closely spaced that they

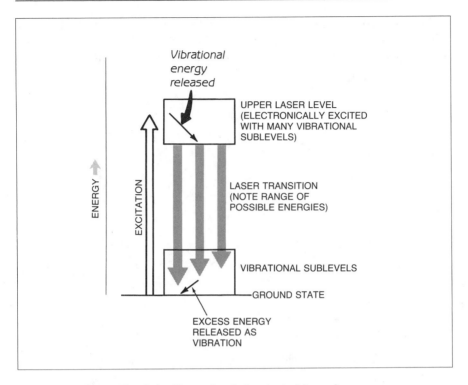

Fig. 9-4   Energy levels in a typical laser dye.

merge together to form a continuum. Instead of tuning a dye laser's wavelength in steps, you can adjust it continuously.

Light excites a dye laser, and much of the absorbed energy ultimately emerges as light. After dye molecules are excited from their ground state, they drop down to the lowest vibrational sublevel of the electronically excited state. The laser transition takes the molecule to a vibrationally excited level of the electronic ground state. The remaining energy is dissipated as heat. As you can see in Fig. 9-4, the laser transition can be to various levels within a range defined by the vibrationally excited sublevels of the ground state, giving it broad gain bandwidth.

The wavelengths emitted by dyes depend on their chemical composition. Most dyes belong to several families of organic compounds, and seemingly countless variations are possible just by adding an atom or two to different places on the molecule. Developing laser dyes is a serious exercise in organic chemistry, and you don't want to worry about the chemical names or the details. Figure 9-5 gives you an idea of what you want to miss—it's the basic structure of the coumarin family of laser dyes. The question marks indicate where different atoms can be added to make new molecules. There are about 100 coumarin dyes.

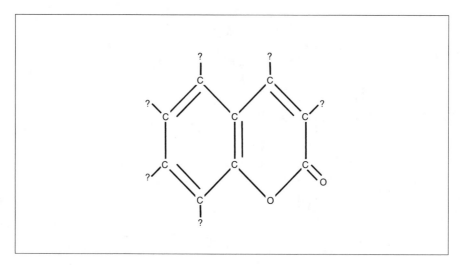

Fig. 9-5   Basic chemical structure of coumarin laser dyes.

Most laser dyes have gain bandwidths of 10 to 70 nm, with typical ranges of 20 to 40 nm. Gain is highest at the center of the range, and drops at longer and shorter wavelengths. Although individual dyes fall far short of covering the entire visible range, groups of dyes can combine to provide continuous coverage. Once you move to the edge of the tuning range of one dye, you switch to another dye. In practice, single dyes have enough range for many purposes, but some lasers are made with groups of dye cells that can be switched into position successively, one after another, to give a broader tuning range.

Output wavelengths depend on the wavelength, power, and pulse characteristics of the pump source. Often the differences are small, but certain pump sources can work much better than others for specific dyes. For example, many dyes will not lase when driven by a continuous-wave laser, but can lase when pumped by a pulsed light source.

Organic dye molecules are more fragile than other laser media, and most degrade after tens or hundreds of hours of use. The bright light needed to excite them can also break up the molecules. This makes it important to keep dye solutions cool, especially if the laser is operated at high power. Like other chemical reactions, dye decomposition occurs faster at warmer temperatures. For this reason, all but the lowest-power dye lasers normally have systems that pump the dye solution through the light-emitting zone. In continuous-wave lasers, the dye flows through in a continuous jet.

Like other complex organic molecules, most dyes do not dissolve readily in water. Dye solutions normally use organic solvents like methanol, ethanol, and other liquids that may be both toxic and flammable. The dyes themselves can be highly toxic or carcinogenic, and most dye solutions are classed as hazardous waste.

Flashlamp-Pumped Dye Lasers

Many laser dyes can be pumped by intense pulses of broadband "white" light from flashlamps. There are two basic types of flashlamp-pumped dye lasers. In Fig. 9-6A, a linear flashlamp transfers energy to a parallel linear dye cell, much like a linear flashlamp pumps a solid-state laser rod. In Fig. 9-6B, the dye flows through the center of a coaxial flashlamp, which transfers its energy directly to the dye inside. For simplicity, the dye flow system is not shown here.

Pulse length of a flashlamp-pumped dye laser depends on the duration of the flashlamp pulses, which are typically about 1 μs, but last up to 500 μs. Flashlamp-pumped dye lasers can produce a few powerful pulses a minute, a few hundred low-energy pulses per second, or some intermediate value.

The main advantages of the flashlamp-pumped dye laser are its ability to produce pulses with higher energy and higher average power than other dye lasers and its modest capital cost. It is used in research and for medical treatment.

Pulsed Laser Pumping

Pulsed lasers can also pump dye lasers. They offer high peak power at specific wavelengths—which can be matched to dye absorption bands. Another advantage is the ability to split the pump beam between an oscillator and an amplifier stage, as shown in Fig. 9-7, to generate higher dye laser power than is possible from an oscillator alone. The dye beam need not be at right angles to the pump beam, as in Fig. 9-7, but pulsed lasers normally do not pump along the length of the dye laser cavity.

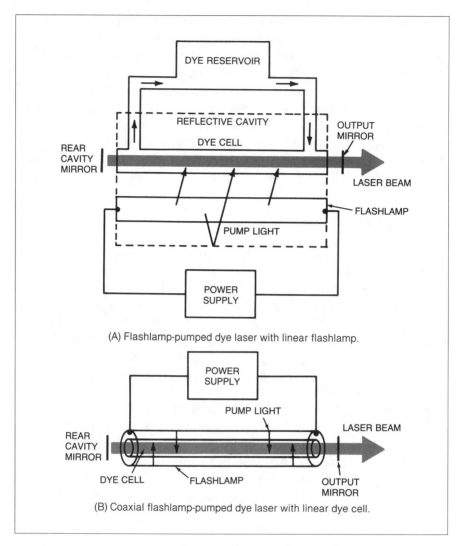

Fig. 9-6   Linear and coaxial flashlamp-pumped dye lasers.

   Although many pulsed lasers are available, the choices for pumping dyes are limited by a fact we mentioned before: the pump wavelength must be shorter than the dye emission wavelength. Because most dye lasers are operated at visible wavelengths, this limits pump wavelengths to the ultraviolet and the short-wavelength end of the visible spectrum. Table 9-1 lists the major pulsed pump lasers and their wavelengths. Conceptually, they work in similar ways, but there are important differences, as described next.

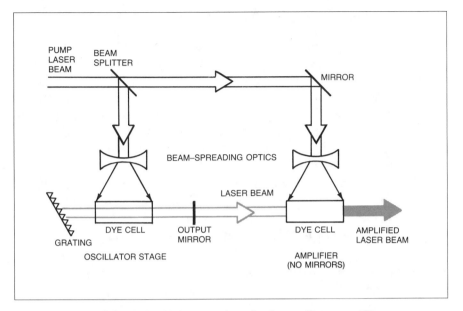

Fig. 9-7   Pulsed laser pumping of a dye oscillator-amplifier.

TABLE 9-1 Pulsed pump lasers for dyes and their wavelengths

Laser	Wavelength
Krypton-fluoride excimer	249 nm
Frequency-quadrupled neodymium	266 nm
Xenon-chloride excimer	308 nm
Nitrogen	337 nm
Xenon-fluoride excimer	351 nm
Frequency-tripled neodymium	355 nm
Copper-vapor	510 and 578 nm
Frequency-doubled neodymium	532 nm

Excimer Laser Pumping

Excimer laser pumps can generate high-power ultraviolet pulses at reasonable repetition rates. Their wavelengths are short enough to pump dyes throughout the visible spectrum, but shorter excimer wavelengths tend to shorten dye lifetimes. Capital costs are higher than for flashlamp-pumped lasers, but excimer lasers are more versatile and easier to operate.

Pulsed Neodymium Laser Pumping

The 1.06-μm fundamental wavelength of neodymium lasers is too long to pump most dyes, except a handful that generate longer infrared wavelengths. However, the second, third, and fourth harmonics have wavelengths short enough to pump visible dyes. The 532-nm green second harmonic can pump dyes at wavelengths as short as 539 nm, and the third harmonic can cover the rest of the visible and part of the ultraviolet. The fourth harmonic is used only to pump a few short-wavelength ultraviolet dyes.

The functional characteristics of pulsed neodymium-laser pumps and excimer pump lasers are similar. Both have similar repetition rates and output powers. Both have higher capital costs and lower operating costs than flashlamp-pumped dye lasers, and both can be used for other purposes than pumping dyes. However, as we saw in Chapter 7, neodymium lasers can emit continuous beams, while excimer lasers cannot.

Nitrogen Laser Pumping

The 337-nm nitrogen laser was the first pulsed ultraviolet laser used to pump dye lasers. However, its pulse energy is limited, and it fell out of favor for many applications after the introduction of more powerful excimer and pulsed neodymium laser pumps. Yet nitrogen lasers have not been forgotten. The simplicity and low cost of small nitrogen lasers has led to the introduction of pulsed dye lasers pumped with small nitrogen lasers. A complete package with pump laser and dye costs only a few thousand dollars, making it attractive for use in instruments or in laboratories with very limited budgets.

Copper-Vapor Laser Pumping

The repetitively pulsed green and yellow lines of the copper-vapor laser can effectively pump dyes at wavelengths beyond about 530 nm. The rapid sequence of pulses can generate high average power less expensively than an argon laser can generate high continuous-wave powers, but the repetitive pulses are not suited for all applications.

Continuous-Wave Laser Pumping

Dye lasers can generate continuous-wave beams when pumped by continuous-wave lasers—argon, krypton, or neodymium. Designs of continuous-wave dye lasers differ markedly from those of pulsed dye lasers. Because the available pump power is modest, the pump beam must be tightly focused to generate a high power in a very small area. It pumps a thin flowing jet of dye solution, which is not housed in a tube that could suffer optical damage from the concentrated pump laser power. The rapid flow of the dye solution also avoids heating to excessive temperatures.

We showed two simple designs for continuous-wave dye lasers in Fig. 9-2. We should warn you that those are simple designs, without the "bells and whistles" that give extremely narrow linewidth in high-performance models. Even the more elaborate ring laser design in Fig. 9-3 is a simplified version of the types used in commercial lasers.

Some measurement techniques require continuous-wave beams. Another important attraction of continuous-wave dye lasers is their potential to emit a very narrow range of frequencies for ultra-precise measurements, because of the transform limit described earlier:

$$\text{Bandwidth} = 0.441/\text{Pulse Length}$$

where bandwidth is in hertz and pulse length in seconds. Thus, the longer the laser emits light, the narrower the minimum bandwidth. Linewidths below 1 MHz (about one part per billion in frequency) are possible in commercial continuous-wave dye lasers, and much lower figures have been recorded in the laboratory.

Ultra-Short Pulses

We also saw, in Chapter 4, that broad linewidth makes it possible to generate ultra-short pulses. Although many dyes have gain bandwidths larger than 10 nm, the dynamics of laser emission limit emission bandwidth to a much narrower range, no more than a few nanometers and usually less. Modelocked lasers can take advantage of that bandwidth to generate pulses in the picosecond (10^{-12} s) range or below, but the shortest pulses directly produced from dye lasers are limited to about 100 femtoseconds (10^{-13} s). Only titanium-sapphire lasers, which have broader bandwidth than dyes, can generate shorter pulses directly.

Special techniques can squeeze pulses from a dye laser down to even shorter durations if they spread the light over a broader range of wavelengths than in the dye laser's output. The approach that has produced the shortest pulses is to pass very short pulses from a dye laser through a short length of optical fiber or other material, which spreads them over a wider range of wavelengths. Then the pulses are reflected back and forth between gratings or prisms to squeeze them down in time. This has yielded pulses as short as 6 femtoseconds $(6 \times 10^{-15} \text{ s})$.

Harmonic Generation

Nonlinear crystals can generate harmonics of dye laser light, just as they do for other lasers. Dye-laser harmonic wavelengths are tunable, as is the fundamental wavelength. Normally, harmonics are generated with pulsed lasers because they produce the high power levels needed for efficient harmonic generation. However, the efficiency of harmonic generation is limited, so it is normally used only to generate wavelengths that dyes cannot produce efficiently; in practice, this means ultraviolet wavelengths.

FREE-ELECTRON LASERS

The free-electron laser differs in many ways from the other types of lasers we have learned about. The active medium is a beam of "free" electrons—unattached to any atom—which is passing through a special type of magnetic field that varies periodically in space. The easiest way to view the magnetic field is by assuming it was created by an array of permanent magnets arranged with alternating polarity, as shown in Fig. 9-8. If the north pole of the first magnet is above the electron beam, as shown in the figure, the south pole of the second magnet must be above the electron beam, then the north pole of the third, and so on.

The electron beam enters the magnet array at a slight angle in Fig. 9-8. The magnetic field from the first magnet bends the electron beam in one direction, and then the opposite-polarity field from the second magnet bends it back in the other direction. This process repeats until the electron beam passes out the other end of the magnet array, which is called a "wiggler" or "undulator" because of its effect on the electron beam.

Electrons wiggling through the wiggler magnet field release some energy as light. Much of that light energy is, in essence, reabsorbed

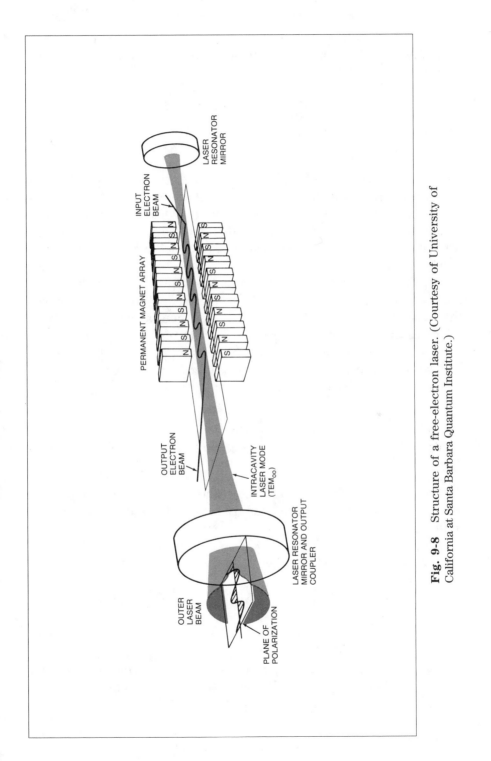

Fig. 9-8 Structure of a free-electron laser. (Courtesy of University of California at Santa Barbara Quantum Institute.)

when the electrons wiggle back the other way. However, light can be amplified if its wavelength λ meets an approximate resonance criterion:

$$\lambda = \frac{p}{2\left(1 - \frac{v^2}{c^2}\right)}$$

where p is the length of a wiggler-magnet "period" (north-to-south-to-north), v is the electron velocity (along the axis of the laser), and c is the speed of light in vacuum. If that $(1 - [v^2/c^2])$ reminds you of formulas you've seen describing the theory of special relativity, there's a reason. Electrons move so fast in a free-electron laser that their behavior is described in relativistic terms.

While that version of the formula is helpful in highlighting the basic physics of a free-electron laser, a more practical version uses the ratio of the accelerated electron's mass and acceleration energy (E, measured in million electronvolts) to its rest mass, 0.511 million electronvolts. The acceleration energy is important because it is the normal way to measure the output of electron-beam generators. With a bit of simplification, the formula becomes

$$\lambda = \frac{0.131p}{\left(0.511 + E\right)^2}$$

As the formula indicates, the free-electron laser wavelength gets shorter as the magnet period decreases and the electron energy increases.

Tunability of Free-Electron Lasers

We have already said that electrons in a free-electron laser are not bound to atoms and thus don't have fixed energy levels. Look carefully at our formula for free-electron laser wavelength, and you can see that there are two ways to adjust it—by changing the wiggler-magnet period, or by changing the electron energy (which depends on velocity). In practice, it's usually easier to change the electron energy than the structure of the wiggler field, but the important fact is that free-electron lasers are inherently tunable in wavelength.

Even more exciting is the fact that the wavelength range is, in principle, extremely broad, ranging from the microwave region at the long-wave end to soft X-rays on the short-wave end. This might make it seem as if you need only turn a knob to adjust electron energy and get the whole range of wavelengths. That is not the case, because practical design constraints limit any one free-electron laser to a much more limited range of wavelengths. Nonetheless, that range is broad compared with other tunable lasers.

Types of Wigglers

Figure 9-8 shows the wiggler as a stack of magnets of alternating polarity. This is the simplest way to view a wiggler, but it can be misleading. The key element of a wiggler is not the magnet *per se*, but the magnetic field it generates. It is the magnetic field that bends the paths of the electrons passing through it. The field direction passes through a complete cycle as you move from north-to-south to south-to-north magnets, then back to a north-to-south magnet. So you can also view the magnetic field as a sinusoidal wave, which oscillates as you move along the wiggler.

This has led to another approach to wiggler design—using a strong electromagnetic wave to provide the sinusoidally varying magnetic field. This may not seem as straightforward as using a magnet array, but it has its own advantages. Recall that free-electron laser wavelength decreases with the period of the wiggler magnet. If the period becomes too small, it gets hard to build arrays of physical magnets. However, electromagnetic waves can easily generate those short periods, making it possible to generate shorter free-electron laser wavelengths without the need for more energetic electrons (which, as we will soon see, require more powerful accelerators).

Electromagnetic-field wigglers also open up some intriguing ways to bootstrap free-electron lasers to shorter wavelengths. Present free-electron lasers can generate more power at longer wavelengths than at shorter wavelengths. Suppose you start with a permanent magnet wiggler that generates a powerful beam at 100 μm in the far-infrared, then use the magnetic field of that laser beam as the

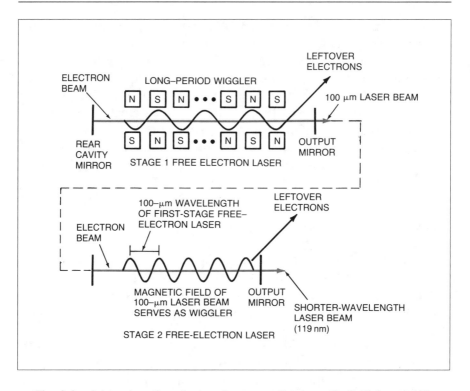

Fig. 9-9 A two-stage free-electron laser uses the magnetic field from a 100-µm free-electron laser as the wiggler of a second stage that generates a 119-nm beam. (The 100-µm beam is bent for artistic convenience—the lasers wouldn't fit on the page in a straight line.)

"wiggler field" to generate a shorter-wavelength laser beam. We show this idea in Fig. 9-9. With a comparatively modest electron energy of 10,000,000 eV, you could generate a second-stage free-electron laser wavelength of 119 nm in the ultraviolet—nearly a factor of 1000 shorter than the first-stage wavelength. Two-stage free-electron lasers are not that simple in practice, but they nonetheless offer an intriguing way to produce short-wavelength tunable laser beams.

In drawing our simple picture of the free-electron laser, we assumed that the magnetic field had uniform amplitude along the entire wiggler. This case is the simplest to analyze, but not necessarily the best. Energy extraction efficiency from the electron beam increases if the period gradually changes along the wiggler. The reason is that the electron beam loses energy as it passes through the wiggler, making it slip out of resonance with a fixed-period field. Changing the period by the proper amount keeps the electron beam resonant through the entire wiggler, so more laser energy can be extracted from the electrons.

Electron Accelerators

One strength of free-electron lasers is that they draw upon a well-developed technology outside of the normal realm of laser physics: charged particle accelerators. That technology had its roots in early atom smashers, and still plays a vital role in particle physics research around the world.

All charged particle accelerators share the same basic principle. An electric potential accelerates an electrically charged particle, thus giving it more energy. For example, a positive potential attracts a negatively charged electron. Accelerators also work with protons, ionized atoms, or charged subatomic particles such as muons. The particle energy is normally measured in electronvolts, where 1 eV is the amount of energy that an electron acquires when falling through a potential of 1 V. The electron accelerators used for free-electron lasers generate particles with energies of millions of electronvolts.

Accelerators generate electron beam pulses with lengths ranging from hundreds of microseconds to picoseconds. The longer pulses typically contain much shorter subpulses. Pulses may be single or repeated. Pulse characteristics strongly affect operation of a free-electron laser, as do other characteristics of the electron beam, including the parallelism of the paths of the electrons and their spread in energies. As with a laser beam, the broader the spread in beam energy and direction, the poorer the quality of an electron beam.

We won't cover details of accelerator design, but we will briefly describe three major types used with free-electron lasers. (Many others are used for other purposes.) The first type used in a free-electron laser was the linear accelerator, in which all the elements lie in a straight line, and the electron travels a long linear path through the accelerator. After the electron beam passes through the undulator, it can be directed to an energy-recovery system or just dumped to a beam dump, as shown in Fig. 9-10A.

One alternative is the storage ring, in which electrons accelerated to a high velocity travel around a closed loop, confined by strong magnets. If one arm of this loop contains an undulator, as shown in

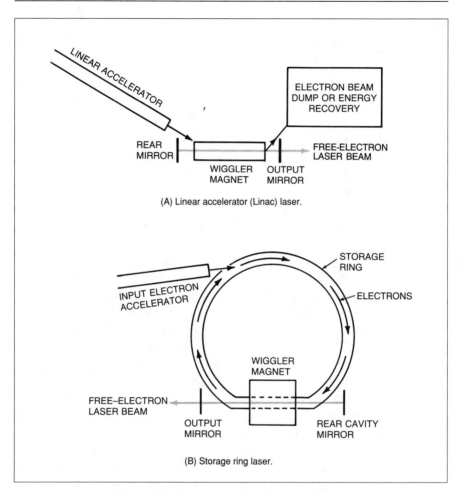

(A) Linear accelerator (Linac) laser.

(B) Storage ring laser.

Fig. 9-10 Two types of free-electron lasers.

Fig. 9-10B, it can drive a free-electron laser. Repeated passes of the electrons through the wiggler should increase efficiency. Unfortunately, passage through the wiggler removes some electron energy, degrading beam quality and taking the electrons out of resonance with the wiggler field. Extra beam-handling equipment is needed to take care of this problem.

Another alternative is the electrostatic or Van de graff generator, which accumulates a large static charge and then discharges it as a pulse of electrons that pass through a wiggler. This approach is attractive because it recovers most energy that the electrons retain after passing through the magnet. It works best for long-wavelength free-electron lasers.

Operating Regimes

The characteristics of free-electron lasers vary over such a wide range that specialists say they have two distinct operating regimes. One is the "Compton" regime, where the physics is best described as interactions between individual particles. The other is the "Raman" regime, best described by collective multiparticle effects. Lasers in the two regimes have quite different characteristics.

Free-electron lasers operate in the Compton regime if the electron energy is high (20,000,000 eV or more), the current flow low, and the wavelengths shorter than about 20 μm (in the infrared, visible, and ultraviolet). In this regime, there is an optimum wiggler length for maximum gain; gain will drop if the wiggler is too long or too short. Gain is typically low, and present only when light waves and electrons are travelling in the same direction. Gain is absent when the light passes through the cavity in the opposite direction.

Free-electron lasers that operate in the collective regime have lower electron energy (typically under 5,000,000 eV) and higher current density. They also operate at wavelengths longer than about 100 μm. Gain is higher in the collective regime and can be exponential with distance travelled through the wiggler. Thus, output powers have been higher.

Cavity Designs

As we saw earlier, ordinary low- and high-gain lasers use different cavity designs. The same is true for free-electron lasers operating in the Compton and Raman regimes.

The low gain in the Compton regime makes low cavity loss essential. This requires a rear-cavity mirror with high reflectivity and an output mirror that lets only a small fraction of the light out of the cavity. The technology for making high-reflectivity visible mirrors is well-developed, but those mirrors don't stand up well under high powers or when exposed to the intense ultraviolet light also generated in a visible-wavelength free-electron laser.

The high gain of the Raman regime makes cavity losses a minimal concern, so the output mirror can transmit much of the light from inside the laser cavity. This makes the optics problem much more manageable, and permits generation of very high-power output pulses. Unfortunately, the far-infrared wavelengths generated by Raman-regime free-electron lasers have limited applications.

Large and Small Free-Electron Lasers

Large and small free-electron lasers have quite different attractions for potential applications. The primary attraction of small free-electron lasers is their versatility and ability to generate reasonable powers (to about 100 W) for research in science and medicine at wavelengths not otherwise available. The main attraction of large free-electron lasers is their potential to generate truly awesome powers with high efficiency.

Potential medical applications have attracted some developers, including John M. J. Madey, the Duke University physicist who built the first free-electron laser at Stanford in the 1970s. Their goal is to build compact lasers that can be tuned between about 1 to 10 μm in the infrared. Few lasers now available can generate reasonable powers between those wavelengths, which look attractive for medical treatment. Other researchers are working on free-electron lasers for longer infrared wavelengths—also hard to generate by other means—for general laboratory use. Shorter-wavelength free-electron lasers are sought for their tunability at wavelengths not available from other lasers.

A potential for power-conversion efficiency of 20% to 50% or more—very high by laser standards—stimulated much interest in very high-power lasers. Developers hope to draw on well-developed technology for generating energetic particle beams, although current levels usually fall below those needed for extremely high-power free-electron lasers. The Strategic Defense Initiative spent hundreds of millions of dollars trying to develop massive high-power free-electron lasers as weapons for nuclear defense. They envisioned a giant laser based on a mountain top, which would send its light to a relay satellite, which would pass the beam on to other relay mirrors, which would eventually deliver it to its target. However, that program was cut sharply and plans to build a large demonstration laser were cancelled.

Promises and Problems for Free-Electron Lasers

Free-electron lasers have exciting promise in many areas. They have impressive potential for efficient, high-power output. Their tunability can open new parts of the spectrum to laser research, especially short wavelengths and the far-infrared, where laser sources have been scarce or unavailable. Their combination of tunability and power makes them promising for many applications in medicine and other areas that may require reasonable power levels at specific wavelengths not available from other sources.

However, any realistic assessment must be tempered by looking at potential problems. So far, the highest powers have been demonstrated in the long-wavelength regime, not at more useful shorter wavelengths. There are big problems with the short-wavelength optics needed to make a good laser resonator. Perhaps most important, progress has been slow and the technology of free-electron lasers remains immature. The main reason is that most experiments are large in scale and thus particularly time-consuming.

X-RAY LASERS

Laser researchers have sought ever-shorter wavelengths. Townes started out with the microwave maser, and then the laser jumped to visible wavelengths—roughly a factor of 10,000 shorter. For two decades after the first ruby laser, researchers seeking shorter wavelengths made only limited progress into the ultraviolet. Only since about 1980 has real progress been made on lasers at extremely short wavelengths in the "soft" X-ray part of the spectrum.

X-ray lasers differ greatly from the other types we've described so far. Differences between X rays and longer-wavelength electromagnetic radiation lead to important differences in structure and operation. Some X-ray laser research is classified, because it depends on nuclear explosions. We will briefly explore this research after we take an initial look at X-ray physics.

X-Ray Physics

Earlier, we learned that visible and ultraviolet lasers operate on electronic transitions. X rays are also produced by electronic transitions, but of a different type. The transitions that produce visible or ultraviolet light involve the outermost electrons attached to an atom or molecule. Those electrons experience only a modest attraction from the atomic nucleus. For light elements such as hydrogen and helium, where only the outer electrons are present, the nuclear charge is low. In heavier elements such as iron or argon, the outer electrons are largely shielded from the much larger nuclear charges by complete shells of inner electrons. As a result, outer-shell transition energies are usually in the 1- to 10-eV range. (We use the electronvolt energy scale because it is a handy way to measure energy in the X-ray range. To convert to wavelength λ, you can use the formula:

$$\lambda \text{ (in m)} = 1.24 \times 10^{-6}/\text{Energy (in eV)}$$

A 1-eV transition has a wavelength of 1.24 μm; a 10-eV transition is at 124 nm.)

Inner-shell electrons are much closer to the positively charged nucleus, and without much shielding they experience a strong pull from the nuclear charge. This means that it takes a lot of energy—100 eV or more—to pluck an electron out of an inner shell and raise it to an outer energy level. Conversely, if an electron drops from an outer shell into a vacancy in an inner shell, as shown in Fig. 9-11, it will release a highly energetic photon. The electromagnetic radiation emitted and absorbed by such transitions are called X rays.

No rigid boundary separates X rays from ultraviolet light; the two parts of the spectrum blend together, and boundary definitions often vary. Some specialists call wavelengths longer than 10 nm "soft X rays," but others prefer to call them the "extreme ultraviolet." (In my more cynical moments, I sometimes thought the choice depended on the desire to claim an "X-ray laser" versus the security restrictions imposed on reporting "X-ray laser" research.) The important point to remember is that there is a gradation of energy levels, some clearly ultraviolet, some clearly X-ray, and some on the hazy middle ground.

X rays affect what they strike much more than visible light because they have much higher photon energy. From a biomedical standpoint, X rays (and short ultraviolet wavelengths) are called "ionizing radiation," because they can strip electrons from atoms or molecules they hit. That has a direct and dramatic effect on biomolecules. X rays are used in cancer therapy, because they can damage cancer cells. However, they can also increase the risk of cancer originating from healthy cells.

Superman's X-ray vision may lead you to the mistaken assumption that X rays can penetrate anything. X rays can penetrate tissue better than bone or teeth because the light elements that make up tissue, such as hydrogen, carbon, and oxygen, do not absorb X rays as strongly as heavier elements, such as calcium in bones and teeth. However, the lighter elements absorb some X rays, and even ordinary air will prevent X rays from going too far. (That's why hospitals and dentists observe X-ray precautions only in the room where the X-ray

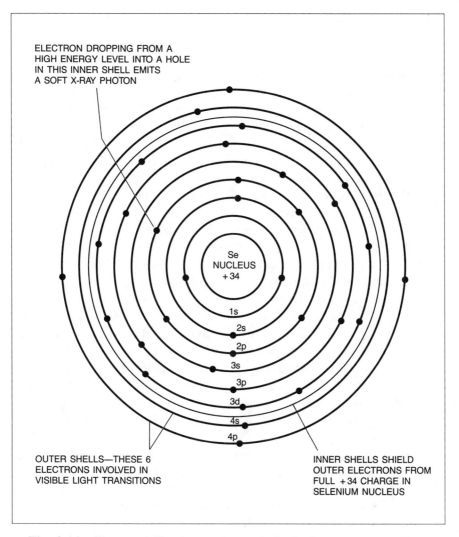

Fig. 9-11   Electron falling into an inner shell of selenium emits an X-ray photon.

is being taken—the X rays don't penetrate farther.) If Superman really had only X-ray vision, his world would be a very dark one.

This strong absorption is not the only problem in dealing with X rays. At X-ray wavelengths, most materials have refractive indexes very close to one, so they can neither reflect nor refract light well, putting some serious limitations on X-ray optics. As we will see later, there has been progress in dealing with this problem, but it is not fully solved.

X-Ray Laser Concepts

As with other types of lasers, an X-ray laser requires a population inversion. However, an X-ray population inversion is very hard to produce. Potential upper laser levels have extremely short lifetimes and drop back to the lower level very quickly. Moreover, it takes a lot of pump energy to raise each atom to the excited level, so it takes a very high energy (or peak power) to produce a population inversion. This requires either a brute force source of tremendous energy or a way to put the energy in precisely the right place at exactly the right time. So far, the emphasis has been on brute force.

The brute force approach is the epitome of what one laser researcher called the "telephone pole" theory of lasing. In the early days of laser research, he recalled, some people believed you could make anything lase—even a telephone pole—if you hit it hard enough. It took a long time before anyone learned how to apply enough brute force to make an X-ray laser.

To produce a population inversion at longer wavelengths, you can get away with raising one electron in each atom to the right excited state. However, to produce an X-ray population inversion you must do a lot more. You must strip many electrons from atoms, and then wait (but not very long) for the atoms to recapture the electrons. The population inversion occurs as the electrons are being recaptured, before they lose all their energy and drop down to the inner electron shells.

Depositing this much energy vaporizes the X-ray laser material. However, for a brief instant before it dissipates, it forms a hot plasma that can emit an X-ray laser pulse. Researchers so far have made X-ray lasers using two major variations on this approach, differing in the nature and scale of energy deposition.

Bomb-Driven X-Ray Lasers

It's hard to imagine more brute force than that provided by a nuclear explosion. That's what researchers from the Lawrence Livermore National Laboratory in California used in a series of secret and highly controversial X-ray laser experiments at the Department of Energy's Nevada Nuclear Test Site during the 1980s. The intense burst of X rays from the nuclear fireball excited the X-ray laser medium, generating intense X-ray laser pulses. Some observers have claimed that the process isn't really a nuclear "explosion," but that seems a largely academic distinction.

The bomb-driven X-ray laser was an early central part of the Strategic Defense Initiative, or "Star Wars" program. Its attraction was the potential to concentrate energy from the nuclear blast and direct it long distances, instead of letting it spread out into space unimpeded, as it would from a conventional bomb. Early experiments led researchers to believe that the energy could be concentrated over long distances, but later tests showed it could not, and the effort was largely abandoned.

Laser-Driven X-Ray Lasers

A parallel unclassified program conducted by Livermore uses a different pumping approach, shown in Fig. 9-12. They zap a thin metal foil with a short, high-energy pulse from a massive Nd-glass laser built for fusion research. The fusion laser pulse contains trillions of watts but only lasts about a nanosecond (10^{-9} s). The intense laser light strips electrons from atoms in a linear segment of the metal foil, creating a linear plasma. As the electrons recombine with the atoms, they emit a pulse of coherent X rays. In a series of experiments, Livermore has generated laser pulses at wavelengths as short as 3.5 nm from several metals.

The Livermore experiments use a single high-energy pulse to both ionize the laser material and excite the ions that emit the X rays. Santanu Basu and Peter Hagelstein took a different approach at the Massachusetts Institute of Technology. They used two comparatively low energy pulses, the first to ionize the atoms to form a plasma, and the second to excite the ions to produce X rays. This let them generate 20.6-nm X rays using pump pulses with less than a joule of energy— less than 1/1000th the energy used in the Livermore experiments. Their X-ray power was much lower than Livermore's, but the Livermore group is working on its own compact laser-driven X-ray laser, which—while not as compact—would produce much more X-ray power.

Another approach is pursued by Syzmon Suckewar at Princeton University, who confines laser-generated plasmas with magnetic fields to reduce the laser pump power required. His group has demonstrated laser gain at 18 nm.

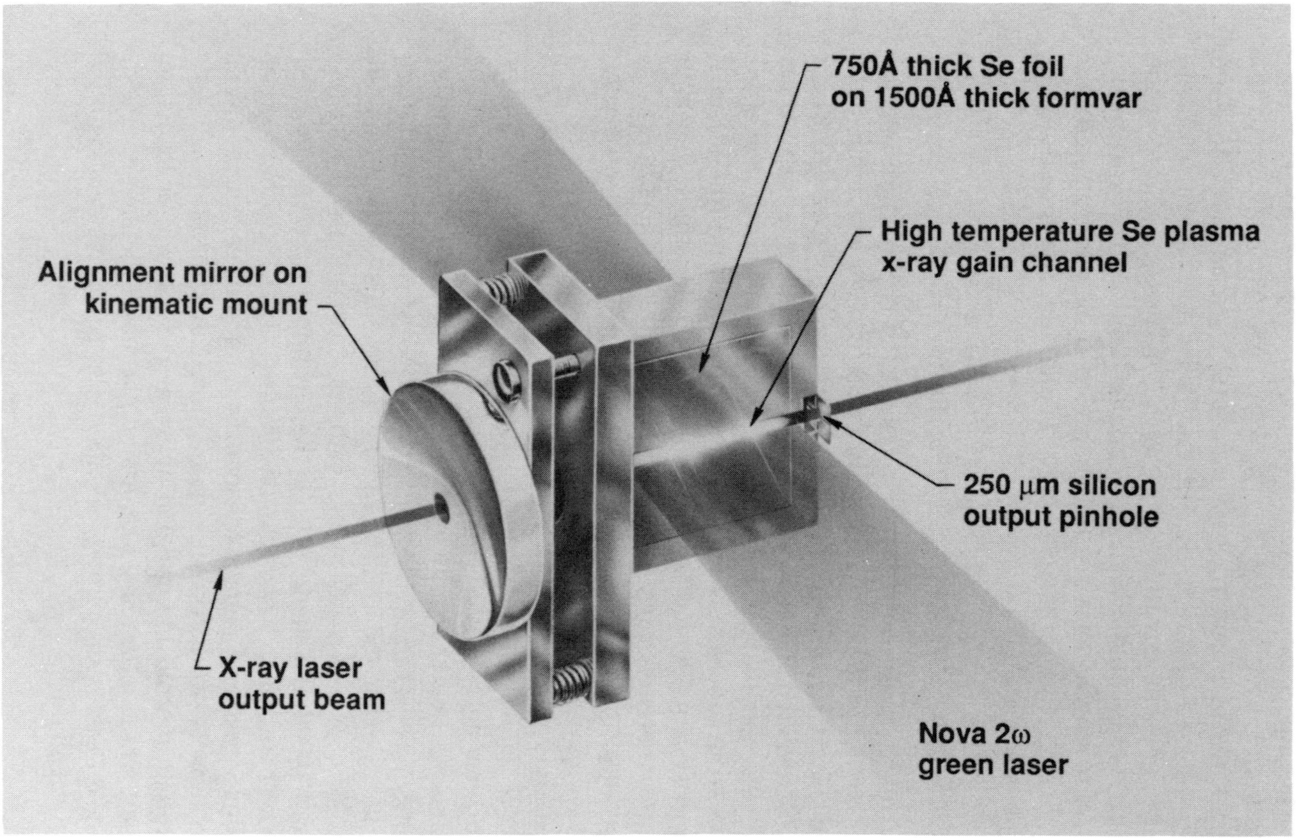

750Å thick Se foil on 1500Å thick formvar

High temperature Se plasma x-ray gain channel

250 μm silicon output pinhole

Nova 2ω green laser

Alignment mirror on kinematic mount

X-ray laser output beam

Fig. 9-12 Intense pulses of green light produce a high-temperature selenium plasma which generates an X-ray laser pulse. (Courtesy of Lawrence Livermore National Laboratory.)

The X-Ray Mirror Problem

As we mentioned earlier, X-ray mirrors are a problem. Fortunately, gain is high enough that they aren't essential for X-ray laser emission. However, some specialists would quibble with our definitions, pointing out that the X-ray "lasers" described so far are not resonant-cavity oscillators but only provide X-ray amplification by stimulated emission, and thus shouldn't be called lasers.

More important, usable X-ray resonator mirrors might help improve the directionality and output power of the X-ray beam, particularly from the laser-pumped unclassified devices. Progress is being made on multilayer X-ray reflectors, similar to the multilayer reflective coatings used for visible-light optics. They are made by depositing extremely thin alternating layers of two materials with slightly different X-ray refractive indexes. Although reflection at each boundary is small, the sum over many layers adds up to reasonable levels.

X-Ray Laser Applications

What could one do with an X-ray laser? The U.S. government spent about a billion dollars learning that it's very difficult to make an X-ray laser that could be used as a weapon. However, development of a comparatively small "laboratory" X-ray laser could lead to many applications in research and industry. Coherent, collimated beams of X rays could be invaluable in studying the structure of matter and even living organisms. X-ray holograms would permit three-dimensional structural studies of living cells—perhaps even of images recorded while the cells were alive. (The X-ray laser pulse would kill them, however.) Coherent X-ray beams might also be useful for other applications, including the processing of semiconductors. The more compact the X-ray laser, the more widespread the applications are likely to be.

OTHER NOVEL LASER CONCEPTS

A few other laser ideas also do not fit into our standard laser categories. We'll mention them briefly here, to show some other ideas being studied.

Gamma-Ray Lasers

Gamma rays have even shorter wavelengths than X rays. Physicists usually define gamma rays as photons produced by transitions between energy levels of the atomic nucleus. These transitions are more energetic than X-ray transitions, so gamma rays have shorter wavelengths.

The shorter wavelengths might make gamma-ray lasers seem even harder to make than X-ray lasers. However, gamma-ray transitions have one big potential advantage—some upper levels have quite long lifetimes as metastable nuclei. This avoids the need for tremendous pump energies to produce a population inversion, as in an X-ray laser.

The story isn't that simple, of course, and some truly formidable obstacles remain. Nonetheless, gamma-ray lasers are an interesting theoretical possibility.

Nuclear-Pumped Lasers

The bomb-driven X-ray laser we described earlier is the most dramatic example of a nuclear-powered laser. There has also been a longer-running and less classified effort to tap the energy of more controlled nuclear reactions.

The essential idea is to excite the laser medium with energy from a nuclear chain reaction. Neutrons or fission fragments produced by a pulsed fission reactor could transfer some energy to a gas laser medium, producing a population inversion and laser action. Two 1974 experiments proved that the concept was feasible. However, power levels have never reached the levels that developers hoped for. Research is continuing, with one goal the use of reactor-driven lasers in space.

WHAT HAVE WE LEARNED

- Optically pumped organic dyes in liquid solutions are the basis of tunable dye lasers. Their major use is to generate tunable-wavelength visible light for scientific research.
- Dye lasers are versatile but complex.

- The broad gain bandwidth of dye lasers lets them generate exceptionally short pulses.
- Most dye lasers have room for accessories inside and outside the cavity.
- Dye-laser cavities include a prism, diffraction grating, or other wavelength-selective element to allow wavelength tuning.
- Dyes have gain bandwidths of 10 to 70 nm, but even without wavelength selection, natural line-narrowing limits linewidth to a few nanometers.
- A ring cavity limits dye lasers to a narrower bandwidth than other cavities.
- Wavelengths emitted by dyes depend on their chemical composition, the pump wavelength, and the duration of pumping, as well as on cavity tuning.
- Dyes decompose gradually during use. Most must be dissolved in organic solvents.
- Dye lasers can be pumped by linear or coaxial flashlamps; pulsed excimer, neodymium, copper-vapor, or nitrogen lasers; or continuous argon, krypton, and neodymium lasers.
- Continuous-wave dye lasers can emit light with an extremely narrow bandwidth.
- Synchronous modelocking can produce extremely short dye-laser pulses.
- Dye-laser harmonics can be tuned in wavelength by tuning the laser's output.
- Free-electron lasers extract energy from a beam of electrons passing through a wiggler magnet. They can be tuned across a broad range of wavelengths by changing the period of the wiggler field or the electron energy.
- The wiggler-magnet field must vary sinusoidally, but it need not come from permanent magnets. It can come from an electromagnetic wave.
- Tapering the wiggler field can increase free-electron laser efficiency.
- Free-electron lasers draw on well-established technology for particle accelerators.
- Free-electron lasers operate in two regimes, one where interactions are between individual particles and the other where

interactions are collective. Different cavities are needed for high-gain Raman and low-gain Compton free-electron lasers.

- All the potential of free-electron lasers has yet to be realized. Important obstacles remain to producing high powers at short wavelengths.

- X rays are produced by inner-shell electronic transitions. They are ionizing radiation that can strip electrons from atoms or molecules.

- It takes extremely high peak powers to produce an X-ray population inversion—either from a conventional laser or a nuclear explosion.

- Reflectivity is much lower at X-ray wavelengths than in the visible. The best mirrors reflect only about 20% of incident X rays.

- Gamma-ray lasers would operate on transitions of atomic nuclei.

- Energy from controlled nuclear fission can power a laser.

WHAT'S NEXT

In the next two chapters, we will explore the applications of lasers. Chapter 10 covers the uses of low-power lasers. Chapter 11 covers the uses of high-power lasers.

Quiz for Chapter 9

1. What is the shortest pulse that can be generated from a dye laser with a bandwidth of 10 GHz?
   a. 44.1 ns
   b. 4.41 ns
   c. 441 picoseconds
   d. 44.1 picoseconds
   e. 4.41 picoseconds

2. How does a diffraction grating select a particular wavelength to oscillate in a laser cavity?
   a. It diffracts only one wavelength back in the right direction to oscillate with the other cavity mirror.

b. It increases losses at other wavelengths by diffracting them out of the laser cavity.

c. It reflects only one wavelength and absorbs the rest.

d. A and B

e. All of the above

3. What property gives organic dye lasers their wavelength tunability?

a. The many vibrational sublevels of electronic transitions

b. The use of organic liquids as solvents

c. The high gain of the optically pumped dyes

d. Photodissociation of the dyes under intense pump light

e. The liquid nature of the laser medium

4. Which of the following pump sources is used in the lowest-cost dye lasers?

a. Excimer lasers

b. Nitrogen lasers

c. Continuous-wave argon lasers

d. Third harmonic of Nd-YAG

e. Electric discharge

5. An excimer-pumped dye laser generates pulses 10 ns long. What is the minimum bandwidth the pulses can have?

a. 4.41 GHz

b. 441 MHz

c. 44.1 MHz

d. 0.4 nm

e. None of the above

6. What is the wavelength of a free-electron laser powered by 20-MeV electrons with a wiggler having a 5-cm period?

a. 15.6 μm

b. 56.1 μm

c. 0.156 mm

d. 1.56 cm

e. 5 cm

7. The first stage of a two-stage free-electron laser is powered by a 5-MeV electron beam and has a wiggler with a 5-cm period. What is its wavelength?

a. 25.6 μm

b. 56.1 μm

c. 0.196 mm

d. 0.216 mm

e. 2.16 mm

8. The second stage of a two-stage free-electron laser uses the electromagnetic field from the first stage in Problem 7 as its wiggler-magnet field. What is its wavelength if it uses the same electron beam?

a. 216 μm

b. 15 μm

c. 930 nm

d. 515 nm

e. None of the above

9. What types of transitions produce X rays?
   a. Vibronic transitions
   b. Outer-shell electronic transitions
   c. Nuclear transitions
   d. Inner-shell electronic transitions
   e. Nuclear explosions

10. What is the wavelength of 1000-eV X rays?
   a. 1000 nm
   b. 100 nm
   c. 12.4 nm
   d. 1.24 nm
   e. 0.124 nm

CHAPTER **10**

Low-Power Laser Applications

ABOUT THIS CHAPTER

Lasers are used for so many applications that a single chapter can't do them justice. In this chapter, we look at applications that require low levels of laser power, typically well under a watt. Such little power does not have a dramatic effect on the objects it strikes, but it can make minor changes, such as exposing photographic film. As we will see, there are many types of low-power laser applications. In some, the laser is little more than a high-performance light bulb, but for others, special features of laser light—such as coherence or tight beam collimation—are essential.

THE ATTRACTIONS OF LASERS

Theodore Maiman's first laser made headlines in the summer of 1960. As word of the laser spread, it seemed that every scientist and engineer wanted his or her own laser. Many had no clear purpose in mind, but the laser seemed like a neat toy, and curiosity lurks deep inside the souls of scientists and engineers. They built their own lasers, or bought lasers from the handful of little companies that sprang up to make them. Then they started zapping just about everything that couldn't get away. They informally measured laser power in "gillettes"—how many razor blades a laser pulse could pierce. It was a fertile time for experiments, new ideas, and accidental discoveries. But like any boom, it busted, and left behind the joke that the laser is "a solution looking for a problem."

Much has changed since the first wave of laser enthusiasm. Today, plenty of jobs demand lasers. In Chapter 11, we will talk about those which use laser power to change something in a major way. In this chapter, we will concentrate on the many uses of low-power lasers.

What can low-power lasers do that other light sources can't? For some jobs, they may merely be serving as well-behaved, long-lived light bulbs. However, other jobs may require special properties of laser light, such as coherence or the tight alignment of laser beams. Let's look at the major differences between low-power lasers and other light sources:

- Lasers produce well-controlled light that can be focused precisely onto a small spot.
- Low-power laser beams, focused tightly in a small spot, can reach a high power density.
- Laser beams can define a straight line.
- Laser light covers only a narrow wavelength range.
- Most laser light is coherent.
- Lasers can generate extremely short pulses.
- Diode and helium-neon lasers produce steady powers for tens of thousands of hours or more.
- Diode lasers are very compact and inexpensive.
- Diode lasers can be modulated directly at high speeds by changing drive current.

On the other hand, lasers do have some significant limitations that affect how they can be used:

- Lasers don't emit white light.
- The least expensive lasers are infrared diode lasers.
- The least costly visible lasers emit red light; yellow, green, and blue lasers are much more expensive.
- The concentration and collimation of laser light makes it hazardous to look directly into a laser beam.

In the rest of this chapter, we will show how these characteristics combine with other factors to determine the uses of low-power lasers. For starters, you might mentally compare a laser to a light bulb. A light bulb is very good for illuminating a room, but not for illuminating one tiny spot far from the bulb. On the other hand, a laser is virtually useless for illuminating a room, but it's very good for delivering light to a tiny spot far away.

READING WITH LASERS

Low-power lasers often read printed symbols, such as the striped product codes on food packages and similar codes on magazines and other products. However, they do not read in the same way that your eye reads. The laser beam illuminates the object being read, but a light detector and an electronic receiver actually decode whatever is being read.

Figure 10-1 is a simplified diagram of the workings of a laser bar-code reader of the type used at supermarket checkout counters.

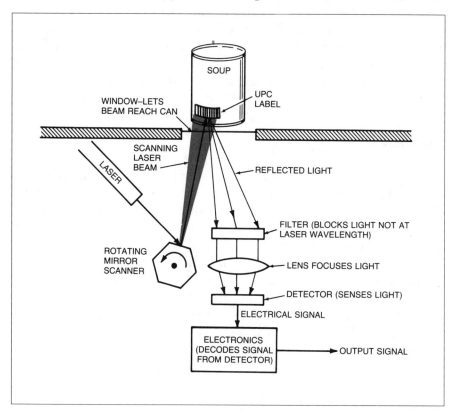

Fig. 10-1   A laser bar-code reader.

A beam deflector—here a rotating mirror—scans the laser beam in a pattern across the zone where the bar code will appear. Laser light reflected from that area is focused onto a detector, which generates an electrical signal proportional to the light reaching it. The amount of reflected light varies with time as the beam scans across dark and light zones. The detector and its electronic circuits see those dark and light areas as periods when reflected light is absent or present, respectively, and decodes them as zeros and ones. Thus, the scanner converts a symbol on a piece of paper into an electronic signal.

Our diagram does not show other lighting, but we can't ignore it in the real world. In a supermarket, for example, bright overhead lights shine down onto the checkout counter, illuminating packages at the same time as the laser. However, the laser light is concentrated at one wavelength (632.8 nm for a helium-neon gas laser), while room light is distributed throughout the spectrum. You can keep the room light from confusing the detector with a filter that transmits only the laser wavelength, which we put in front of the detector in our diagram. A little room light does get through at the laser wavelength, but it is overwhelmed by the laser light and does not interfere with reading. Without the filter, most of the light reaching the detector would be from the room, making it impossible to read the bar codes accurately.

Supermarket Scanning

The place you are most likely to see a laser beam is in the supermarket. The laser's presence is not advertised, but it rests under the checkout counter. Its beam comes up through a window to read the bar codes that identify food packages, and you can see the low-power red beam scanning labels if you look carefully.

Supermarket scanners work like the generalized laser scanner in Fig. 10-1, but there are a few important refinements. The beam-scanning pattern and the bar code are both designed so that packages can be moved rapidly across the scanning window without bothering to align them. The scanner can read the code front-to-back or back-to-front; the code itself tells it which is which. It can even read the code when the package is tilted at an angle, as long as the scanning window is clean.

The scanner does not read the price directly from packages supplied by outside companies, such as cans of soup or boxes of cereal.

It reads a bar code that identifies a specific product and then looks up the price of that product in a table stored in the store's computer. That price is added to your bill. Many stores print special bar codes for produce and meat that they package. Those bar codes indicate price and weight, which can be read directly at the checkout.

The symbols used on food packages are the Universal Product Code (UPC), which was designed to be read with a 632.8-nm helium-neon laser. The wavelength is important because it affects specifications for printing UPC symbols. For the laser to read the bar code correctly, the dark and light stripes must reflect different fractions of light at the red wavelength, which is all that the detector "sees." It took some food processors a while to learn that it was only the laser wavelength that mattered. For a while, one Boston-area milk company printed red-striped bar codes on its white-paper milk cartons. Both the red stripes and the white paper reflected the red laser light strongly, so the symbols were invisible to laser scanners.

The red wavelength was chosen in the mid-1970s because the helium-neon laser was the most economical visible laser. The designers wanted a code readable in visible light so that packages could be visually inspected at the printer. The need for red light, and the need for a well-collimated beam, kept designers from switching to near-infrared semiconductor lasers to read supermarket packages. The advent of red semiconductor lasers does not completely overcome the wavelength problem, because the optics that filter out background light are fine-tuned for the 633-nm helium-neon line, but most red diode lasers emit at 670 to 680 nm. Helium-neon lasers remain the standard choice for supermarket bar-code scanners, but probably will be replaced eventually by 635-nm diode lasers.

Similar bar codes have been adapted for many other purposes. Magazines use similar symbols on their covers to speed processing by wholesale distributors. Other retailers use similar codes. The government uses a similar code to identify materials in its inventory. Libraries paste bar codes on books and library cards and use readers to feed data on borrowed books directly into their computers.

Hand-Held Bar-Code Readers

The biggest growth in bar-code readers recently has been in hand-held devices. These readers are much simpler and less expensive than checkout scanners, but they have to be passed directly over the symbol they are reading. This avoids the need for scanning optics, because the person holding the reader moves the beam. By placing the reader close to the symbol, it reduces the amount of stray light reaching it—avoiding the need for expensive filters.

Putting the reader close to the bar code also eliminates the need for a highly directional beam. Many hand-held "wands" actually contain red LEDs rather than lasers; other compact scanners use 670-nm red semiconductor lasers. (The reduction in background light makes wavelength much less critical.) This further reduces cost, leading to their widespread use.

OPTICAL DISKS AND DATA STORAGE

Most of the world's lasers read a different type of information— digitized sound encoded on reflective audio compact discs (spelled that way to frustrate copy editors who think all disks contain k's). The lasers are tiny inexpensive semiconductor lasers, and they're buried deep inside compact disc players, where you would never find them unless you disassembled the players. Many millions of CD players are sold each year, making them impressively successful even by the standards of consumer electronics. Optical disk technology has also found other applications, including computer data storage and video playback.

Optical Disk Technology

The basic concept of optical disk technology is shown in Fig. 10-2. Light from a semiconductor laser is focused onto the surface of a rapidly spinning disk. The surface of the disk is covered with tiny spots, which record bits of data and have different reflectivity than blank areas on the disk. Laser light reflected from the disk is focused onto a detector, which generates an electrical output. The result is a series of pulses corresponding to the data recorded on the disk.

Optical storage can cram tremendous amounts of data onto a small surface area. You can focus a diode laser beam to a spot roughly one wavelength across. For the 780-nm diode lasers used in CD players, that means spots roughly 1 μm across. For the computer data-storage equivalent of a CD—called a "CD-ROM" for Compact Disc Read-Only Memory—that means a storage capacity of 600 million

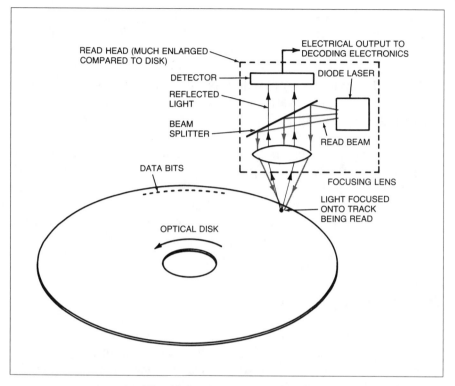

READ HEAD (MUCH ENLARGED
COMPARED TO DISK)

ELECTRICAL OUTPUT TO
DECODING ELECTRONICS

DETECTOR

DIODE LASER

REFLECTED
LIGHT

BEAM
SPLITTER

READ BEAM

DATA BITS

FOCUSING LENS

LIGHT FOCUSED
ONTO TRACK
BEING READ

OPTICAL DISK

Fig. 10-2   Optical disk system.

bytes on a 12-cm disk. Lasers with shorter wavelengths could read smaller spots and allow denser data storage.

Figure 10-2 shows only the reading of optical data, as from prerecorded CDs and CD-ROMs. However, other optical disks have surfaces that respond to light, making it possible to write data by modulating the laser beam to store data bits. We'll describe the details below.

Videodisks

Ironically, the idea that originally stimulated interest in optical disks has never been a great success—the home videodisk player. In the 1970s, the Dutch electronics giant N.V. Philips and the American entertainment firm MCA Inc. both decided to develop home players of prerecorded video programs. They settled on optical disk players built around mass-produced helium-neon lasers. Video programs were encoded in analog form on 30-cm (12-inch) disks.

Their technology worked, although it got to market a bit later than planned. Unfortunately, videocassette recorders got to market first, and the public preferred the tape system because it let them record as well as play back programs. To further confuse the market, RCA introduced an incompatible type of videodisk system based on capacitive disks. The laserless RCA videodisk made a big but short-lived market splash and quickly sank to the bottom, a $500 million fiasco.

Laser videodisk players remain on the market more than a dozen years after their original introduction. Current models use semiconductor lasers and have added other refinements, but are otherwise similar to the earlier videodisks. Prices of both prerecorded disks and of the players have come down, but they remain more expensive than VCRs. However, their better quality is earning them acceptance among videophiles.

Compact Disc Players

If you don't own a compact disc player, you probably have at least looked longingly at one. The crisp, clean digital sound is an audiophile's delight. The compact disc is a straightforward spinoff of videodisk technology. It was jointly developed by Philips and Sony, who converted the recording format from analog to digital and moved to a smaller 12-cm (4.75-inch) disk, which could hold up to 72 minutes of music. It has largely supplanted phonograph records, which are becoming rarities.

By the time Philips and Sony developed the CD format, diode lasers had matured enough to be a better choice for audio players than helium-neon lasers. Diode lasers are much smaller, so they can be mounted directly in the playing head. They also require much less power than gas lasers. They have proved easy to make at low cost in the large quantities required. The GaAlAs lasers used for CD players, which emit a few milliwatts continuous-wave at about 780 nm, sell for only a few dollars each in large quantity. The mass production of CD lasers and other optical components has made them available inexpensively for other purposes.

Compact discs, like phonograph records, are mass-produced by pressing plastic against a master, which itself is recorded with a He-Cd laser. The master pattern transfers a spiral of tiny pits to the copy. Extremely high precision is required because the pits are only about a micrometer across, but once that precision is attained, reproduction costs are low.

CD-ROMs, CD-I, CD-V and Other Spinoffs

The impressive success of the audio CD encouraged Philips and Sony to modify the format for other applications. The first such modification, and still the most important, is the CD-ROM. Like an audio CD, it uses 12-cm disks, but it digitally encodes 600 Mbytes of computer data instead of music.

The CD-ROM is a very attractive format for publishing large data-bases and other reference compilations for computer retrieval. One reason is the disk's extremely high capacity; the text from a standard 24-volume encyclopedia fills only about a third of a CD-ROM. Another is the fast electronic access to information anywhere on the disk available with suitable software. Enter the word "laser," and the electronic encyclopedia will display its main entry and cross-references to other entries that mention lasers.

CD-ROM players are similar to those used with conventional CDs, but they don't need to convert the digital data bits into audio format. (However, the need to directly interpret digital data bits makes error correction more important for CD-ROMs than for audio disks, where a single bad bit is normally not detectable.) Likewise, the same technology can produce CD-ROMs as well as CDs, although, again, minimization of errors is more critical for CD-ROMs than for CDs.

The original CD-ROM format is limited to prerecorded data only. The CD-I (Compact Disc-Interactive) format can handle simple inter-active graphics as well as text. The CD-V (Compact Disc-Video) format can show short sequences of full-motion video. The use of data compression can increase disk capacity. Other variations are also in development.

Writing on Optical Storage Media

You can read from standard CDs and CD-ROMs, but you can't write new information onto them. The surfaces of the prerecorded disks are not sensitive to light. If you want to store data on optical disks, you need disks coated with materials that change in some way when exposed to light. Then you can store data bits at selected places on the disk by turning the beam off and on at a high enough power level.

Several types of optical storage systems are in commercial use, and many more have been developed. The differences reflect the way that recordable media are used. Read-only systems like CDs and CD-ROMs can work only if information (computer data or music) is published in the proper format. This requires standards, in which many companies agree to use the same format for disks and players so their products can be interchanged. The formats are usually compromises, which may not be ideal for any one purpose but can be used by many different people.

Writable optical disks are used in a different way—for local storage. It helps to have the blank disks a standard size, so they can be mass-produced economically, but it is not essential that they all have the same format. That has led to a few different formats.

The first writable disks to reach the market were used for computer data storage. They used materials that changed permanently when illuminated, like photographic film rather than magnetic disks. Called "write-once, read-many" or "WORM" disks, they are read with a low-power beam at one wavelength, which does not affect the disk. Writing is with a beam of higher power and/or shorter wavelength, which forms holes or spots on the disk surface. Like CD-ROMs, WORM disks can store large quantities of data. Some consider the permanent recording a problem because it prevents reuse of the disk; others consider it an advantage because it prevents unauthorized alterations.

A different approach is used in erasable optical disks, called "magneto-optic" recording because it relies on a combination of magnetic and optical phenomena. These disks use different materials and recording techniques that make it possible to erase and reuse the disks as you would magnetic disks for computer storage.

It takes a combination of light from a diode laser and a magnetic field to record a data bit on a magneto-optic disk. The disk is coated with a partly transparent material that can be magnetized if heated beyond a certain temperature. At lower temperatures it keeps the magnetic field direction it took when heated. A beam from a diode laser heats the coating to the critical temperature, as shown in Fig. 10-3. Once the coating exceeds that temperature, it becomes magnetized in a direction set by a magnet on the other side of the disk. Heating the material again when it is exposed to a different magnetic field can change the field direction in the coating again, making the data erasable.

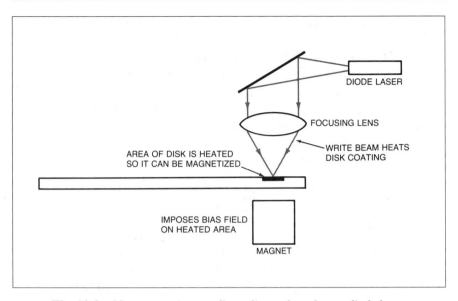

DIODE LASER

FOCUSING LENS

WRITE BEAM HEATS
DISK COATING

AREA OF DISK IS HEATED
SO IT CAN BE MAGNETIZED

IMPOSES BIAS FIELD
ON HEATED AREA

MAGNET

Fig. 10-3   Magneto-optic recording relies on heat from a diode laser.

Reading is with a lower-power beam from a semiconductor laser, which passes back and forth through the partly transparent disk coating. If the material is magnetized, it affects the polarization of the transmitted light in a way that can be detected by looking at the laser pulse reflected from the disk. Thus, although diode lasers write and read the data on the disk, the information is actually stored in magnetic form. Magneto-optic data storage has found wider acceptance than write-once drives, and is widely available on the commercial market.

Floptical disks are a different hybrid of magnetic and optical technology. They use lasers to write precision optical tracks on magnetic disks the same size as 3.5-inch floppy disks, but the data is written and read magnetically. This allows storage of 21 Mbytes on a 3.5-inch magnetic disk.

Other applications for recordable optical disks are emerging, although none has yet been widely accepted. Eastman Kodak has developed the Photo-CD system, which records images electronically and then transfers them to a CD-ROM-format disk as color images. Special centers do the transfer by writing onto write-once disks.

Writable disks can also be used to record music, although the development of consumer audio products has lagged behind that of computer disks. Sony uses magneto-optical recording in its Mini Disc system, which records on a 6.4-cm disk coated with terbium ferrite cobalt. It is able to squeeze as much data onto the smaller disk as on a larger CD by compressing the digital data to require less storage space. The disks can be prerecorded or sold blank for consumer recording.

A competing approach is the CD-R, or Compact Disc-Recordable format. This uses write-once disks that are written in the same 12-cm format as prerecorded audio disks and can be played in a standard CD player. It and minidiscs must compete with the digital compact cassette format for consumer acceptance.

LASER WRITING

Laser beams can write as well as read. The concept is shown schematically in Fig. 10-4. The laser beam is turned on and off as it moves across a light-sensitive surface. In Fig. 10-4, when the beam is on, it writes a dark spot; when it is off, it does not change the light-sensitive surface. After the laser scans a single line across the surface, either the beam or the surface can be moved slightly so the laser can write another row of spots next to the first.

This basic concept appears in many variations. The most common is the laser printer of computer output, now used with many personal computers. Others include laser systems for reproducing printing plates and for setting printing type.

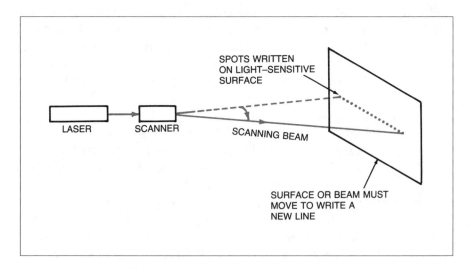

Fig. 10-4   Scanning laser beam writes a series of spots on a light-sensitive surface.

Laser Computer Printers

Look closely at a laser printer attached to a personal computer, and you'll see a strong resemblance to an office photocopier. There's a good deal of similar technology in the two. Some printers use the same type of rechargable cartridges used by inexpensive photocopiers; others use customized versions.

In an ordinary plain-paper copier, a bright light shines onto the page to be copied, and light reflected from the page is imaged onto a rotating drum coated with light-sensitive material. The coating initially carries a static electric charge that is discharged by light reflected from the page. The drum continues rotating after the light strikes it, and regions that were not illuminated and still carry an electric charge pick up a dark material called a "toner." The drum then transfers the toner to plain paper, producing a copy of the original page. (The dark toner adheres to the parts of the drum where the dark areas of the original page were imaged.)

A laser printer contains the same drum arrangement, as shown in Fig. 10-5, but there is no original. Instead, a modulated laser beam scans across the drum's coated surface, writing a pattern of charged and uncharged areas that corresponds to the printed page. The rotating drum then picks up the toner, and transfers the image to plain paper as in a conventional copier. The image is made not by copying an original, but by scanning with a laser beam.

With typical resolution of 300 to 600 dots/inch, the output of a laser printer looks almost as good as a typeset page. In fact, many times the copy for whole books—including all text and art—are produced using a laser printer.

We should note, however, that some "laser printers" don't really contain lasers. Instead of a single laser and a moving mirror, they contain a row of tiny LEDs spaced very closely together. The LEDs are placed close to the rotating drum, and the output of each one writes a dot on the surface, corresponding to the dots a scanning laser beam writes on one pass across the page.

Laser "Platemakers"

Another use of laser writing is to make printing plates for newspapers. The newspaper staff enters articles and other material to be printed into a computer, where the material is laid out into pages in

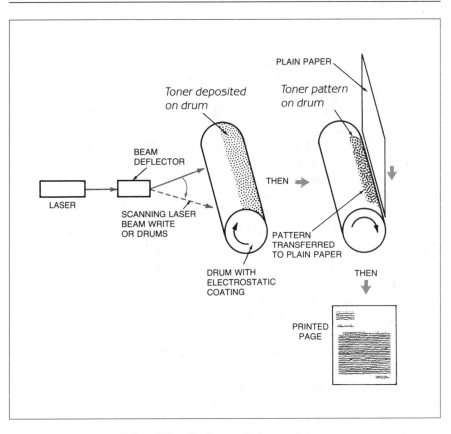

Fig. 10-5   Workings of a laser printer.

the proper form for publication. Then a high-resolution laser beam scans a light-sensitive printing plate, recording images of the words and pictures to be printed. (Original copies of advertisements are typically inserted by other means.) After the plate is processed, newspaper pages can be printed from it.

This platemaking technique can be extended to make copies at satellite printing plants away from the newspaper's main offices. The electronic information used to write the first printing plate is encoded and stored, then transmitted to a distant printing plant, where a laser platemaker uses that information to write the printing plates used at that plant. This technique is used by nationally distributed newspapers such as *The Wall Street Journal* and *USA Today* in the United States, and by national papers in Europe.

FIBER-OPTIC COMMUNICATIONS

Fiber optics have become the dominant medium for long-distance telecommunications around the world. Optical fibers link telephone company switching offices and provide the backbone for the long-distance telephone networks operated by AT&T, Sprint, and MCI Telecommunications. Optical fibers in submarine cables carry telephone calls under the Atlantic and Pacific Oceans. Eventually, optical fibers may bring telephone and other telecommunications services all the way to homes. LEDs can drive fiber-optic systems in a building or on a small campus, but semiconductor lasers are the standard light sources for longer-distance fiber-optic systems.

Figure 10-6 shows key elements of a long-distance fiber-optic system. The signal that reaches the transmitter generates a current, which passes through a semiconductor laser. The signal is a series of digital pulses, and it generates a series of current pulses, which in turn generate light pulses from the laser. The laser emits light pulses directly into the core of the fiber, which transmits them—over distances up to tens of miles or kilometers—to a distant receiver. There, a light detector converts the light pulses back into a series of electrical pulses.

The semiconductor laser is well matched to fiber-optic communication requirements. Its emitting area is only a few micrometers wide, small enough to couple light well into the tiny light-carrying cores of

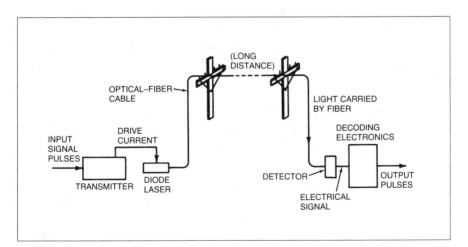

Fig. 10-6   Long-distance fiber-optic system.

optical fibers, which are only about 10 μm across. It is very compact, and operates on the same voltage and current levels used by conventional semiconductor electronics.

Interest in fiber-optic communications has driven much work on semiconductor lasers. Early fiber optics were designed to work at the 800- to 900-nm wavelengths of GaAlAs semiconductor lasers. The discovery that glass optical fibers had lower loss at 1300 and 1550 nm led to the development of InGaAsP lasers for use at those wavelengths. Output powers of a few milliwatts are adequate for long-distance transmission through optical fibers with typical loss in the 0.4- to 0.5-dB/km range at 1300 nm and 0.2 to 0.3 dB/km at 1550 nm.

New long-distance fiber-optic systems use the fiber amplifiers described in Chapter 7. Other laser technology is being developed for local distribution networks, which would replace current cable television and telephone systems. (For more about fiber optics, see the author's book, *Understanding Fiber Optics*, published by Sams Publishing.)

OPTICAL COMPUTING

Optical devices can manipulate light in ways more complex than merely focusing or bending light rays. They can switch light rays in different directions, block or transmit light, or convert the information carried by a light wave into different forms. This whole range of processes is sometimes called "optical computing" or "optical processing." Some require lasers, some work with laser light or incoherent light, and some only work with incoherent light. Most remain developmental. While we can't go into much detail here, we can list some key concepts and applications:

- *Optical switching*: Like electronic switching, it redirects optical signals, often through optical fibers. It is being developed for applications including fiber-optic networks.
- *Optical interconnection*: Making alternative paths among electronic devices, especially those which could not be made by electronic conductors. While electronic interconnections typically must be in a circuit plane, optical interconnections can lie outside the circuit plane, so they can be made between points on the surfaces of two electronic chips. They are particularly attractive for linking complex chips, where the space along the sides for electronic connections is limited.

- *Analog optical transforms*: Conversion of information presented as a function of time in a time-varying optical signal $f(t)$ into a function of frequency $F(v)$ that contains the same information. This operation is called a Fourier transform, and is essential in analyzing the frequencies that make up a signal that varies in time.

- *Special analog transforms*, such as conversion of raw data from synthetic-aperture (side-looking) radars into usable images. Developed in the 1950s, this was the first practical use of optical computing. Military officials considered it so important that they kept it classified for many years.

- *Operations on vectors and matrices*, which are difficult to perform with digital electronic computers because they contain many separate elements, but are important in computer modelling and signal processing. Optical systems promise to do them faster by handling operations on different matrix or vector elements in parallel with arrays of optical elements.

- *Massively parallel computing* other than matrix-vector manipulation, based on the inherent ability of optical devices to process many inputs in parallel without interference. (You can think of this as analogous to the ability to pass many rays through a single lens.)

- *"Neural network" models*, which try to use the inherently high interconnection possibilities of optics to reproduce some functions of the human brain.

The possibilities of optical computing are exciting and nearly endless. Some enthusiasts envision optical computers a thousand times more powerful than today's best supercomputers. To be fair, however, we should warn you that many of these promises have been lurking just out of reach for many years. Nor are optical computers good for simple arithmetic. As one veteran researcher quipped, "Optical computers can do things you would never dream of, but they can't balance your checkbook."

LINEAR MEASUREMENTS AND STRAIGHT LINES

Laser beams have proved invaluable for many types of measurement. Some of the most important use laser beams as straight lines, and laser wavelengths as units of length. Others simply use laser beams as pointers.

Laser Beams for Alignment

The use of a laser beam to draw a straight line may sound trivial or even mundane, but it is important in the construction industry, surveying, and even agriculture. Laser instruments for this purpose use a visible red beam from a low-power helium-neon laser or a red diode laser. They have simplified some surveying tasks, and also aid in construction and agriculture.

One of the more important construction applications is in defining a plane surface to aid in lining up mounts for suspended ceilings or partitions. A laser is mounted in a tripod, with the beam directed up into a prism that bends the beam so it emerges out the side. The prism rotates in a full circle, sweeping the beam around the walls. Adjust this laser plane generator properly, and it sweeps the laser spot around the walls at the same height. Construction workers mount the hangers for suspended ceilings at this level, so they are all even around a room. Turn the laser plane generator 90 degrees, and it can mark places for partitions.

In agriculture, laser beams define the gradients of irrigated fields. The slope should be large enough that the water doesn't form puddles, but slow enough that the water doesn't run off too fast. A tripod-mounted laser can draw a straight line at the desired angle (for example, at 1 degree from the horizontal). Then the farmer can mount a sensor on his grading equipment to automatically keep the blade at the right height as it moves around the field.

Measuring Distance by Counting Waves

The wavelength of light is a very convenient unit for measuring small distances. The trick is to use the interference of light waves that we described much earlier to count the number of wavelengths. Interferometric techniques can make very exact measurements of the distance between two points.

To understand the basic idea of interferometry, we'll look at a simple example of how to measure a change in distance. Suppose we start with the arrangement shown at the top of Fig. 10-7A. Light from a laser is split into two beams with a beamsplitter and then directed at a pair of mirrors. The mirrors reflect the laser light back to inter-

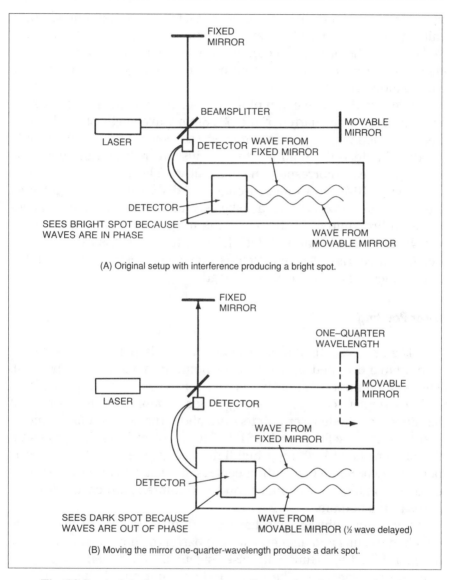

(A) Original setup with interference producing a bright spot.

(B) Moving the mirror one-quarter-wavelength produces a dark spot.

Fig. 10-7   An interferometer measures distance in units of wavelength.

fere with each other at a detector near the laser. We assume that the reflected waves are initially in phase when they meet at the detector, so the detector sees a bright spot.

Figure 10-7B shows what happens if we move one mirror just one-quarter wavelength farther from the laser. This makes that reflected wave's round-trip path one-half-wave longer than it was, so it is out of phase with the other reflected wave. The two wave amplitudes add to zero, producing a dark spot at the detector. If we keep moving the mirror, the detector will see another light spot, then a dark spot, and so on. If we hook the detector up to electronics that can count the light and dark cycles, it can count how many quarter-wavelengths the mirror moves.

Because the wavelength of light is small, these measurements are quite precise. One-quarter of the 632.8-nm wavelength of the helium-neon laser is 158.2 nm, so if the mirror was moved 1 μm—a thousandth of a millimeter—it would go through just over three full light-dark-light cycles, each corresponding to a half-wavelength.

Figure 10-7 shows measurement of distance in a straight line. However, interferometry can also be used to measure distance changes across a broad area. In that case, the interference pattern becomes visible as a set of light and dark fringes, tracing paths over the area being studied. Each fringe indicates a change in distance between the two surfaces of one-quarter-wavelength.

Laser Pointing

The simplest thing a laser beam can do is point, and as visible lasers have dropped in price, laser pointers have become popular. The well-equipped executive giving a presentation may carry his or her own semiconductor laser pointer. Complete with a couple of batteries, red diode laser pointers are about the size of a fat ballpoint pen, and some sell for under $100. They project bright red spots on the screen, ideal high-tech highlights for whatever information is being shown. Some convention centers still use an earlier generation of laser pointers—fat helium-neon laser pointers, attached by power cords to the podium.

Red laser pointers also have a very different use. They can be attached to guns and aligned with the barrel to show the point where a bullet would hit. Guns with laser sights are used mainly by police and security agencies. The bright red spot not only helps them aim the gun, but can help persuade the potential target to surrender before the bullet hits its mark.

RANGEFINDING AND LASER RADAR

The same principle used in radar measurements of distance can be used for laser distance measurements and finding the range to potential targets. You aim the laser at a target and then fire a short pulse of light. Then you measure the time it takes the reflected light to return to a receiver at the laser. The measured time t lets you calculate the distance d from the equation:

$$d = ct/2$$

where c is the speed of light. You need to divide by two because the light actually travels twice the distance to the target—once on its way there and the second time on its way back.

The laser pulse should be short to measure distance accurately. The uncertainty in distance measurements is given by the same formula, but this time substituting the pulse length for t. If the pulse lasts 1 μs, the distance uncertainty is

$$\text{Distance Uncertainty} = 3 \times 10^5 \text{ km/s} \times 10^{-6} \text{ s/2}$$

$$= 0.15 \text{ km}$$

Reduce the pulse length to 10 ns, and you get a much smaller uncertainty:

$$\text{Distance Uncertainty} = 3 \times 10^5 \text{ km/s} \times 10^{-8} \text{ s/2}$$

$$= 1.5 \text{ m}$$

In practice, the distance uncertainty is strongly affected by the accuracy of pulse-timing and measurement electronics.

Laser rangefinders are used in surveying and a variety of research and geophysical measurements. Short-pulse laser rangefinders have measured the distance from the earth to the moon. They can also measure the precise distance from points on earth to satellites, making it possible to detect very small motions of the planet's surface. Such laser measurements can detect the motion of continents and changes along faults that might be subject to earthquakes.

The biggest users of laser rangefinders are the armed forces. They are standard equipment on the battlefield to measure the distance to potential targets. In modern systems, the rangefinder may be hooked directly into a gunnery computer to tell the gunner information he needs to hit the target. One special type of military laser rangefinder is a proximity sensor, used in certain anti-aircraft missiles. As the missile approaches its target, the laser fires a short pulse and times how long it takes the reflected light to return. The shorter the return time, the closer the target. The proximity sensor waits for the missile to come near enough to the target that detonation of the missile's warhead would destroy it—then the proximity sensor detonates the warhead.

Solid-state neodymium lasers are the usual choice for battlefield rangefinders. Ruby or semiconductor lasers have been used for other purposes. Cheap semiconductor lasers are the obvious choice for proximity sensors, where the lasers can only be used once.

OTHER MILITARY TARGETING AIDS

Rangefinders are not the only low-power military lasers on the battlefield. Other laser systems are used to mark targets for precision-guided munitions, or in battle simulation games that were invented years before Laser Tag reached the toy stores.

Laser Target Designators

A laser target designator is an extension on the rangefinding and pointing laser systems we mentioned earlier. A designator fires a series of coded pulses at a battlefield target. This series of pulses serves as a unique "mark," which identifies the target so that a "smart" bomb or missile can home in on it.

The smart bomb or missile contains a sensor that looks for the characteristic series of pulses emitted by a target designator. (The coding is used to prevent the bomb from being misguided by bright lights, fires, reflections of the sun, or other sources of light.) The simplest such sensors focus light from the target onto detectors divided into four quadrants. As long as equal amounts of light fall onto each quadrant, the bomb is known to be on course. If one quadrant starts getting more light than the others, the bomb corrects its course to balance the light reaching all the quadrants.

First used in the Vietnam War, laser-guided bombs have proved much more accurate than conventional unguided bombs. They were particularly impressive against Iraqi targets in the Persian Gulf War. However, the use of laser target designators is not without its risks. The soldier holding the designator must keep the target "marked" until the bomb hits it. That means that he must keep the target in sight. Unfortunately, if he can see the target, the target can see him, which can make life rather unpleasant—if not short—if the target contains enemy soldiers who learn they are being marked by a laser designator and can find the soldier with the designator.

Another hazard is eye damage from the laser pulses. Military agencies are not concerned about enemy soldiers who they're trying to put out of action, but with the safety of their own soldiers who use laser designators in training exercises. This is an important concern, because modern soldiers spend more time in training than in real battles. Efforts to reduce the hazard have led to the development of rangefinders and target designators that operate at wavelengths of 1.5 μm and longer, which are too long to penetrate to the retina, the most sensitive part of the eye. (See Appendix A for an overview of laser safety.)

Laser Battle Simulation

In the 1970s, the Naval Training Equipment Center in Orlando worked with military contractors to develop the Modular Integrated Laser Engagement System, called "Miles." Miles is a training system that equips soldiers with pulsed semiconductor lasers and sensors. The lasers are attached to all kinds of weapons, and each fires a characteristic sequence of pulses. One code indicates that the pulses come from a rifle, another a bazooka, and a third heavy artillery. Sensors are strapped on trucks and tanks as well as to soldiers.

When the war games start, the soldiers fire laser pulses at each other, and the sensors keep score. A laser-simulated rifle shot can "kill" a soldier. Tanks, however, can only be knocked out by certain types of weapons. (To keep things honest, when the sensors on a tank detect a "kill," they turn off the controls and fire a plume of purple smoke to indicate to everyone on the battlefield that the tank is dead.)

Similar laser battle simulation systems are used today by armies around the world. Doubtless it was someone who had seen one of them who invented Laser Tag games, which are based on much the same principle but don't actually use lasers.

SPECTROSCOPIC MEASUREMENTS

In describing the physical basis for lasers in Chapter 2, we told how materials absorb and emit light at characteristic wavelengths that depend on internal energy levels. The development of lasers has helped scientists study these energy levels and wavelengths in much more detail than was previously possible. Many of the most sophisticated uses of lasers involve measuring wavelengths of light absorbed and emitted by various materials, a discipline called "spectroscopy." Such observations can identify materials, tell the quantities present, and help in studying their properties. There are several different types of laser spectroscopy, but we will concentrate on those of the most practical importance.

Fluorescence Spectroscopy

Fluorescence spectroscopy is the measurement of wavelengths at which materials "fluoresce" when they are illuminated by a short-wavelength light source. The wavelengths emitted depend on both the nature of the material and the excitation wavelength.

Figure 10-8 gives a simple example of how fluorescence spectroscopy works. Suppose we want to check newly made clothing to see if it is contaminated with oil that would make the clothes unsellable. We know the oil fluoresces at 450 nm when illuminated with 308-nm

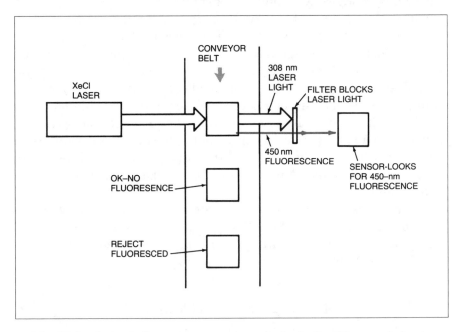

Fig. 10-8 A simple fluorescence spectrometer looks for 450-nm emission to detect contamination.

pulses from a xenon-chloride laser. We can run the clothes along a conveyor belt, hitting each one with one or more excimer pulses. We look at them through a narrow-band filter that transmits only light at 450 nm. If more than a certain amount of light passes through the filter, we know that the oil has contaminated the clothing, and it should be rejected.

The same idea can be used to detect other contaminants, but we may have to use other lasers and look for other fluorescence wavelengths. We can also do more with fluorescence spectroscopy than our simple example indicates. We can, for example, measure the amount of fluorescence to measure how much material is present. That could let us do a quick test for water pollution by certain chemicals.

It's possible to be misled by measurements at only one wavelength, because other compounds might fluoresce at the same wavelength. To avoid such problems, we can look simultaneously at several wavelengths. In that case, we could spread out a spectrum across an array of detectors, with each element in the detector array picking up light at a different wavelength. Then we could feed outputs from those detector elements into a computer, which would compare the measured spectrum with the spectra of various compounds. For example, fluorescence at 450, 462, 501, and 512 nm might indicate that compound A was present, while fluorescence at 450, 480, and 630 nm might indicate that compound B was present. Measurement of how much light was emitted at each wavelength would indicate how much of each compound was present.

One of the most interesting uses of fluorescence spectroscopy is to reveal otherwise invisible fingerprints. Some oils from human skin that are present in fingerprints fluoresce when illuminated with light near 500 nm. These oils linger long after other evidence of the fingerprint is gone and are visible on surfaces such as paper, where fingerprints cannot be recovered otherwise. Thus, some police agencies examine articles with argon or copper-vapor lasers in hope of finding otherwise invisible latent fingerprints. The fluorescent fingerprints can be photographed and identified—and have sent some criminals to jail.

Absorption Spectroscopy

Absorption spectroscopy is analogous to emission spectroscopy, except that it identifies materials by the characteristic wavelengths that they *absorb*. Absorption spectroscopy can be thought of as observing which for what wavelengths a sample has subtracted from the light shined through it.

You don't have to have a laser for absorption spectroscopy. You can start with an ordinary light source that emits the full spectrum of visible light, and then spread out the transmitted spectrum to see what wavelengths have been absorbed. However, the laser gives much finer resolution, so you can identify the precise wavelengths that have been absorbed. With an ordinary light source, you might know only that light was absorbed near 650 nm. A laser could tell you that light was absorbed at two wavelengths in that range, 649.87 and 650.13 nm. Because the laser can concentrate light at a single wavelength, it can also spot absorption lines too weak to see using other techniques.

Tunable lasers are the most common types used in absorption spectroscopy. Typically, they are "swept" in wavelength by tuning throughout their emission range. For example, a dye laser could be tuned from 640 to 670 nm, as shown in Fig. 10-9. (The laser output could be pulsed or continuous-wave, but different measurement instruments would

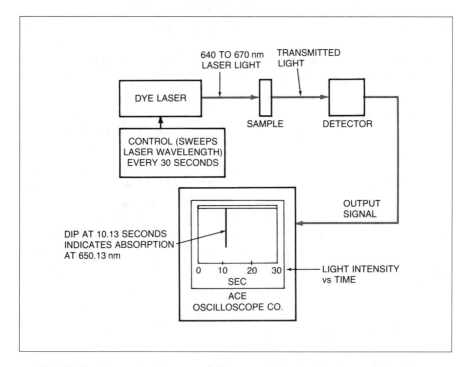

Fig. 10-9 Absorption is measured by sweeping dye-laser wavelength from 640 to 670 nm every 30 s.

be used for each type of laser.) To measure absorption as a function of wavelength, you would measure the amount of light transmitted throughout the sweep, keeping track of the time. If the entire sweep took 30 s, and the tuning rate was uniform in wavelength from the short to the long end of the spectrum, you would know that absorption 10.13 s after the start of the sweep was at a wavelength of 650.13 nm.

Absorption spectroscopy is useful in the infrared as well as the visible. The infrared is often preferred for studies of organic compounds, because the absorption bands most useful for identification—arising from molecular vibration and rotation—lie in the infrared. While some lasers have tunable output at those wavelengths, most infrared absorption spectroscopy is still done with nonlaser sources, because the extremely high resolution possible with lasers is not essential.

Biomedical Diagnostics

A major application of spectroscopic measurements is in instruments for biomedical diagnostics. The instruments can use either fluorescence or absorption spectroscopy to look for abnormalities, count cells, or make other measurements.

Suppose, for example, that you could stain cancer cells with a dye that fluoresced brightly at 480 nm when illuminated by a 308-nm XeCl laser. You could put the cells into a carrier liquid and flow them past a laser and a detector. If the detector saw 480-nm fluorescence, it would indicate the presence of cancer cells.

An even simpler use of absorption spectroscopy is counting cells in a blood sample. The blood could be made to flow through a tube so thin that it lets only one cell through at a time. Every time a cell passed, it would prevent light from a low-power laser from reaching a detector. The detector electronics would register the interruption of the light beam and count that as the passage of a blood cell.

The attraction of these spectroscopic laser systems is their ability to automate and speed measurements. Many biomedical measurements have traditionally been made manually by technicians, who must look for abnormal cells or count the number of cells present. Technicians can get bored and make mistakes, but the laser system won't. The automated measurement systems help reduce costs as well as improve reliability.

Pollution Measurements

Another important use of spectroscopic laser systems is to detect air or water pollution. For example, you could aim a carbon-dioxide laser at the plume of smoke emitted by a smokestack and measure the reflected wavelengths. You could tell the quantities of pollutants emitted by measuring how much light is reflected at certain wavelengths. Thus, environmental agencies can measure concentrations of harmful gases such as sulfur dioxide and nitrogen oxides without having to climb to the top of a smokestack.

A combination of laser techniques can measure, from the ground, the concentration of ozone (O_3) high in the atmosphere. Ozone is important because it keeps the sun's potentially harmful short-wavelength ultraviolet radiation from reaching the surface of the earth. Scientists are concerned that certain pollutants are destroying ozone, especially above the North and South Poles, allowing too much ultraviolet light to reach the ground. However, it has been difficult to measure ozone concentrations precisely so far above the ground.

Lasers can solve that problem. Scientists start with a laser that emits short pulses at a wavelength absorbed or scattered by ozone. (In practice, they use the Raman shifting technique described in Chapter 5 to move output from an excimer laser to the desired wavelength.) They fire a pulse up into the atmosphere, and watch for a return signal. Then they may fire a second pulse at a different wavelength, and likewise watch for a return signal. By comparing the time it takes the signal to return, they can measure altitude like a rangefinder measures distance. By comparing the strengths of the two wavelengths—one affected by ozone, one not affected—they can measure the levels of ozone at those altitudes.

Basic Research

Many uses of spectroscopy remain in the realm of basic research. Laser light, carefully controlled in wavelength, direction, and timing, is an exquisitely sensitive probe of the inner workings of atoms and molecules. Lasers can measure very slight differences between atomic energy levels that are important in evaluating theories of atomic structure. They can measure the very slight differences in energy levels caused by the differing weights of natural and synthetic isotopes. They can measure the very slight shifts in wavelength caused by the motion of atoms and molecules themselves. They have been able to detect the presence of single atoms.

The importance of lasers in exploring the physics of atoms and molecules was recognized by the Nobel Committee in 1981, when it awarded the 1981 Nobel Prize in Physics to Arthur L. Schawlow of Stanford University and Nicolaas Bloembergen of Harvard for their work on laser spectroscopy. (The third person to share in the 1981 Physics Prize, Kai Seigbahn, was cited for research in other types of spectroscopy.)

TIME MEASUREMENTS

The fastest man-made events are pulses of light. The current speed record is six femtoseconds—6×10^{-15} second. Those ultrashort pulses of visible light, generated at AT&T Bell Laboratories, were only about three wavelengths long. Generating them required a complex optical system, but recently researchers at Washington State University produced 11-femtosecond pulses directly from a titanium-sapphire laser. Femtosecond pulses have become important tools in research and development.

What can you do with femtosecond pulses? They can help measure how fast chemical and physical processes occur. By combining time-resolved measurements and absorption spectroscopy, you can follow the progress of a chemical reaction by taking "snapshots" of composition every 10 or 100 femtoseconds. That technique can give valuable information on the speed of chemical reactions and can identify intermediate compounds that are formed during a reaction.

Fast laser pulses can also measure electrical processes, such as changes in conduction in semiconductor devices. For example, one ultrafast laser technique measures the electric field on the surface of integrated circuits a thousand times faster than the best sampling oscilloscopes. This lets scientists time events in electronic circuits that are too fast to measure with conventional techniques.

HOLOGRAPHY

Virtually all holograms are made with lasers, but few people are aware that holography was invented in 1948, a dozen years before the laser. Hungarian-born engineer Dennis Gabor, working in Britain after

World War II, originally conceived of holography as a way to improve the resolution of the electron microscope. His goal was to record and reconstruct not just the amplitude of a wave (as in a photograph), but also its phase—information lost in normal photography. That information should allow reconstruction of the entire light wave.

Gabor wasn't thinking of recording three-dimensional images. His original holograms were of flat two-dimensional transparencies. It wasn't until 15 years later, when Emmett Leith and Juris Upatienks started making laser holograms, that people realized the possibility for three-dimensional holograms.

Coherence is the property of laser light that is important for making holograms. If you're going to make a hologram of a three-dimensional object, the light you use must have a coherence length at least equal to the dimensions of the object. (That is, the waves that make up the light must march along in phase for at least that distance.) Other light isn't coherent over such distances, but laser light is. This lets you record both the amplitude and phase of the light reflected from the entire object. Being able to record that wavefront, in turn, lets you reconstruct a three-dimensional image of the original object.

The process of recording a hologram is shown in Fig. 10-10. Light from a laser (usually a low-power helium-neon laser) is split into two beams by a beamsplitter. Half forms the object beam, which is reflected from the object. The other half forms the reference beam, which follows a separate path and is combined with the reflected object beam to record the hologram.

Figure 10-10 shows the hologram being recorded on a photographic plate. The plate actually records an intensity pattern—the interference pattern generated by combining the object beam and the reference beam. This interference pattern contains information on the phase of the object beam. All you have to do is create another reference beam (possible because you know its characteristics) and shine it at the hologram. Scattering of the reference beam from the interference pattern recorded on the hologram generates a light wave with the same phase and intensity distribution as the original object beam. The result is a three-dimensional image that seems to float in space, roughly the same size as the original object, as we show at the bottom of Fig. 10-10.

The first holograms had to be viewed as well as created by laser light, as we show in our figure. However, later advances made it possible to view laser-created holograms in white light. Careful choice of the way in which the hologram is recorded and viewed let white light

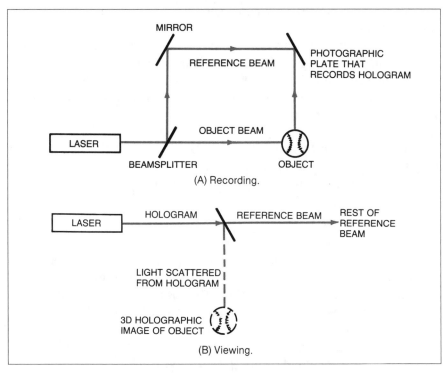

Fig. 10-10 Recording and viewing a hologram.

serve as the reference beam. You can view so-called rainbow holograms in any color of the spectrum; the appearance depends on the viewing angle. These rainbow holograms can be mass-produced by stamping in metallized plastic, so they can be made cheap enough to put on millions of credit cards and copies of *National Geographic*, and to use in special postage stamps.

Holographic Interferometry

Holography can be combined with interferometry to make sensitive measurements of how objects deform under stress. The basic idea is fairly simple, and requires a pulsed laser to take holograms. First, you take a hologram without applying any stress to the object. Then you apply an increment of stress and record another hologram on the same photographic plate.

When you view the combined hologram, you see a single image of the object crossed with bright and dark lines. Those are the interference fringes that we described earlier. They measure how far the object's surface moved between exposures. The amount and uniformity of change is important. Suppose you tested an aircraft tire before and after applying pressure. If the changes are uniform, they indicate that the tire expanded slightly over its entire surface when air pressure increased. That is what should happen if the tire is sound. On the other hand, if the changes are concentrated in a small area, they usually betray a flaw less able to withstand strong force than the rest of the tire.

Holographic interferometry has become the standard way to assess the quality of aircraft tires, and it is also used to assess the quality of many other high-performance components.

Holographic Optical Elements

We have seen how holograms can reconstruct three-dimensional images. However, there's another way to view holograms—as optical elements. From that viewpoint, the diffraction pattern recorded in the hologram scatters or bends light in a controllable way. By using this approach, you can design a hologram that functions as a sophisticated lens, redirecting light by scattering it. Typically, holographic optical elements are designed by computer programs.

Holographic optical elements are not as efficient in redirecting light as lenses or mirrors, but they can do things that are difficult or impossible with other types of optics. One example is a large holographic lens used in military aircraft displays. The lens projects an image of the control panel in the air in front of the pilot, forming what is called a "head up" display. This lets the pilot avoid looking down at his dials and controls in the midst of a battle. The large area and short focal length make it impractical to design the lens as anything but a holographic optical element.

ART AND ENTERTAINMENT

We have yet to cover the most striking uses of low-power visible lasers—in displays and light shows. Millions of people around the world have seen brightly colored laser beams light up the night sky, a theatre, or a planetarium.

The concept is fairly simple. Just bounce beams from visible lasers off moving mirrors. The mirrors scan the laser beams in patterns made visible by scattering from air particles, and reflections from clouds, nearby buildings, or a screen. The scanning patterns can be controlled by music, by a computer program, or by someone operating the lasers (who becomes the optical analog of a musician). It is fascinating to watch the patterns unfold, and the display is safe as long as the beam is kept out of people's eyes. In practice, the key to a good laser display is imaginative control of the scanning system.

Many types of visible lasers have been used in displays. A few light shows have used dye lasers, but the typical types are argon- and krypton-ion lasers, which generate powers of 1 W or more at several visible wavelengths. The argon-ion lines generate green and blue lines. Krypton has a strong red line, and can also produce yellow— which is a hard color to find in laser light. The different wavelengths can come from separate lasers, or from a single "mixed gas" laser that contains both argon and krypton and emits on lines of both.

WHAT HAVE WE LEARNED

- Laser light illuminates words and codes so they can be read by special detectors and decoding electronics.
- UPC symbols used on food packages are designed to be read at the 632.8-nm helium-neon laser wavelength.
- Optical disk technology was initially developed for home video playback. Today, most lasers in the world are used in compact disc (CD) audio players, which play digital sound from 12-cm disks with diode lasers.
- The mass production of CD lasers and other optical components has made inexpensive optical components available for other purposes.
- Other formats have been adapted from CDs. The CD-ROM format can store 600 Mbytes of digital data for computer use. CD-I is an interactive video format; CD-V is a moving video format.
- Lasers can store data by writing on light-sensitive disks. Some are write-once materials, on which data cannot be changed once it is written. Most erasable optical disks use magneto-optic materials, on which data is read and written optically, but stored magnetically.

- A scanning laser beam can write text or images on a light-sensitive surface. Laser computer printers write with a diode laser on a photocopier drum.
- Laser platemakers reproduce copies of newspaper printing plates at remote locations.
- Diode lasers are the light sources used in fiber-optic systems for long-distance telephone communications.
- Optics can switch and manipulate light to process signals and perform computing functions.
- Lasers can draw straight lines and planes to align machinery and construction equipment.
- Interferometry uses interference effects for precise measurements of short distances in units of the wavelength of light.
- Laser radars measure distance by timing the round trip of a reflected pulse.
- A major use of laser rangefinders is pinpointing targets on the battlefield.
- Laser target designators fire coded pulses to identify a target to sensors on a homing bomb or missile.
- Laser pulses can simulate weapons fire in war games.
- Spectroscopy is the study of the wavelengths absorbed and emitted by materials. It can identify some materials by the wavelengths of light they absorb or emit.
- Fluorescence is the emission of light at a longer wavelength after illumination by a shorter wavelength.
- Biomedical diagnostics are a major use of spectroscopic measurements.
- Spectroscopic laser systems can detect air or water pollution and monitor ozone concentrations in the atmosphere.
- Laser spectroscopy is particularly important for basic research.
- 2-D holograms were made before lasers were available. The long coherence length of lasers made 3-D holograms possible.
- Holographic interferometry can measure deformation of objects under stress.
- Holograms can serve as optical elements.
- Bright scanning laser beams make striking displays and works of art.

WHAT'S NEXT

In our final chapter, we will study the uses of high-power laser beams.

Quiz for Chapter 10

1. Which of the following are differences between laser light and light from other sources?
   a. Laser light has much longer coherence length.
   b. Laser light can be focused onto a small spot.
   c. Laser light contains only a narrow range of wavelengths.
   d. Lasers can generate extremely short pulses.
   e. All of the above

2. Which of the following is an important limitation of lasers?
   a. Laser light is invisible.
   b. Light rays from a laser are parallel.
   c. Inexpensive lasers do not have wavelengths shorter than the red end of the visible spectrum.
   d. Lasers do not have long lifetimes.
   e. Laser light interferes with itself.

3. What type of light sources must be used to read the Universal Product Code on food packages?
   a. Red light sources
   b. Infrared diode lasers
   c. Fluorescent lights
   d. Any laser
   e. Argon lasers

4. What was the first type of optical disk developed?
   a. CD-ROM
   b. Videodisk
   c. Compact disc
   d. Write-once, read-many (WORM)
   e. Erasable

5. What are the differences between WORM and CD-ROM disks?
   a. Data can be written once on WORM disks but not at all on CD-ROMs.
   b. WORM disks are 13 cm in diameter, CD-ROMs are 12 cm.
   c. Standard recording formats are different.
   d. A and B
   e. A, B, and C

6. Which of the following is not a major application for semiconductor diode lasers?
   a. Fiber-optic communication systems
   b. Optical disk data storage
   c. Compact disc players

d. Holography

e. Laser printers

7. Which of the following is not a promising application for optical computing and processing?

a. Interconnection of chips

b. Balancing your checkbook

c. Matrix-vector operations

d. Fourier transforms

e. Processing synthetic-aperture radar signals

8. An interferometer that uses the 632.8-nm light from a helium-neon laser has measured motion of a mirror in an arrangement like Fig. 10-7 as 15 pairs of dark-light fringes. What distance does this equal?

a. 632.8 nm

b. 2.37 μm

c. 4.75 μm

d. 7.12 μm

e. 9.49 μm

9. A pulse of light returns to a laser rangefinder 100 μs after it is emitted. How far away is the object being measured?

a. 30 m

b. 1 km

c. 15 km

d. 30 km

e. None of the above

10. How is laser spectroscopy used?

a. In basic research

b. To measure concentrations of pollutants

c. To count blood cells

d. All of the above

e. None of the above

CHAPTER 11

High-Power Laser Applications

ABOUT THIS CHAPTER

In Chapter 10, we described important uses of low-power lasers, loosely defined as those which don't make major changes to objects. This chapter covers the major applications of high-power lasers. The applications practical today include cutting, welding, drilling, and marking in industry, some specialized tasks in the manufacture of electronic components, and a variety of medical treatments. Several other applications for high-power lasers remain in research and development, including causing and controlling chemical reactions, purification of isotopes, laser weapons, and thermonuclear fusion.

HIGH- VS. LOW-POWER LASER APPLICATIONS

The line we have drawn between high- and low-power laser applications can be a hazy one. We call high-power applications those which cause major changes to the object the laser beam strikes—but what do we mean by "major"? Cutting and welding clearly are major changes, whether the cutting is of sheet steel or the surgical removal of tissue. The marking of serial numbers and product codes on components is closer to the line, but we tend to consider that a subcategory of industrial materials working.

On the other hand, we put applications that cause changes in light-sensitive materials on the other side of the line. Examples include laser printers and optical data storage. Part of our rationale is that ordinary light can make similar changes—like the exposure of photographic film; part is that some of these applications are closely related to other low-power applications—for example, optical data storage by writing on disks is one of a family of optical disk applications that logically belong together. A final element of the rationale is the need to make some separation lest these final two chapters merge into a single monstrous and unmanageable whole.

In reading the previous chapter, you may have noted the diversity of low-power laser applications. There is less evident diversity in the use of high-power lasers. Many are variations on the basic theme of laser materials working—heating an object to temperatures sufficient to vaporize and remove its surface layer. From this standpoint, many medical applications are merely specialized materials working (a technically accurate view, if a bit callous to the patient), and laser weapons are merely intended to perform materials working on unfriendly objects.

You will also note that some interesting high-power laser applications remain in development, even after many years of research. For example, the prospects for the practical use of high-power laser weapons and laser fusion remain unclear, despite research dating back to the 1960s and the expenditure of billions of dollars. The problem is that both are massive programs, which require developing a complex set of supporting technology as well as the big lasers needed.

Finally, all uses of high-power lasers have benefitted in some way from military support. Military basic-research programs helped develop lasers of the power levels needed for materials working and medicine. As we will see later, other programs have been more or less directly supported by military programs. Most of that military support has been stimulated by interest in lasers as weapons.

ATTRACTIONS OF HIGH-POWER LASERS

In Chapter 10, we listed the attractions of low-power lasers. Let's take another look at the attractions of lasers, this time concentrating on those important for high-power applications:

- Lasers produce well-controlled light that can be focused precisely onto a small spot.
- High-power laser beams can produce very high power densities when focused onto small areas.
- Laser beams do not move or distort the objects they strike.

- Lasers can generate light in short pulses, thus producing very high power levels for short intervals.
- Lasers can be controlled directly by robotic systems.
- Different laser wavelengths are absorbed by different materials.
- Laser light travels a well-defined straight line.

This list is not identical with the one in Chapter 10, but some points are quite similar. On the other hand, the list of disadvantages of high-power lasers is rather different:

- Lasers have high capital cost.
- Many materials strongly reflect the wavelengths emitted by the most powerful lasers.
- Laser light cannot cut deeply into materials.
- In some cases, laser light can be hard to control completely.
- Extremely powerful beams are hard to deliver to targets.

In the rest of this chapter, we will look systematically at the major uses of high-power lasers and their limitations.

MATERIALS WORKING

Materials working is the broad field of cutting, welding, drilling, and otherwise modifying industrial materials, including both metals and nonmetals. The 1960s engineers who played at measuring laser power in gillettes were—at least in some primitive sense—exploring one aspect of materials working (drilling holes in razor blades).

It's instructive to look at some of the earliest successes of lasers in materials working because they highlight how to take advantage of the properties of laser beams. We'll look at two that represent interesting extremes before moving on to look at materials working in general—drilling holes in diamond dies for drawing wire, and drilling holes in baby-bottle nipples.

Drilling Diamond Dies

One way to make thin metal wires is by pulling the metal through tiny holes in diamond "dies." Diamond is an excellent die material because it is the hardest substance known. It also conducts heat well, and thus can dissipate the heat generated by friction when metal is pulled through the hole.

The problem is that the hardness of diamond also makes it an extremely difficult material to drill. Because diamond is the hardest material known, it takes diamond bits to drill holes in diamonds, and those diamond bits get dull as they drill. Laser light, however, does not get dull. If a laser beam is focused onto a tiny spot in the diamond, it can remove the carbon atoms—heating them so that they combine with oxygen in air or merely evaporate. Deposit enough laser energy, and you can make a hole through the diamond.

Engineers discovered that lasers could drill holes in diamond dies during the 1960s, and ever since this has been a standard example of laser applications. It is not a high-volume application, but it is one where there are no attractive alternatives.

Drilling Baby-Bottle Nipples

The problems in drilling holes in baby-bottle nipples are quite different from those in drilling diamond. The rubber used for nipples is very soft and flexible, which makes it hard to drill holes in the material. For milk to flow properly, the holes in the nipples must be small. The best way to make such holes mechanically is with tiny wire pins, but the rubber bends, and the fine pins can be caught and broken by the flexible nipple material. The task is something like trying to punch holes in Jell-O with a thin soda straw.

A laser beam is an excellent alternative because nothing solid contacts the nipple to push it out of shape. The laser pulse simply burns a tiny hole through the rubber, without any chance for anything to get stuck. Like drilling holes in diamond dies, this laser application has been around for many years and doesn't require many lasers—but it is nonetheless the best solution to the problem.

Noncontact Processing

As we have indicated, an important factor in drilling both diamond dies and baby-bottle nipples is that laser drilling is a noncontact process. There's no physical laser "bit" that can get dull while drilling diamond. Nor can the laser beam apply any force to distort the baby-bottle nipple or get caught while drilling the hole. Avoiding the need to bring a drill bit against the object also speeds processing.

The noncontact nature of laser processing is important for many types of laser drilling, cutting, and welding. Interestingly, many materials most often machined by lasers tend to fall into categories similar to diamond and rubber—either very soft or very hard. Titanium, a light metal used in military aircraft, is too hard to cut easily with conventional saws, but can be cut readily with a carbon-dioxide laser. Other advanced alloys and hard materials are also often cut with lasers. On the other extreme are some rubbers, plastics and other soft materials that are usually cut with lasers.

Wavelength Effects and Energy Deposition

Our examples haven't told us much about the fundamental interaction that drives laser materials working—the transfer of energy from the laser beam to the object. The actual process is complex, but you should have at least a general understanding of what happens to appreciate the nature of laser materials working.

We saw earlier that all surfaces reflect, transmit, and absorb some energy. If we want laser energy to do something to an object, it is energy absorption that is most important. Reflected and transmitted energy doesn't do any good, because it is not transferred to the object. As we indicated earlier, the amount of energy reflected or absorbed depends on wavelength. Differences in absorption at different wavelengths can be quite large, and can make certain laser applications impractical.

Table 11-1 lists the reflectivity of some common metals and a few other materials at various laser wavelengths. Metals tend to become more reflective at longer wavelengths, while skin and paint absorb more light at longer wavelengths. Transmission is very low in reasonable thicknesses of these materials, so we assume that the remaining light is absorbed. The differences are dramatic for some metals, which absorb reasonably well at the 1.06-μm neodymium wavelength but are highly reflective at the 10.6-μm carbon-dioxide wavelength. Metals that reflect strongly at both wavelengths are poor candidates for laser applications; aluminum is particularly poor because it is soft and easy to cut with conventional saws and blades. Titanium, on the other hand, is a good choice for laser materials working because it absorbs more strongly at 10.6 μm than other materials, and it is so hard that it is almost impossible to cut with a conventional saw.

TABLE 11-1 Absorption of different materials at important laser wavelengths: as a function of wavelength

	Laser/Wavelength			
Material	Argon 500 nm	Ruby 694 nm	Nd-YAG 1064 nm	CO_2 10.6 μm
Aluminum	9%	11%	8%	1.9%
Copper	56%	17%	10%	1.5%
Human skin (dark)	88%	65%	60%	95%
Human skin (light)	57%	35%	50%	95%
Iron	68%	64%	–	3.5%
Nickel	40%	32%	26%	3%
Sea water	*	*	*	90%
Titanium	48%	45%	42%	8%
White paint	30%	20%	10%	90%

*At these wavelengths, sea water transmits most light not reflected at the surface

We should note that low absorption at 10.6 μm does not make it impossible to cut metals with carbon-dioxide lasers. Although only a small fraction of the laser light is absorbed, the carbon-dioxide laser can deliver more power than Nd-YAG lasers, offsetting the higher absorption at the shorter wavelength.

Absorption of the solid surface does not tell the whole story, however. Once the surface absorbs some energy, the material starts to melt and then vaporize. For most materials, absorption increases with temperature, so the first bit of heating is the hardest. Moreover, many metals absorb even more light after they have been melted.

However, vaporization can create other problems by forming a layer of plasma or vapor between the laser and the object, as shown in Fig. 11-1. This plasma layer can absorb some laser energy, blocking it from reaching the object. Sometimes, turbulence produced by heating the object can remove some or all of this plasma. However, in other cases the plasma has to be removed to improve energy coupling. The simplest way is often blowing the plasma away.

Another complication comes as the laser beam starts going deep into the material. The deeper the hole, the farther the material vaporized by the laser beam must go to get out of the hole. Depending on the depth of the hole and the temperature of the vaporized material, some of it can be deposited in the hole or around it. At best, this can reduce the efficiency with which the laser removes material. At worst, it can stop the process altogether.

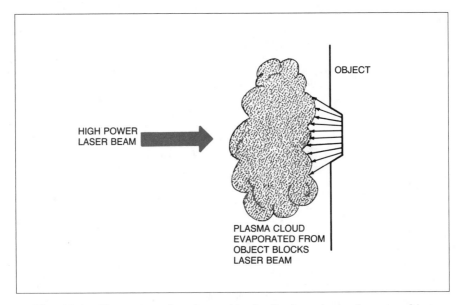

Fig. 11-1   Formation of a plasma blocks the laser beam from reaching the object.

Focusing of the laser beam also influences the maximum depth of the laser hole. Figure 11-2 shows the difference between a lens with short focal length and one with long focal length. The short-focus lens produces a smaller spot on the object, but the beam also spreads out faster beyond the point where the focal spot diameter is smallest. The longer-focus lens does not produce as tight a focal spot, but the beam does not spread out as fast, so it can drill deeper holes and cut thicker sheets of material.

The Drilling Process

When we talk of drilling, we specifically mean making a single hole through an object. Drilling is done with one or a series of short laser pulses, each of which removes some material, making the hole deeper. The process requires high peak power and works well for short repetitive pulses. (Shorter pulses have the high peak power needed to vaporize and remove material from the hole.) The number of pulses required depends on factors including wavelength, peak power, the nature of the material, and repetition rate. For the most efficient drilling, it may be necessary to change the optical focus during drilling to reach deeper into the drilled hole. The laser beam and/or the

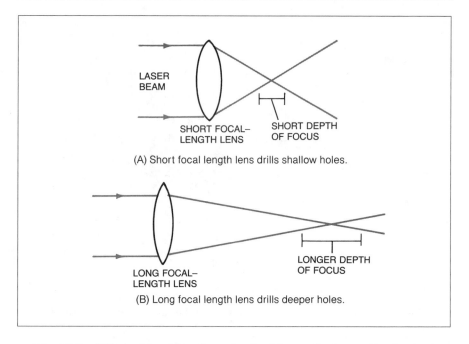

(A) Short focal length lens drills shallow holes.

(B) Long focal length lens drills deeper holes.

Fig. 11-2   Effect of focal length on depth of focus; the longer the depth of focus, the deeper the hole that the beam can drill.

object are moved to change the spot being drilled. It is also important to make sure that the vaporized material gets out of the hole.

It is simplest to envision drilling one hole at a time with a series of laser pulses. However, in a few applications many holes are drilled at once, sometimes with a single pulse that is divided into several lower-power pulses by passing through one or more beamsplitters. One example is the "drilling" of ventilation holes in cigarette papers. A pulse from a carbon-dioxide laser is split into many separate pulses, each of which punches a separate hole in the cigarette paper.

Marking and Scribing

The processes of laser marking and scribing are similar to drilling, but don't require making holes all the way through the material. In marking, the goal is simply to write an indelible message, serial number, or trademark on a component. This can be done by drilling shallow holes without bothering to penetrate deep into the material. The shallower the hole, the faster the job can be done.

Marking is often done by writing a series of dots on the surface to create the desired pattern (e.g., a serial number). Alternatively, the laser beam can be focused through a stencil-like mask to remove material from a larger area. The laser beam can mark a bare surface, but marking efficiency can be enhanced by coating the surface with something that absorbs laser light, such as black paint.

Scribing involves drilling somewhat deeper holes, more or less as a perforated line, in a brittle ceramic material. When the ceramic is bent, it breaks along the weakened line scribed by the laser pulses. This technique is most often used in electronics manufacture, a topic described in more detail below.

Cutting Processes

Laser cutting works in somewhat the same way as drilling, but with some important exceptions. You can think of cutting as drilling a series of holes that overlap. In cutting, the beam and/or the object move continuously, and the beam itself is normally continuous rather than pulsed. You can think of laser cutting as a process something like slicing with a knife, but in this case the knife is a beam of light.

Cutting is typically done with the assistance of a jet of air, oxygen, or dry nitrogen. For nonmetals, the role of the jet is to blow debris away from the cutting zone and improve the quality of the cut. Lasers can cut readily through sheets of wood, paper, and plastic, but thick materials are more difficult, and foods tend to char when cut, leaving an unappetizing black layer. Efforts to slice bread with lasers have produced burnt toast.

Metal cutting works differently and uses a different type of jet. The laser beam heats the metal to a temperature hot enough that it burns as oxygen in the jet passes over it. This process is properly called "laser-assisted cutting," because the oxygen in the jet actually does the cutting.

Welding

Welding may seem similar to cutting, but the two processes differ in fundamental ways. Cutting is the separation of one object into two (or more) pieces. Welding is the joining of two (or more) pieces into a single unit. Both require heating the entire thickness of the material, but the conditions differ.

Special parameters must be considered in welding, as shown in Fig. 11-3. In this example, we see two pieces of metal about to be welded together. Note that they must be fitted together precisely, with very little room between them. The laser beam heats the edges of the two plates to their melting points, causing them to fuse together where they are in contact. If the pieces are not in contact along the entire junction, the weld will be flawed. The laser beam must penetrate the entire depth of the weld (the thickness of the material) to make a proper joint. The heat of welding transforms the metal through the entire depth of the weld and in a zone extending on both sides of the actual junction. The width of that weld zone depends on the nature of the materials being welded, the power delivered by the laser, and how fast the laser beam moves over the surface. The farther from the midpoint of the weld, the less the heat affects the metal.

The compositions of the pieces being joined together are very important in welding. The materials do not have to be identical, but metallurgists have learned that some compositions do not bond together

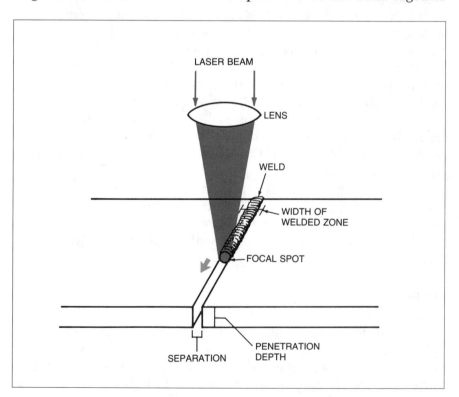

Fig. 11-3   Two metal plates being welded with a laser.

well, no matter how thoroughly they are heated or what technique is used for welding.

Like cutting and drilling, laser welding is limited in the thickness of material it can handle. However, it has gained wide acceptance in manufacturing high-technology products. The cases of heart pacemakers and the blades of Gillette Sensor razors are among products welded by lasers.

Heat Treating

Another materials-working use of high-power lasers is heat treating, a process in which a metal surface is heated to convert the surface layer from one crystalline state to another. Typically, heat treating converts the surface layer to a crystalline state that is harder and more resistant to wear than the rest of the material. That crystalline form of the metal may be too brittle to use in bulk form, but strong enough if it merely forms a layer on another type of metal crystal.

In laser heat treating, the beam—usually from a carbon-dioxide laser—scans across the surface of the metal. The surface may be covered with a coating to help it absorb the laser energy more efficiently. There are alternative ways to heat-treat metal surfaces, but a laser beam can reach into areas that other processes cannot, such as the insides of engine cylinders. Typically, heat treating is done with a high-power continuous laser, with the beam spread out to cover a large area.

Materials-Working Power Levels

The laser power levels chosen for particular processes depend upon the type of process, the type of material, its thickness, the laser wavelength, and other factors. Carbon-dioxide laser powers of under a kilowatt generally suffice for most applications involving nonmetals. Multikilowatt carbon-dioxide lasers may be used for cutting or welding thick metal plates (up to an inch or two thick), and for heat treating.

In some applications, you can trade off the speed with which the beam moves against laser power. That is, you can use a lower-power laser if you are willing to move the beam more slowly. For some processes, the results of moving a 4-kW beam at 2 m/s are functionally identical to those of moving a 2-kW beam at 1 m/s. That is, the same amount of energy is deposited on the area being cut or welded in a given time. However, if processing speed is critical, the higher-power laser may be preferred.

3-D MODELLING

Lasers also can be used to "build" three-dimensional objects, using a technique very different than other types of laser materials working, which requires fairly low laser power. The technique, called "stereo-lithography," relies on the fact that some liquids turn to solids when exposed to ultraviolet light. It is used to build up three-dimensional models from designs stored on computers.

In a stereolithography system, an ultraviolet ion laser beam is focused onto a platform slightly below the surface in a bath of liquid which solidifies when exposed to ultraviolet light. The computer controls the scanning of the beam across the surface of the liquid, forming a thin layer of solid on the platform. Then the platform is lowered farther into the liquid, and the laser beam scans the surface again, following another pattern to build the next layer of the model. After that layer is formed, the platform is lowered again, and the cycle repeats. Layer by layer, the system builds a 3-D model, which engineers can use in prototype development.

ELECTRONICS MANUFACTURE

In our discussion of hole drilling, we mentioned how lasers "scribe" ceramic wafers. This is one of several uses of lasers for materials processing in the electronics industry. Many differ from the more general materials-processing applications described above, mostly in their concentration on precision. Scribing, for example, involves drawing fine lines in brittle ceramic and semiconductor wafers so that they can be broken into precisely sized chips for components. Others rely on fundamentally different processes, in which light alters the composition of materials deposited on a semiconductor wafer.

Scribing, Mask Repair, and Circuit Changes

Wafer scribers and other laser systems built for the electronics industry are customized devices, designed and built for a specific application, not for general-purpose use. They may be built around solid-state neodymium lasers, argon lasers, carbon-dioxide lasers, or, in certain cases, other types.

Another special-purpose laser system for electronics manufacture is a resistor trimmer. This is a laser built to modify thin-film resistors deposited on hybrid circuits. The resistance of these devices depends on their surface area, which can be modified by removing some surface material with a laser beam. This removal fine-tunes resistance to the desired value.

Other laser systems can repair masks used in making integrated circuits (ICs). Suppose, for example, inspection of an IC mask shows a blob of excess material in a critical spot on the circuit. A laser can vaporize the excess material, cleaning up the mask so it can be used to manufacture good electronic components.

Likewise, laser pulses can customize certain types of integrated circuits by making or breaking connections. These special circuits are designed with laser customization in mind, and include extra connections that can be broken or made with a laser beam. If the chip is used for one application, the manufacturer can break one connection; if it is used for another application, the company can break another connection or set of connections. Changing these connections changes the behavior of the integrated circuit. Although certain chips have to be altered, the economics of integrated-circuit production favors mass production of many identical chips over small runs of customized chips.

Photolithography

Lasers can also provide light sources for photolithography, a critical process used in forming surface patterns on semiconductors. In photolithography, the surface of a wafer is coated with a light-sensitive material, called a "photoresist." Selectively exposing certain areas of the resist forms the pattern to be created on the wafer. The exposure changes the resist so part of it can be removed by chemical etching, leaving the desired pattern on the surface.

As integrated circuits shrink in size, it becomes more and more important to make smaller lines on the surface. The minimum width of features depends on the wavelength of the light that exposes the photoresist. The smaller the wavelength, the thinner the features. This has led developers to move from visible light sources to ultraviolet sources, including pulsed excimer lasers, which are in use today, making circuits with features 0.25 to 0.5 μm across.

Even shorter wavelengths are needed to produce smaller features. This has led to extensive efforts to develop X-ray sources for photolithography, including nonlaser as well as laser approaches. One technique that has been demonstrated (and is the basis of a commercial system) is using a short, powerful laser pulse to produce a powerful pulse of X rays that could illuminate the photoresist. A longer-term possibility is the use of X-ray lasers.

Other potential laser applications in semiconductor electronics are being studied in the laboratory. One important example is the deposition of metal films on semiconductors. Recent experiments have shown that lasers can decompose metal-containing gases just above the surfaces of semiconductors or insulators to produce metal films that conduct electricity well. Researchers are working on extending this technique so that they can use lasers to write electrically conductive paths on electronic circuits. Similar laser techniques have been developed to deposit thin films of high-temperature superconductor and diamond.

MEDICAL TREATMENT

Physicians weren't far behind engineers when it came to starting experimenting with lasers. The first to use lasers were dermatologists and ophthalmologists (eye specialists), who were used to working with light. However, it was not long before other specialists were looking at potential uses of lasers. Today, important fundamental work is still being done on the interaction between laser light and tissue, but lasers are in widespread use in medicine, and specialists have a good understanding of the basic processes involved.

Fundamentals of Laser-Tissue Interactions

We can consider the interaction between laser light and tissue as a problem in light absorption as a function of wavelength. Tissue, like any other material, absorbs light differently at different wavelengths, as you can see in Table 11-1 for light and dark skin. The characteristic wavelengths at which tissue absorbs light depend on its composition, and we can get an idea of how tissue should behave if we look at its major components.

The most important component of tissue is water, which absorbs strongly in much of the infrared. The absorption is so strong, and tissue contains so much water, that it isn't a bad approximation to consider tissue as absorbing light like water, especially in the infrared. At the 10.6-μm wavelength of carbon-dioxide lasers, water reflects about 10% of the incident light and absorbs about 80% in the first 20 μm thickness. Absorption is even stronger at 3 and 6 μm, although reflection is somewhat higher at those wavelengths.

The 10.6-μm wavelength is a favorite for laser surgery because it is readily available and absorbed strongly by water and tissue. Focus a carbon-dioxide laser onto tissue at high enough intensity, and the laser energy will vaporize the cells. The absorption is so strong that only the upper layer of cells will be killed; the lower level of cells will survive with little damage. The lower layer suffers even less damage if a laser emitting at 3 μm is used instead, because water absorbs that wavelength even more strongly.

The interaction between tissue and carbon-dioxide laser light has another useful aspect. The laser penetrates just deep enough into tissue to seal small blood vessels and stop bleeding. This makes the CO_2 laser an especially valuable tool for surgery in regions rich in blood vessels, such as the female reproductive tract. The CO_2 laser's ability to remove thin layers of blood-rich tissues gives it a special slot in the surgeon's collection of tools.

One important drawback of CO_2 lasers is that silica glass optical fibers are opaque at 10 μm. Surgeons like fiber-optic beam delivery because easy-to-handle fiber probes give them precise control over where light energy goes. Fibers can deliver 1-μm beams from neodymium lasers, but those beams are not absorbed as strongly and penetrate a millimeter or so into tissue. The resulting cauterizing effect can benefit in some cases, but is often harmful and leaves wounds that are slow to heal. As an alternative, surgeons are beginning to use holmium lasers which emit at 2 μm, a wavelength transmitted by silica fibers and strongly absorbed by water, which does not penetrate far into tissue. Holmium lasers are being adapted for delicate spinal and joint surgery, where they can remove cartilage as well as soft tissue. Water absorption is highest at 3 μm, and the solid-state erbium laser which emits at that wavelength can be used to cut bone. However, silica fibers cannot transmit 3μm, and surgeons continue to prefer conventional instruments for cutting bone.

Other laser wavelengths in the visible and near-infrared can serve other medical purposes. Advances in the state of the art of laser medicine are leading to better criteria for matching lasers to specific treatment needs. If you wanted to target a particular thing in the body—for example, abnormal blood vessels—you would search for wavelengths that the target tissue absorbs much more strongly than surrounding tissue. This would concentrate laser energy in the tissue you want to remove, while causing minimal side effects to surrounding tissue.

As we will soon see, a growing number of laser medical techniques were developed using this approach. The physician picks the best laser for the job, selecting a desired pulse length as well as wavelength to control energy deposition. The desire to control wavelength has led to the development of medical laser systems built around dye lasers, because their wavelengths can be tuned to the best possible value.

Laser Surgery

The surgical removal of tissue, shown conceptually in Fig. 11-4, is the simplest medical application of lasers to describe. The surgeon scans the beam from a carbon-dioxide laser across the tissue to be removed. Water in the cells absorbs the 10-μm beam strongly, vaporizing the top layers of cells but doing little damage to the underlying tissue. The process also seals off (or "cauterizes") blood vessels smaller than about a millimeter in diameter, effectively stopping bleeding.

Use of carbon-dioxide lasers has become routine for some types of surgery. It is particularly valuable for gynecological surgery, because it can remove abnormal tissue from the surface of blood-rich tissues. For example, lasers can remove abnormal and potentially precancerous cells on the surfaces of the vaginas of women whose mothers took the drug DES while they were pregnant. Lasers can also treat a condition called endometriosis, a condition affecting some 6 to 8 million American women which can cause pain and abnormal bleeding and impair fertility.

Another common but rather different use of carbon-dioxide lasers is to perform delicate microsurgery on the larynx without having to open up the throat. Passing the CO_2 beam down the throat in a small tube lets the surgeon remove small cancers or wart-like growths from the vocal cords without damaging them. Some specialists believe there's no effective substitute for the laser for treating some conditions.

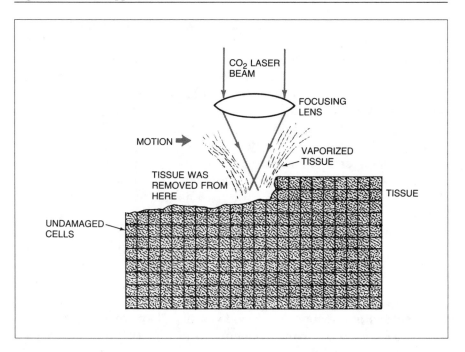

Fig. 11-4 A CO_2 laser removes the top layer of cells with minimal damage to those underneath.

We should emphasize that all surgeons are not about to throw away their scalpels in favor of lasers. The scalpel remains a powerful all-purpose tool, useful in a wide range of surgery, including many types where lasers can't do the job. As in materials working, it is much easier for a laser beam to deliver energy to the surface than to penetrate deep inside a body. (However, optical fibers can deliver laser energy to small selected points in the body, letting them treat conditions like kidney stones without extensive surgery, as described later in this chapter.)

Laser Eye Treatment

The first big successes of medical lasers were in eye treatment, and the laser remains an especially valuable tool for ophthalmologists. The simplest laser eye treatment to explain is a rare condition called a detached retina.

The retina is the light-sensitive layer at the back of the eye. Sometimes the retina can become torn and detached from the back of the eyeball. The damage can spread, and without treatment the entire retina can come loose from the back of the eye, causing blindness. The detachment can be stopped by focusing a laser pulse onto the retina so that it causes a burn, which forms scar tissue. Although this damages a small area of the retina, the scar tissue "welds" the retina down to the back of the eyeball, so it cannot break free. The laser treatment represented a big advance over earlier techniques, which required risky open-eye surgery.

A much more common laser procedure is treatment of a condition called "diabetic retinopathy," which is common in people who suffer from juvenile-onset diabetes. For reasons not fully understood, networks of abnormal blood vessels spread across the surface of the retina in many diabetics. These blood vessels are very fragile, and they can leak blood into the normally clear liquid of the eye—gradually dimming vision.

Ophthalmologists have found that illuminating the diseased retina with under a watt of continuous-wave visible laser light can slow spread of the abnormal blood vessels. They are not sure exactly how the laser works, but it seems to close off the blood vessels before they can do serious damage. The laser treatment is not a sure cure, but careful studies have shown that it can at least forestall blindness, and it has been used to treat well over 20 million diabetic eyes around the world.

Another type of laser eye treatment is more akin to conventional surgery. It uses a short laser pulse to treat a common complication of cataract surgery. Cataracts occur when the natural lens of the eye becomes cloudy and obstructs vision. They usually occur in older people. The standard treatment is to remove the natural lens material, leaving behind the back membrane of the lens, and to implant a plastic lens. In about a third of all cases, the natural membrane itself can become cloudy, again obstructing vision.

A short, intense pulse from a neodymium laser can pass through the lens implant and break the cloudy membrane, as shown in Fig. 11-5. The laser pulse is focused with a lens having very short focal length, so the laser light comes to a very tight focus exactly at the membrane surface. The ophthalmologist fires a series of laser pulses, which produce tiny sparks that break the membrane. Because the beam is focused very tightly, the power density at the implanted plastic lens is too low to damage it. Behind the membrane, the laser light again spreads out over a large area, so it cannot damage the retina.

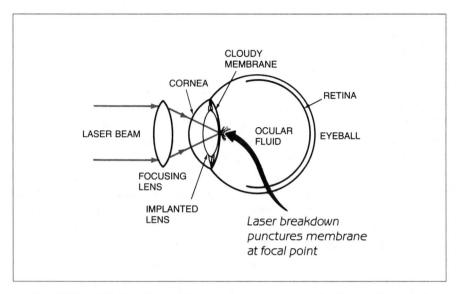

Fig. 11-5 A tightly focused laser pulse punctures a cloudy membrane left behind after cataract surgery.

The operation is done in a physician's office, and the membrane opens up almost instantaneously so the patient can see. One leading ophthalmologist says some patients almost jump for joy. Important advantages of the laser treatment include avoiding the time and expense of hospitalization, and the risk of surgically opening the eye to cut the membrane.

Refractive Surgery

A new, promising, and somewhat controversial laser eye treatment is called *refractive surgery*. It uses an excimer laser to change the shape of the cornea to correct the refractive defects that force people to wear eyeglasses or contact lenses.

The cornea is the outer transparent layer of the eye, shown in Fig. 11-5. It accounts for most of the eye's refractive power (more than the better-known lens) and is easily accessible because it is exposed to the air. Researchers have found that short, intense pulses of ultraviolet light can remove material from the surface of the cornea. This can be used to reshape or sculpt the cornea to correct for the eye's refractive defects. For example, removing material from the center of the cornea can reduce the excess focusing power that causes near-sightedness or myopia.

At this writing, laser refractive surgery is an experimental treatment undergoing clinical tests in the United States and other countries. While the people who developed the technique are enthusiastic, some other eye specialists are skeptical. The surgery is so new that its effects have not been followed for a long time, so it is not certain if there are any long-term effects from removing the surface of the cornea. There have been some reports of cloudy vision after the surgery (which may be cleared by another laser treatment). Not all eyes are corrected perfectly; some refractive errors may remain after the surgery. Refractive surgery also works better for some people than others. People with moderate near-sightedness benefit most. It does not work as well for the severely near-sighted. Nor can it yet treat far-sightedness.

In short, the results aren't in yet. Laser refractive surgery could someday be common, but so far the risks remain uncertain.

Laser Dermatology

Dermatologists, who had long used light to treat skin disorders, were among early medical-laser enthusiasts. The laser has not had as dramatic an impact on skin treatment as it has on eye treatment, but it has given dermatologists an important new tool to treat some conditions.

The biggest success of laser dermatology is in treating dark-red birthmarks called "portwine stains," which often appear on the face or neck. Networks of abnormal blood vessels just under the surface of the skin cause portwine stains; the blood gives the birthmarks their dark color. Because portwine stains are dispersed over the surface of the skin, conventional surgery cannot remove them effectively, and many sufferers had little choice but to cover them with cosmetics.

The standard laser treatment for portwine stains long has been to illuminate them with blue-green light from an argon laser, which is absorbed by the dark-colored blood and heats the blood vessels. The laser beam burns and blisters the skin, but it also causes the blood vessels to close over a number of weeks. As the burns heal, the birthmark bleaches away. While the treatment can be painful, it is usually effective in removing dark birthmarks, although it can cause scarring in some patients, and it cannot be used with young children or people with comparatively pale birthmarks.

A recent alternative is the use of a dye laser tuned to 577 nm, the wavelength at which hemoglobin in blood has its peak absorption. Because this concentrates absorption in the blood vessels, it avoids the blistering of the skin that accompanies argon-laser treatment. It is also more selective, and thus can be used to treat birthmarks too light for the argon-laser treatment. This and the avoidance of burning also allows the treatment of young children, whose birthmarks are normally pale but darken with age.

Laser dermatologists have tried for years to remove tattoos, with mixed success. In the last few years, they have begun using the venerable ruby laser, but tattoos remain tough targets.

Laser Dentistry

After years of neglecting lasers, some dentists have begun to buy and use them to treat patients. Lasers are not very efficient at removing solid enamel, but they can remove decayed areas. More important to both dentist and patient, laser pulses are less threatening than the traditional whirring mechanical drill, and can be less painful as well. Lasers can also treat gum disease.

Fiber-Optic Surgery

One strength of laser medicine is its ability to deliver energy precisely to the right place. Conventional laser surgery works well for parts of the body that are exposed to the laser beam, but not so well for the interior of the body. However, optical fibers can reach inside the body, delivering light to hard-to-reach places and avoiding the need for invasive surgery with a knife.

One example is the "laser lithotriptor," which delivers laser pulses through an optical fiber to shatter stones that can grow in the kidney and ureter. The idea is to thread an optical fiber through the ureter until it is pointed directly at the stone (which can be done using X rays or ultrasound imaging) and then blast away with laser pulses to shatter the stone into pieces small enough to pass through the ureter without pain. Dye lasers have been used because they can be tuned to precisely the right wavelength to destroy the stones. While ultrasound can also be focused to destroy kidney stones, optical fibers can reach some places that ultrasound cannot.

A similar technique can be used to destroy gallstones.

Laser Cancer Treatment

Lasers are an integral part of a new treatment for cancer called "photodynamic therapy." The treatment relies on a dye, called "hematopophyrin derivative" (HpD), which cancer cells retain much longer than healthy cells.

A patient is given an injection of HpD, then waits for three days while healthy cells flush out the dye, leaving it concentrated in cancer cells. Then the physician illuminates the cancer with red light at a wavelength around 631 nm, where HpD absorbs light strongly but healthy cells absorb light only weakly. The absorbed light makes the dye react chemically with oxygen molecules in cancer cells, producing a very reactive form of oxygen that can kill the cancer cells in high enough concentrations.

Although some experiments have produced encouraging results, photodynamic therapy has not been as effective in treating some cancers as some researchers had hoped and has also caused some side effects. So far it has been approved for regular clinical use only in Canada to treat certain bladder cancers.

Laser Artery Clearing

Another experimental laser treatment that got much publicity in the 1980s was laser angioplasty, in which optical fibers delivered laser energy to the inside of arteries. The laser energy was then supposed to remove arterial plaque, opening up the arteries so that blood could flow freely, reducing the risk of strokes and heart attacks.

The idea sounds a little like a laser version of a plumber's snake, but arteries aren't like pipes. The best success has been in the arteries to limbs; coronary arteries, where clots can be most deadly, are so tangled that it is very hard to thread fibers through them. Yet even in limbs, the problem remains that the laser energy that can remove arterial blockages also can remove arterial walls. Serious concerns remain about damage to the arteries, and some observers worry that the problems may not be solvable for the tangled arteries near the heart. Research continues, but with more caution than several years ago.

Cold Laser Treatment

In all the laser treatments we have described so far, the laser beam produces an observable change in something, whether by vaporizing tissue or closing blood vessels. However, there are also a family of "cold" laser treatments that require only a few milliwatts of power from a helium-neon or semiconductor laser. The orthodox medical establishment is skeptical of their effectiveness, but some people have claimed good results. Examples include:

- Laser "acupuncture," in which a low-power laser beam illuminates the acupuncture points originally defined for classical Chinese acupuncture with needles. The effects are said to be similar to needle acupuncture.
- Alleviation of chronic pain by illuminating affected areas with a low-power laser.
- Speeding of wound healing by illumination with a low-power laser beam.

While there has been legitimate controversy over the effects of such cold laser treatments, there also have been some allegations of fraud concerning the use of low-power lasers for "biostimulation." Several years ago, the Federal Trade Commission filed complaints against two Florida practitioners who were making what the FTC considered excessive claims about the benefits of nonsurgical laser "facelifts." The American Society of Plastic and Reconstructive Surgeons made even stronger charges.

PHOTOCHEMISTRY AND ISOTOPE SEPARATION

The energy carried in a light wave or photon can trigger chemical reactions (called *photochemical* reactions because a photon causes a chemical change). Light energy can break apart a complex molecule, remove an electron from the outer shell of an atom or molecule, or excite an atom or molecule into a state where it is more ready to engage in chemical reactions.

We have already covered some types of photochemistry, but we haven't used that label before. Photographic film records images by photochemistry, when a photon of visible light breaks apart a silver-halide molecule, leaving black metallic grains of silver (the exposed part of a black and white negative). Some types of optical data storage use photochemical reactions to record bits of data.

Laser photochemistry is playing an important role in some types of electronic material processing. Laser light causes the reactions that "expose" some photoresists used to write patterns on semiconductor wafers. Lasers evaporate the raw materials deposited to form thin films of high-temperature superconducting ceramics. Laser light breaks down the gaseous compounds that deposit carbon atoms to form diamond films, and other gases that leave metal films on electronic components. Ultraviolet laser light converts a liquid into a solid to form 3-D models in stereolithography.

Another interesting possibility is to use the inherently high selectivity of photochemical reactions. We saw earlier that atoms and molecules have different characteristic absorption wavelengths. This means that light of a specific wavelength can affect one type of molecule, but not others present in the same mixture. Thus, selective photochemistry is possible if the light sources are lasers tuned to certain wavelengths.

Suppose, for example, you have a mixture of three chemically similar molecules and you want to excite one of them so that it will react with another substance to form a drug. If the desired molecule absorbs light at 400 nm—and the other two absorb at 398 and 402 nm—you need only illuminate the mixture with a 400-nm laser with a line width of less than 1 nm. It will excite the desired molecule so it reacts, but it will not affect the other molecules.

Chemists have developed a formidable arsenal of tools, so they don't use selective laser photochemistry for routine chemical synthesis. However, laser photochemistry has made considerable progress in the much tougher problem of isotope separation, which is important in making materials for nuclear reactors and nuclear weapons.

Isotopes are atoms of the same element that contain different numbers of neutrons in their nucleus but the same number of protons. The number of protons in the nucleus (which equals the number of electrons in a neutral atom) defines the atomic number, and hence the identity, of an element. For example, all atoms with 92 protons in their nuclei are uranium atoms. However, uranium atoms can have different numbers of neutrons in their nuclei. The two most common uranium isotopes are uranium-238 (with 146 neutrons) and uranium-235 (with 143 neutrons). U-238 is the most common in nature, but it cannot sustain the fission chain reaction needed to drive a nuclear reactor. To sustain a fission reaction, you need the less common U-235. Conventional fission reactors need a few percent of U-235 to sustain a chain reaction, but the natural U-235 concentration is only about 0.7%. The only way to get those reactors to work is to enrich U-235 concentration above the natural level.

The problem is that isotopes are chemically almost identical. Because they have the same number of protons in the nucleus, they have the same electron configurations. The differences in nuclear mass shift the electronic energy levels slightly, but not enough to make their chemical properties different. Conventional isotope enrichment processes have to rely on small differences in mass between different isotopes that give gases containing them different weights.

Laser photochemistry can make the distinction between the two isotopes because lasers can be tuned precisely enough to excite energy levels of one isotope but not the other. Figure 11-6 shows the concept behind the atomic vapor laser isotope enrichment process for uranium developed at the Lawrence Livermore National Laboratory in California. A dye laser, pumped by a copper-vapor laser, is tuned to a narrow range of wavelengths that excites U-235 atoms but not U-238 in uranium vapor. The excitation process removes an electron from the U-235 atom, leaving it positively charged. That positive ion is attracted to a negatively charged electrode. On the other hand, U-238 atoms retain all their electrons and are not collected by the negative electrode.

Our figure makes it look like the negative electrode only collects U-235, but the process is not that simple—fortunately. We don't really

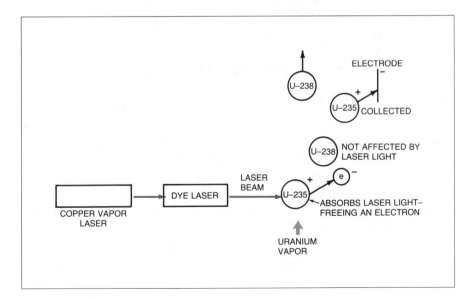

Fig. 11-6 Dye laser is tuned to a wavelength that ionizes U-235 atoms but not U-238, allowing collection of the rarer U-235 isotope.

want it to be too easy to produce pure U-235, because the fissionable isotope can be used for nuclear bombs as well as nuclear reactors. In a single pass, the laser process does not produce 100% pure U-235, but it does "enrich" the product in U-235. The process also has other complications, such as the handling of uranium vapor (truly nasty stuff!). The government has planned to use laser isotope separation of uranium only to produce reactor fuel, but has not yet begun building a plant.

Livermore also developed a similar process for separating isotopes of plutonium, which are produced in nuclear reactors when U-238 atoms absorb neutrons. Plutonium-239, like U-235, is fissionable and can sustain nuclear chain reactions in reactors or bombs. However, reactors also convert U-238 to the heavier plutonium-240 and -241 isotopes, which are not fissionable, and tend to absorb neutrons and slow nuclear chain reactions. Those heavier isotopes must be removed to purify the plutonium-239 for further use.

Scientists working at Livermore have demonstrated laser separation of plutonium isotopes for the Department of Energy. The United States does not use plutonium to fuel reactors, so the main application of that technique would be to make nuclear weapons. Disarmament at the end of the Cold War left the Department of Energy with a stockpile of excess plutonium, so the agency stopped work on the laser approach (called "Special Isotope Separation").

LASER-DRIVEN NUCLEAR FUSION

Another use of lasers in nuclear energy is in research into nuclear fusion. The fission reactors we use today split the nuclei of heavy atoms such as uranium and plutonium to generate energy. Nuclear fusion, which occurs naturally in the sun and stars, releases energy by combining the nuclei of light elements such as hydrogen and helium.

Fusion has long seemed an attractive energy source. In theory, it should produce less radioactive nuclear garbage than fission reactors, which inevitably leave radioactive fragments of heavy nuclei. Another advantage is that the light nuclei that fuel a fusion reaction are far more common on earth than the heavy elements needed to fuel fission reactors.

The bad news is that the right conditions for fusion are very hard to produce. Extremely high temperatures and pressures are needed for light nuclei to overcome their mutual repulsion (because both have positive charge) and combine to release energy. Those pressures and temperatures are readily available in the core of the sun. We can produce them on earth by exploding a nuclear fission bomb (which serves as the trigger for a hydrogen or fusion bomb). However, to generate power, we want to release energy in a controlled manner, and a bomb is hardly a controlled release of energy.

Controlled fusion researchers pursued two approaches. One is magnetic confinement, which tries to keep the hot fusion plasma in a strong magnetic field. The other is inertial confinement, which seeks to compress the fusion fuel into a tiny, hot, dense plasma for a short period, so nuclear fusion can occur and release energy. The two techniques take fundamentally different approaches to producing the combination of confinement time, high temperature, and density needed for fusion to occur. Magnetic confinement seeks to hold the plasma together for a long time, but does not try to maintain the highest possible pressure or temperature. Inertial confinement can produce higher temperatures and pressures, but it sustains them for only a fraction of a second.

Lasers enter the picture because they have been the main approach to inertial confinement fusion for so long that many people call the field "laser fusion," although beams of charged particles can also be used. Figure 11-7 shows the idea of laser fusion. Short, intense pulses of laser light strike a pellet containing hydrogen isotopes that can fuse together to release energy. The laser energy must be distributed uniformly around the surface of the sphere. When the laser energy hits the surface, it produces an implosion, forcing the material inside the tiny sphere inwards and simultaneously heating it to high temperatures. At the peak of the implosion, the nuclei in the pellet fuse together, releasing energy.

The laser power levels needed for fusion are very high—the peak power is measured in trillions of watts (terawatts). Also, the time scale is very short—laser pulses last around a nanosecond, a billionth of a second. So far the quantities of fusion energy generated are small, but the process has produced fusion reactions.

Before you get too enthusiastic about laser fusion, a few words of caution are appropriate. Even the optimists consider laser fusion a 21st-century technology. It's by no means certain that laser fusion will ever generate power commercially, and if it does, the date probably will be well into the 21st century. Of course, magnetic confinement fusion also won't be ready until the 21st century, but most observers consider it more mature.

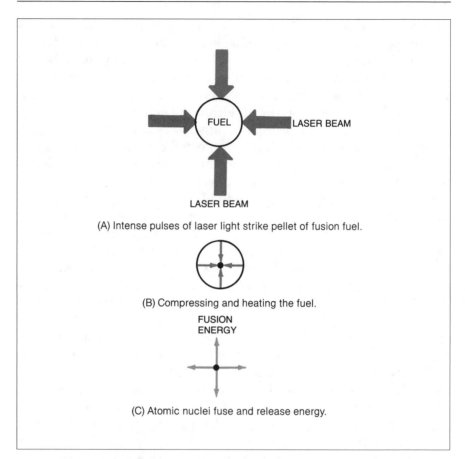

(A) Intense pulses of laser light strike pellet of fusion fuel.

(B) Compressing and heating the fuel.

(C) Atomic nuclei fuse and release energy.

Fig. 11-7   Intense laser pulses implode a pellet containing nuclear fuel to generate fusion reactions.

In the meantime, laser fusion actually serves a somewhat different purpose. The implosion of a pellet of fusion fuel can simulate the physical processes during the explosion of a hydrogen bomb. Fusion microexplosions let bomb designers study the weapon physics on a more convenient scale—and at much less cost than using underground nuclear explosions. These simulations are so important that they have been the main rationale for the U.S. government's support of the laser fusion program, which is run by the Department of Energy's Division of Military Programs. In the event of a total ban on nuclear testing, they would make laser fusion a very important program indeed.

LASER WEAPONS

Lasers themselves have long been investigated as weapons. Military officials have been enthusiastic since the late 1950s. When TRG brought Gordon Gould's laser proposal to the Pentagon, the Department of Defense were so excited they gave the company a $1 million grant, rather than the $300,000 they had requested. They have spent billions on laser weapons since then, and have discovered that they are not easy to make.

Destroying Enemy Targets

The simplest way to think of laser weapons is as materials processing on unfriendly objects. The basic processes involved are similar. A laser weapon must do something to the surface of a target, although the specific goal is to make the target useless as a weapon rather than to drill a hole. Laser weapons have many potential missions. The one that has gotten the most attention is for defense against attack by nuclear missiles. However, high-energy lasers have also been studied for use against satellites, short-distance battlefield missiles, and sensors.

The task of laser weapons is much more complex than materials processing. In materials processing, you try to make objects absorb laser light. In warfare, the other side tries to protect potential targets against laser attack. Military targets are much farther away, making it harder to focus laser energy onto them. They may be coated with materials that deflect laser energy, rather than absorb it. And the targets will be moving in ways that the person (or control system) firing the laser isn't supposed to know. All this adds up to requirements for much higher powers than for materials processing. The most powerful lasers used for materials processing deliver continuous-wave powers of about 45 kW—but 5 to 10 kW is much more common. Precise power levels needed to destroy military targets are classified, but it's clear that average or continuous powers in the megawatts are needed for use against missiles.

Distance is a serious complication for many reasons. The atmosphere is nowhere near as transparent as we think. Air always absorbs a small fraction of the light it transmits. Absorption of only 0.1% of the energy in the beam may seem a small loss, but if the beam is powerful, that tiny fraction can heat the atmosphere significantly. For a 1-MW laser beam, 0.1% is 1 kW, and that's enough to heat the atmosphere and make it turbulent so that it bends the laser beam in unpredictable directions. This turbulence makes it hard both to track the target and to focus the laser beam onto it effectively.

Going into space can overcome atmospheric transmission problems, but it begs the problem of how to get massive hardware into space and provide power for it when it's up there. Moreover, there are hosts of other problems, including the need to identify and track targets and control the entire weapon system effectively. Plus, the distances are much greater—hundreds or thousands of miles—so the beam can spread out considerably. In fact, building a high-energy laser may be one of the lesser problems.

In short, there are truly formidable problems to building laser weapons able to destroy nuclear-armed ballistic missiles. The problems are not necessarily insurmountable, but it will take billions of dollars to investigate them, and there is absolutely no assurance that we will be able to overcome them. The Strategic Defense Initiative spent some $6 billion studying high-energy lasers for nuclear defense, but that program has been scaled back sharply, and most big demonstrations have been cancelled.

Anti-Sensor and Anti-Personnel Lasers

We can make the task of building laser weapons appear more manageable if we redefine it to the more limited goal of disabling targets. If we limit our goal to knocking out sensors—such as the infrared sensors on guided missiles or the sensitive electronic eyes of spy satellites—much lower laser powers will suffice. Satellites are very vulnerable because their orbits can be predicted precisely, and their electronic eyes are very sensitive.

Electro-optical sensors are also widely used on the battlefield, which has led to programs to build anti-sensor lasers to blind military systems. The stated goal is to disable weapon systems. However, the lasers could also blind enemy soldiers, because the human eye is very sensitive to intense pulses of laser light. The laser rangefinders and target designators mentioned in Chapter 10 are powerful enough to cause eye damage in soldiers who are too close to them. That is a serious problem in war games, because it endangers friendly troops. More powerful lasers could be a potent weapon against distant enemy soldiers, especially any unlucky enough to be looking through binoculars when hit by the beam.

An anti-personnel laser would probably not blind a soldier totally, but it could easily destroy the fovea, the part of the retina that gives us detailed vision. You are using the fovea when you read these words; without it, you probably couldn't read or do much skilled labor. The potential use of lasers to intentionally blind enemy troops worries many observers, and proposals have been made to consider anti-personnel lasers inhumane.

FUTURISTIC HIGH-POWER LASERS

What does the future hold for high-power lasers? Some scientists have proposed some way-out ideas that may become possible in the 21st century. Laser beams might transmit power between satellites, from the ground to satellites, or from solar power satellites to the ground.

Powerful laser beams might propel spacecraft by evaporating fuel kept on the satellite. One idea is to nudge satellites into different orbits by evaporating ice they carry, so the force from the evaporating gas would push the satellite in the right direction. People have even proposed using lasers to launch satellites from the ground, but extremely high powers might be needed.

Some people who used to work on high-energy laser weapons for nuclear defense have come up with an idea of their own—using high-energy lasers to deflect or destroy any comets or asteroids that come too close to Earth. However, many astronomers think that's premature, because we don't even know what the asteroid threat is.

WHAT HAVE WE LEARNED

- High-power applications change materials that are not unusually sensitive to light. They are less diverse than lower-power applications.
- High-power lasers can concentrate beams to produce extremely high power densities on small areas. They do not apply physical force to the objects they strike.
- Materials working includes cutting, welding, drilling, and marking both metals and nonmetals.
- Lasers can drill holes in diamond, the hardest substance known, or in flexible materials like baby-bottle nipples.

- Energy transfer from the laser beam depends on wavelength and material absorption. Most metals absorb little light at CO_2 laser wavelengths.

- Plasma released from an illuminated object's surface can block the laser beam.

- Lenses with longer focal length have greater depth of focus so they can cut or drill deeper.

- Drilling requires pulsed lasers and depends on peak power.

- Lasers can mark or scribe materials without drilling holes all the way through.

- Lasers cut with a continuous beam. Metal cutting usually depends on oxygen to remove material.

- Welding joins two pieces into a single unit.

- Heat treating transforms the surface of a metal to a harder form.

- Power levels for materials working range up to kilowatts.

- Lasers are widely used in making semiconductor electronics. They can generate ultraviolet light for photolithography, and are one potential source of X rays for future X-ray lithography systems.

- Laser-tissue interactions depend on power, wavelength, and pulse duration.

- The 10-μm CO_2 laser can vaporize tissue and prevent bleeding. Visible and near-infrared lasers have other medical uses.

- Lasers can treat detached retinas and diabetes-related blindness.

- Lasers can treat complications from cataract surgery and are being tested to correct defective vision.

- Laser light can trigger chemical reactions, and deposit materials such as thin films of diamond and high-temperature superconducting ceramics.

- Laser photochemistry can separate isotopes of uranium and plutonium for nuclear reactors and bombs.

- In laser fusion, intense laser pulses heat and compress a fusion-fuel target. Fusion has been demonstrated, but commercial reactors are a long way away.

- High-power lasers seemed attractive as military weapons, but the technology has proved very difficult. Anti-personnel and anti-sensor laser weapons are much easier to build than anti-missile lasers.

- Anti-personnel lasers may be a threat to future soldiers.

Quiz for Chapter 11

1. Which of the following is not an attraction of high-power lasers?
   a. Ability to generate extremely high powers in very short pulses
   b. Amenable to robotic control
   c. Do not apply physical force to objects
   d. Are extremely efficient
   e. Can focus light energy tightly onto a small spot to generate very high powers

2. Which of the following laser properties is most important for laser drilling of baby-bottle nipples?
   a. Ability to generate extremely high powers in very short pulses
   b. Amenable to robotic control
   c. Do not apply physical force to objects
   d. Are extremely efficient
   e. Can focus light energy tightly onto a small spot to generate very high powers

3. A 1000-W continuous-wave carbon-dioxide laser illuminates a titanium sheet for 1 s. How much energy does the titanium absorb (assuming that absorption does not change with heating)?
   a. 100 W
   b. 80 joules
   c. 120 joules
   d. 800 joules
   e. None of the above

4. Which of the following are desirable for drilling holes in a thick (2-cm) slab of material?
   a. Short laser pulses
   b. Short focal-length lens
   c. Long focal-length lens
   d. A and B
   e. A and C

5. Which of the following are not important in producing a good laser weld?
   a. Short laser pulses
   b. Penetration of the laser beam into the joint
   c. Close fitting of the pieces being welded
   d. Compositions of the pieces being welded
   e. Scanning the laser beam along the joint at the right speed

6. Which of the following lasers produces a beam that is absorbed most efficiently by tissue?
  a. Argon-ion
  b. Neodymium-YAG
  c. Ruby
  d. Carbon-dioxide
  e. Semiconductor diode (GaAlAs)

7. What is the most important medical use of the carbon-dioxide laser?
  a. Eye examinations
  b. Treatment of bleeding ulcers
  c. Removal of thin layers from blood-rich tissue
  d. Cutting bone
  e. Removal of portwine-stain birthmarks

8. Lasers are useful in isotope separation because they do which of the following?
  a. Provide extremely narrow linewidth light to excite one isotope but not another
  b. Can deliver very short pulses
  c. Concentrate power onto one atom but not adjacent atoms
  d. Heat uranium metal until it is vaporized
  e. Produce thermonuclear fusion

9. What must laser beams do to cause fusion reactions?
  a. Provide extremely narrow linewidth light to excite one isotope but not another
  b. Compress fusion fuel to very high densities while heating it to high temperatures
  c. Push atomic nuclei together
  d. Heat uranium isotopes
  e. All of the above

10. What types of targets are likely to be threatened by laser weapons first?
  a. Nuclear-armed ballistic missiles
  b. Extraterrestrial space ships carrying hostile aliens
  c. Eyes of enemy soldiers
  d. Space stations
  e. Helicopters on the battlefield

Appendix A

LASER SAFETY

There are two main potential hazards associated with lasers: the beam and the power supply. Even low-power laser beams can pose some dangers to the eye, while high-power laser beams can burn the skin. The power supply, like any source of high voltage—or even wall current—can kill.

We have seen that a laser beam is made up of parallel light rays. The concentration of parallel light rays in a small beam makes laser powers as low as the milliwatt range a potential hazard to your eyes. The reason is that the eye focuses those rays, concentrating their optical power in a small enough area that they can burn the retina. The hazard of a few-milliwatt visible laser beam is comparable to that from looking directly at the sun. An accidental glance past the sun may dazzle your eyes, but staring directly at the sun can cause permanent eye damage. The same is true for lasers. The safest rule is never to look directly into any laser beam, even if you think it has very low power.

Higher laser powers pose more serious eye hazards. As continuous-wave power levels rise, it takes less time for the light to do lasting damage to the eye. If the power is concentrated in short high-power pulses, even a single shot can do lasting damage. One laser scientist has described how a direct hit from a stray pulse made blood erupt inside his eye. Such an event is not likely to blind you totally, but it can leave a blind spot that impairs your vision. The impairment could be severe if the blind spot is in the fovea, the part of the retina which provides detail vision.

Both visible and invisible laser light pose hazards to your eyes. Invisible near-infrared and near-ultraviolet light can enter your eye, and near-infrared light (in particular) can reach your retina without you sensing it. Hazards become less at wavelengths longer than about 1.55 µm, because the light cannot reach your retina, but you should nonetheless exercise care when working with any laser. Some near-infrared lasers can be very deceptive, because they seem to be emitting extremely weak red beams but are actually emitting significant powers at wavelengths that your eyes sense very inefficiently.

Laser safety goggles are made to block the wavelengths emitted by specific lasers and transmit other light. For example, goggles made to block the 1.06-µm neodymium laser wavelength transmit visible light; those made to block the argon-laser lines at 511 and 488 nm transmit red light. Good goggles should block light from reaching your eye via any side paths, such as accidental reflections from the side. Remember that it is vital to use goggles made for the type of laser you're using. Goggles made for the red helium-neon wavelength do not block light from an argon laser.

In nearly two decades of writing about lasers, I have never known anyone to be killed by a laser beam. However, several people have been killed when they accidentally touched high-voltage sources within a laser. It is much too easy to take electricity for granted. Even when the power is off, components such as capacitors in high-voltage power supplies can retain a lethal dose of electricity for a long time. Watch out. (This homily is dedicated to the memory of Gordon Brill, a laser-industry veteran who was killed while servicing a copper-vapor laser.)

If you are going to work with lasers, do yourself a favor and become familiar with how to use them safely. Information is available from the U.S. government agency charged with regulating laser safety: The Center for Devices and Radiological Health, 5600 Fishers Lane, Rockville, MD 20857. The Lawrence Livermore National Laboratory has prepared a helpful booklet, *A Guide to Eyewear for Protection from Laser Light*, for people using lasers. You can request a copy from Corrine Burgin, Scientific Editor, L-399, Lawrence Livermore National Laboratory, P.O. Box 808, Livermore, CA 94550.

Appendix B

CONSTANTS

Speed of light in vacuum (c): 2.99792458×10^8 m/s
Planck's constant (h): $6.6260755 \times 10^{-34}$ joule-s
Planck's constant (h): $4.1356692 \times 10^{-15}$ eV-s
Boltzmann constant (k): 1.380658×10^{-23} joule/Kelvin
Boltzmann constant (k): 8.617385×10^{-5} eV/K
Rydberg constant (R): 1.0973731534×10^7/m
Rydberg constant (R): 13.6056981 eV

CONVERSIONS

Electronvolts to joules: 1 eV = $1.60217733 \times 10^{-19}$ joule
A 1-eV photon has frequency: $2.41798836 \times 10^{14}$ Hz
A 1-eV photon has wavelength: 1.23984 μm
Mass of electron: $9.1093897 \times 10^{-31}$ kg
Mass of proton: $1.6726231 \times 10^{-27}$ kg

SYMBOLS TO REMEMBER

λ (Greek lambda): wavelength
n: refractive index
ν (Greek nu): frequency
ω (Greek omega): angular frequency, $= 2\pi\nu$
c: speed of light
h: Planck's constant

Laser Glossary

Acousto-Optic—Interaction between an acoustic wave and a light wave, used in beam deflectors, modulators, and Q switches.

Active Medium—The light-emitting material in a laser, sometimes used to identify a specific atom in a crystal.

Alexandrite—A synthetic crystal doped with chromium to form a tunable solid-state laser that emits near-infrared light.

Angstrom ($\mathring{A}$)—A unit of length, 0.1 nm or 10^{-10} m, abbreviated $\mathring{A}$. It is not a standard SI unit, but is often used to measure wavelength in the visible spectrum.

Arc Lamp—A high-intensity lamp in which an electric discharge continuously produces light.

Attenuation—Reduction of light intensity, or loss. It can be measured in optical density or decibels as well as percentage or fraction of light lost.

Attenuator—An optical element which transmits only a given fraction of incident light.

Average Power—The average level of power in a series of pulses. It equals pulse energy times the number of pulses divided by the time interval.

Beamsplitter—A device that divides incident light into two separate beams, one reflected and one transmitted.

Birefringent—Has a refractive index that differs for light of different polarization.

Bistable Optics—Optical devices with two stable transmission states.

Brewster's Angle—The angle at which a surface does not reflect light of one linear polarization.

Chemical Laser—A laser that is excited by a chemical reaction. The most common type produces hydrogen fluoride or deuterium fluoride.

Circular Polarization—Light polarized so the polarization vector rotates periodically without changing in length, describing a circle. Circularly polarized light can be considered as two equal-intensity linearly polarized beams, one 90 degrees in phase ahead of the other.

Coating—Material applied in one or more layers to the surface of an optical element to change the way it reflects or transmits light.

Coherence—Alignment of the phase and wavelength of light waves with respect to each other. If the waves are perfectly aligned, the light is coherent.

Collimate—Make light rays parallel.

Concave—Curving inwards, so the central parts are deeper than the outside, like the inside of a bowl.

Confocal—Having the focal point of two mirrors at the same place. Confocal resonators have concave end mirrors which have the same focal point in the middle of the laser cavity.

Continuous-wave—Emitting a steady beam.

Convex—Curving like the outside of a ball so the outer parts are lower than the center.

Cycles per Second—Number of oscillations a wave makes, the frequency. 1 Hz = 1 cycle per second.

Decibel—A logarithmic comparison of power levels, abbreviated dB, and defined as the value: $10 \log(P_2/P_1)$—i.e., 10 times the base-10 logarithm of the ratio of the two power levels.

Detector—A device that generates an electric signal when illuminated by light.

Difference Frequency—An output wave with frequency the difference of the frequencies of two input waves.

Diffraction—Scattering or spreading of light waves when they pass an edge.

Diode—An electronic device that preferentially conducts current in one direction but not in the other. Semiconductor diodes contain a p-n junction between regions of different doping, which lets current flow in one direction but not the other. Diodes can emit light (e.g., laser diodes) or detect it (photodiodes).

Diode Laser—A semiconductor laser in which a current flowing through the device causes electrons and holes to recombine at the junction of p- and n-doped regions, where stimulated emission (laser action) takes place.

Divergence—The angular spreading of a laser beam with distance.

Electro-Optic—The interaction of light and electric fields, typically changing the light wave. Used in some modulators, Q switches, and beam deflectors.

Electromagnetic Radiation—Waves made up of oscillating electrical and magnetic fields perpendicular to one another, travelling at the speed of light, which can also be viewed as being photons or quanta of energy. Electromagnetic radiation includes radio waves, microwaves, infrared, visible light, ultraviolet radiation, X rays, and gamma rays.

Electronic Transition—Change in energy level of an electron.

Etalon—An optical resonator, typically with a short distance between reflective surfaces, placed inside laser resonators to restrict the range of wavelengths.

Excimer Laser—A pulsed ultraviolet laser in which the active medium is a short-lived molecule containing a rare gas such as xenon and a halogen such as chlorine.

Far-Infrared Laser—One of a family of gas lasers emitting light at the far infrared wavelengths of 30 to 1000 μm.

Free-Electron Laser—A laser in which stimulated emission comes from electrons passing through a magnetic field that varies in space.

Frequency—For light waves, the number of wave peaks per second passing a point. Measured in hertz, or cycles per second.

Fused Silica—Synthetic silica (SiO_2) formed from highly purified materials.

Gallium Aluminum Arsenide—A semiconductor used in LEDs, diode lasers, and certain detectors. Chemically, $Ga_{1-x}Al_xAs$, where x is a number less than one.

Gallium Arsenide—A semiconductor used in LEDs, laser diodes, detectors, and electronic components; chemically, GaAs.

Glass—An amorphous solid, typically made mostly of silica (SiO_2) unless otherwise identified. Silica glasses transmit visible light.

Harmonic Generation—Multiplication of the frequency of a light wave.

Hertz—Frequency in cycles per second.

Heterojunction—A boundary between semiconductors that differ in composition, such as GaAs and GaAlAs.

Index of Refraction—The ratio of the speed of light in vacuum to the speed of light in a material, a crucial measure of a material's optical characteristics. Usually abbreviated n.

Indium Gallium Arsenide Phosphide—A semiconductor used in lasers, LEDs, and detectors. The band gap—and hence the wavelength emitted by light sources and detected by detectors—depends on the mixture of the four elements. Abbreviated InGaAsP.

Infrared—Invisible wavelengths longer than 700 nm and shorter than about 1 mm. Wavelengths from about 700 nm to 2000 nm are called the near-infrared.

InGaAsP—Indium gallium arsenide phosphide, a semiconductor compound used in light sources and detectors. The band gap, depends on composition, often written $In_{1-x}Ga_xAs_{1-y}P_y$, where x and y are numbers less than one.

Injection Laser—Another name for semiconductor laser, coming from the fact that current carriers are injected to produce light.

Integrated Optics—Optical elements analogous to integrated electronic circuits, with multiple devices on one substrate.

Intensity—Power per unit solid angle.

Interference—The addition of the amplitudes of light waves. In destructive interference the waves cancel; in constructive interference they combine to make more intense light.

Irradiance—Power per unit area.

Joule—Unit of energy equal to 1 W of power delivered for 1 s.

Junction Laser—A semiconductor diode laser.

Laser—From Light Amplification by Stimulated Emission of Radiation, one of the wide range of devices that generate light by that principle. Laser light is directional, covers a narrow range of wavelengths, and is more coherent than ordinary light.

LED—Light-emitting diode, a semiconductor diode that emits incoherent light by spontaneous emission.

Light—Strictly speaking, light is electromagnetic radiation visible to the human eye. Commonly, the term is applied to electromagnetic radiation close to the visible spectrum that acts similarly, including the near-infrared and near-ultraviolet.

Light-Emitting Diode—LED, a semiconductor diode that produces incoherent light by spontaneous emission.

Longitudinal Modes—Oscillation modes of a laser along the length of its cavity, so twice the length of the cavity equals an integral number of wavelengths. Distinct from transverse modes, which are across the width of the cavity.

Maser—Microwave analog of a laser, an acronym for Microwave Amplification by Stimulated Emission of Radiation.

Metastable—An excited energy level that has an unusually long lifetime.

Micrometer—One millionth of a meter, abbreviated μm. The μ is a Greek mu.

Mode—A manner of oscillation in a laser. Modes can be longitudinal or transverse.

Modelock—The locking together of many longitudinal modes in a laser cavity, producing a series of short laser pulses. Modelocking can be visualized as clumping together a group of photons that bounce back and forth in the laser cavity.

Monochromatic—Containing only a single wavelength or frequency.

Multimode—Containing multiple modes of light. Typically refers to lasers that operate in two or more transverse modes.

n Region—Part of a semiconductor doped so that it has an excess of electrons as current carriers.

Nanometer—A unit of length equal to 10^{-9} m. Its most common use is to measure visible wavelengths.

Nanosecond—A billionth of a second, 10^{-9} s.

Near-Infrared—The part of the infrared nearest the visible spectrum, typically 700 to 1500 or 2000 nm, but not rigidly defined.

Normal (Angle)—Perpendicular to a surface.

Orthogonal—Perpendicular.

Oscillator—A laser cavity with mirrors so stimulated emission can oscillate within it. To some purists, only an oscillator can be a laser.

p Region—Part of a semiconductor doped with electron acceptors in which "holes" (vacancies in the valence electron level) are the dominant current carriers.

Peak Power—Highest instantaneous power level in a pulse.

Phase—Position of a wave in its oscillation cycle.

Photodetector—A light detector.

Photodiode—Usually, a semiconductor diode that produces an electrical signal proportional to light falling upon it. There also are vacuum photodiodes that can detect light.

Photometer—An instrument for measuring the amount of light visible to the human eye.

Photons—Quanta of electromagnetic radiation. Light can be viewed as either a wave or a series of photons.

Polarization—Alignment of the electric and magnetic fields that make up an electromagnetic wave. Normally, this refers to the electric field. If light waves all have a particular polarization pattern, they are called polarized.

Polarization Vector—A vector indicating the direction of the electric field in an electromagnetic wave.

Polarizer—A device that transmits light of only one polarization.

Population Inversion—The condition when more atoms are in an upper energy level than in a lower one. A population inversion is needed for laser action.

Pumping—The way a laser gets the energy to produce a population inversion.

Q Switch—A device that changes the Q (quality factor) of a laser cavity to produce a short, powerful pulse.

Quantum—Divided into discrete pieces or levels. A photon is a quantum of light energy. Quantum levels are discrete states with specific values of energy.

Quartz—A natural crystalline form of silica (SiO_2).

Quaternary—A compound made of four elements, e.g., InGaAsP.

Radian—A unit of angular measure; 2π radians equals 360 degrees, a circle.

Radiant Flux—Instantaneous power level in watts.

Radiometer—An instrument to measure power (watts) in electromagnetic radiation; distinct from a photometer, which measures light perceived by the human eye.

Rays—Straight lines that represent the path taken by light.

Recombination—The dropping of an electron with enough energy to circulate freely into an atomic state with lower energy. Semiconductor lasers emit light when a conduction-band electron is captured by a "hole" in a semiconductor, causing the conduction band to drop to a vacancy in the valence band.

Refraction—The bending of light as it passes between materials of different refractive indexes.

Refractive Index—The ratio of the speed of light in vacuum to the speed of light in a material, a crucial measure of a material's optical characteristics. Abbreviated n.

Resonator—A region with mirrors on the ends and a laser medium in the middle. Stimulated emission from the laser medium resonates between the mirrors, one of which lets some light emerge as a laser beam.

Retroreflector—An optical device that reflects incident light back in precisely the direction from which it came. It is typically a prism or a cluster of three mirrors forming the corner of a cube.

Semiconductor Diode Laser—A laser in which recombination of current carriers at a p-n junction generates stimulated emission.

Silica—Silicon dioxide, SiO_2, the major constituent of ordinary glass.

Silica Glass—Glass in which the main constituent is silica.

Single-Frequency Laser—A laser that emits only a very narrow range of wavelengths, nominally a single frequency but actually a very narrow range small enough to be considered a single frequency.

Single-Mode—Containing only a single mode. Beware of ambiguities because of the difference between transverse and longitudinal modes. A laser operating in a single transverse mode often does not operate in a single longitudinal mode.

Speckle—Coherent noise produced by laser light. It gives a mottled appearance to holograms viewed in laser light.

Spontaneous Emission—Emission of a photon without outside stimulation when an atom or molecule drops from a high energy state to a lower one.

Stimulated Emission—Emission of a photon that is stimulated by another photon of the same energy; the process that makes laser light.

Ternary—Compound made of three elements, e.g., GaAlAs.

III-V Semiconductor—A semiconductor compound made of one (or more) elements from the IIIA column of the periodic table (Al, Ga, and In) and one (or more) elements from the VA column (N, P, As, or Sb). Used in LEDs, diode lasers, and detectors.

Threshold—The excitation level at which laser emission starts.

Threshold Current—The minimum current needed to sustain laser action, usually in a semiconductor laser.

Time Response—The time it takes to react to a change in signal level.

Total Internal Reflection—Total reflection of light back into a material when it strikes the interface with a material having a lower refractive index at a glancing angle.

Transition—Shift between energy levels.

Transverse Modes—Modes across the width of a laser. Distinct from longitudinal modes, which are along the length.

Tunable—Adjustable in wavelength.

Ultraviolet—Part of the electromagnetic spectrum at wavelengths shorter than 400 nm, to about 10 nm, invisible to the human eye.

Vibronic—A electronic transition accompanied by a change in vibrational energy level.

Visible Light—Electromagnetic radiation visible to the human eye, at wavelengths of 400 to 700 nm.

Waveguide—A structure that guides electromagnetic waves along its length. An optical fiber is an optical waveguide.

Wavelength—The distance an electromagnetic wave travels during one cycle of oscillation. Wavelengths of light are usually measured in nanometers (10^{-9} m) or micrometers (10^{-6} m). The standard symbol is λ.

X ray—A photon emitted when an electron drops into an inner shell of an atom. Typically, wavelengths are 0.03 to 10 nm, but the boundaries are not well defined.

YAG—Yttrium aluminum garnet, a crystalline host for neodymium lasers.

Index

Answers to Quiz Questions

Chapter 1:

1. c	2. b	3. a
4. e	5. b	6. d
7. a	8. b	9. d
10. b		

Chapter 2:

1. b	2. c	3. a
4. d	5. c	6. d
7. a	8. b	9. d
10. c	11. d	12. c

Chapter 3:

1. b	2. e	3. b
4. c	5. d	6. e
7. a	8. d	9. c
10. a		

Chapter 4:

1. b	2. e	3. a
4. e	5. c	6. e
7. b	8. d	9. b
10. a		

Chapter 5:

1. a	2. d	3. d
4. b	5. c	6. e
7. c	8. a	9. c
10. e		

Chapter 6:

1. c	2. a	3. b
4. d	5. d	6. a
7. e	8. c	9. e
10. d		

Chapter 7:

1. e	2. b	3. a
4. d	5. b	6. e
7. c	8. c	9. a
10. d		

Chapter 8:

1. c	2. a	3. c
4. b	5. c	6. b
7. d	8. b	9. a
10. e		

Chapter 9:

1. d	2. d	3. a
4. b	5. b	6. a
7. d	8. c	9. d
10. d		

Chapter 10:

1. e	2. c	3. a
4. b	5. e	6. d
7. b	8. c	9. c
10. d		

Chapter 11:

1. d	2. c	3. b
4. e	5. a	6. d
7. c	8. a	9. b
10. c		